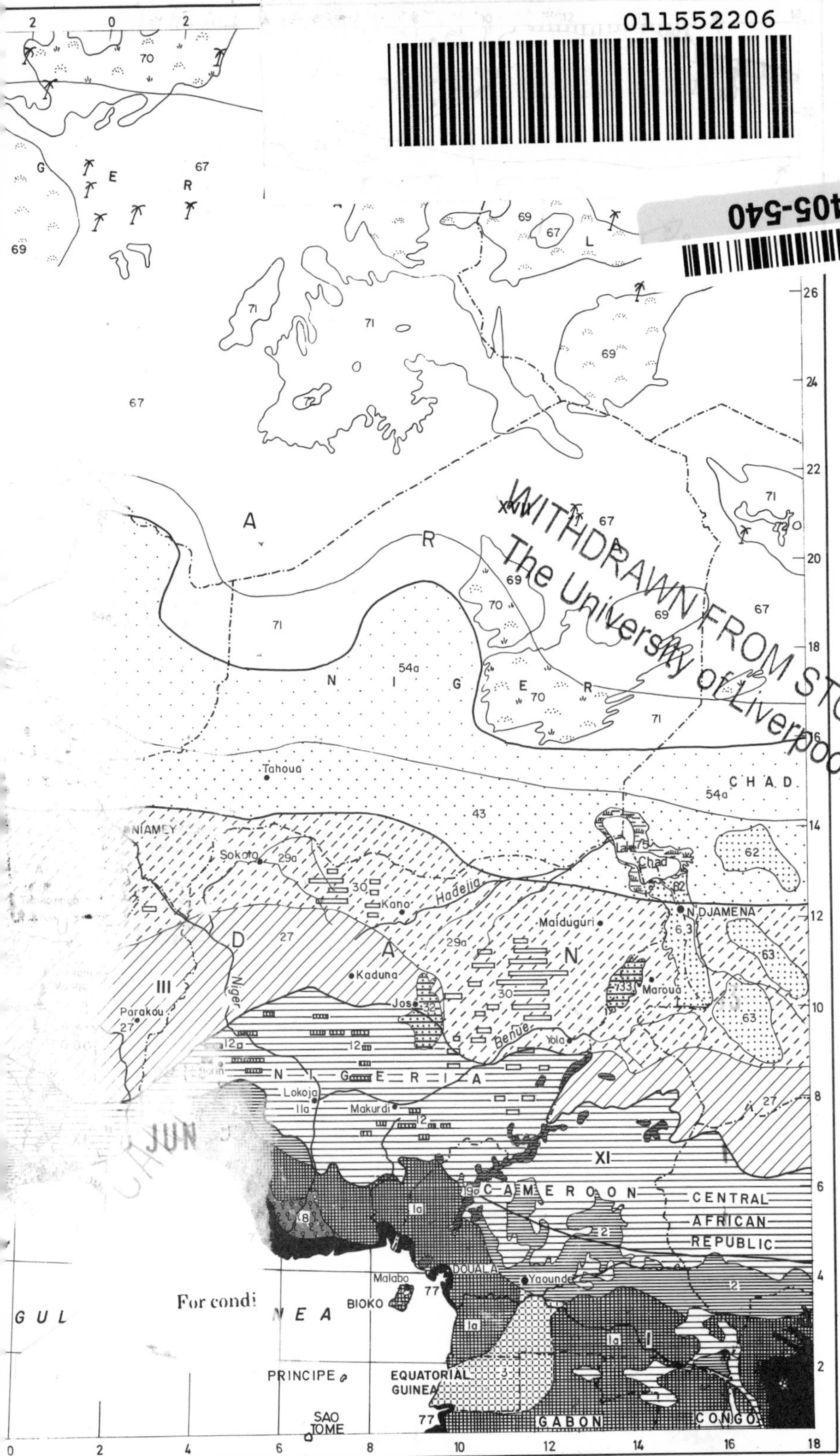

GUL

For condi'

SAO TOME

PRINCIPE

EQUATORIAL GUINEA

GABON

CONGO

BIOKO

Malabo

Douala

Yaoundé

CENTRAL AFRICAN REPUBLIC

CAMEROON

N I G E R I A

Makurdi

Lokoja

Benue

Yola

Jos

Kaduna

Kano

Maiduguri

Hadejia

Sokoto

N'DJAMENA

L. Chad

C H A D

Tahoua

Parakou

Niger

NIAMEY

N I G E R

Zinder

Agadez

Tamanrasset

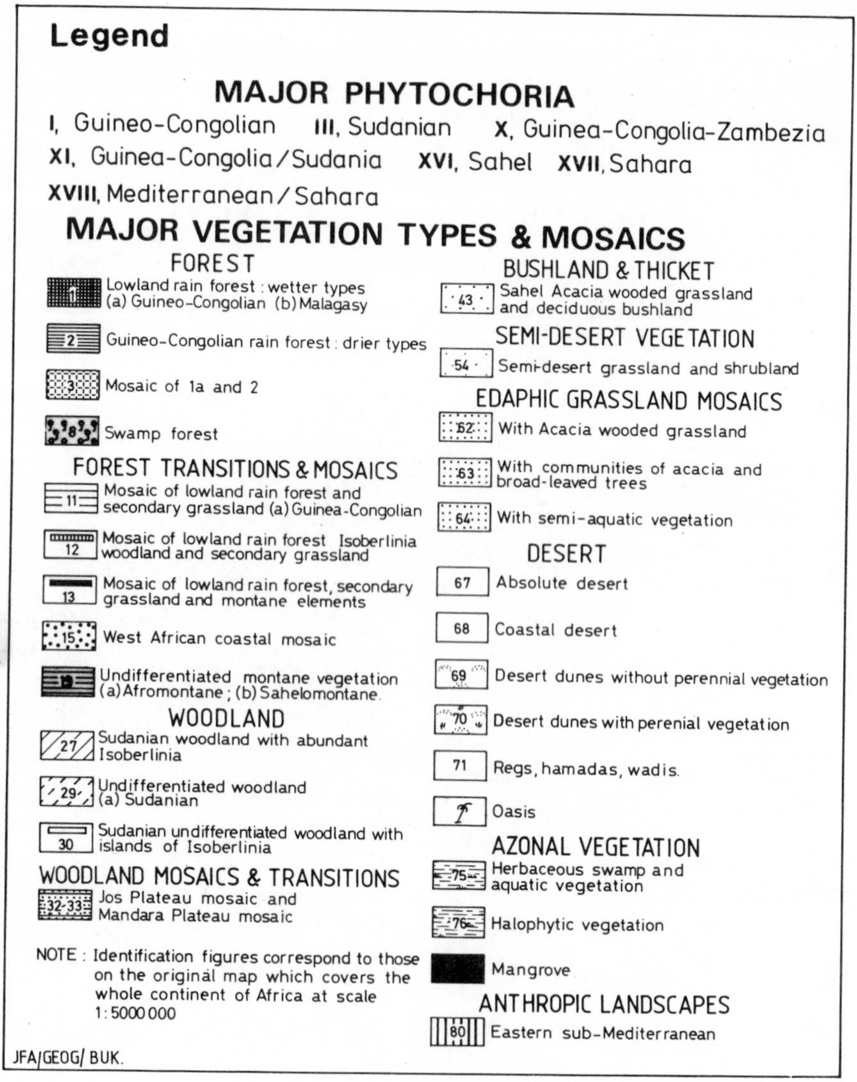

Fig. 1.1   Vegetation map of West Africa (White, 1983), based upon the *Unesco Vegetation Map of Africa*, © Unesco 1981.

Plant Ecology
in
West Africa

Systems and Processes

Edited by

George W. Lawson

Bayero University, Nigeria

JOHN WILEY & SONS
Chichester · New York · Brisbane · Toronto · Singapore

Library of Congress Cataloging in Publishing Data:
Main entry under title:

Plant ecology in West Africa.

Includes bibliographies and index.
   1.   Botany—Africa, West—Ecology. I. Lawson, George W.
QK393.P58   1985   581.5'0966   84–17347

ISBN 0 471 90364 7

British Library Cataloguing in Publication Data:

Plant ecology in West Africa.
   1. Botany—Africa, West—Ecology
I. Lawson, George W.
   581.5'0966   QK393

ISBN 0 471 90364 7

Printed and bound in Great Britain

Contributors

H. G. BAKER — Department of Botany, University of California, Berkeley, California 94720, USA

H. GILLET — Laboratoire d'Ethnobotanique, 57 rue Cuvier, Paris, France

(the late) J. B. HALL — Formerly of Department of Botany, University of Ghana

I. HEDBERG — Institute of Systematic Botany, University of Uppsala, Sweden

A. O. ISICHEI — Department of Botany, University of Ife, Nigeria

D. M. JOHN — Department of Botany, British Museum (Natural History), London

G. W. LAWSON — Department of Biological Sciences, Bayero University, Kano, Nigeria

J. M. LOCK — Cambridge, England

J. K. MORTON — Department of Biology, University of Waterloo, Ontario, Canada

W. W. SANFORD — Department of Botany, University of Ife, Nigeria

M. A. SOWUNMI — Department of Archaeology, University of Ibadan, Nigeria

M. D. SWAINE — Department of Botany, University of Aberdeen, Scotland

About the Contributors

HERBERT GEORGE BAKER received his bachelor's and doctor's degrees from the Universityof London. From 1945 till 1954 he was on the faculty of the University of Leeds, moving from there to the University of the Gold Coast (now the University of Ghana) where he was Professor of Botany till 1957. Then he joined the University of California, Berkeley, as a professor (till the present) and Director of the Botanical Garden (till 1969). His current research is on the chemical constitution of nectar and pollen rewards to flower-visitors. For the last 20 years his tropical work has been mostly in Costa Rica, with some time in Australia and Sri Lanka.

HUBERT GILLET has undertaken many journeys through the African Sahel region since 1958, especially in Chad, Niger, and Senegal. During the course of eight visits to Ennedi in northern Chad he made an intensive study of the flora and vegetation of this massif, and was able to explore areas completely untouched by human interference. He has also made a detailed investigation of the pasturage of the Quadi Rimé Ranch in central Chad and catalogued the plants of Aïr in northern Niger. Professor Gillet, whose present interest is in the food plants of the large wild herbivores of the Sahel, works in the Department of Ethnobotany and Ethnozoology at the National Museum of Natural History, Paris, France.

J. B. HALL worked in Ghana from 1955 to 1980, first as biology master at Mfantsipim School, Cape Coast, and subsequently as a lecturer in the botany departments of Cape Coast University and the University of Ghana. From 1971 to 1980 he travelled extensively through the forests of Ghana, Ivory Coast, and Nigeria and, with M. D., Swaine, published a new floristically based classification of Ghana forest, and numerous papers on the ecology and taxonomy of West African forest plants. He was also joint author and editor of a series of A–level biology textbooks for use in tropical African schools. John Hall's knowledge of and contribution to West African botany were immense. He died suddenly in May 1984.

INGA M. M. HEDBERG works at the Institute of Systematic Botany at Uppsala University, Sweden. Her interests are in cytotaxonomy, the flora of Ethopia project — of which she has been co-ordinator since 1980 — and in the problems of conservation in less developed countries. In 1966 as Deputy Secretary-General of AETFAT (Association pour Étude Taxonomique de la Flore d'Afrique Tropical) she collaborated with her husband Olov Hedberg, the well-known specialist on East African montane vegetation, in organizing and running an international conference at Uppsala on Conservation of Vegetation in Africa South of the Sahara (published in 1968). Dr. Hedberg is secretary of the Swedish National Committee of Biology and a member of the IUBS working group on medicinal plants.

AUGUSTINE ONWUEGBUKIWE ISICHEI took his first degree and doctorate at the University of Ife, Nigeria, and is Lecturer in Ecology in the Department of Botany at Ife University. His special interest is in nitrogen cycling in savanna and he has worked extensively on problems concerned with the elucidation of stocks and flows of nitrogen in Nigerian savanna, and, in addition, on cyanobacteria and rhizobia as contributors of nitrogen to savanna ecosystems.

DAVID M. JOHN. After completing his postgraduate studies in 1968 on tidal and subtidal ecosystems in the British Isles, Dr John joined the staff of the Botany Department at the University of Ghana, Legon. The next 12 years were spent lecturing and researching into the taxonomy and ecology of marine and freshwater algae. In pursuit of his research he visited most of the coastal countries of West Africa (western Sahara south to Angola) and the islands of Ascension (mid-Atlantic ridge) and Antigua (West Indies). From 1974 he was Senior Lecturer in Botany and since 1980 has been Head of the Freshwater Algae Section in the Department of Botany in the British Museum (Natural History), London.

GEORGE W. LAWSON has been connected with tropical African botany for over 30 years and has worked on problems of marine, brackish-water, freshwater, and terrestrial ecology in the region. He has held chairs in botany at the universities of Ghana, Nairobi, and Lagos, and a visiting professorship at the University of Dar es Salaam, Tanzania. At present he is Head of the Department of Biological Sciences at Bayero University, Kano, and was recently awarded a higher doctorate by the University of London for his work on African botany.

J. MICHAEL LOCK. After graduating in botany from Cambridge University, Dr Lock went to Uganda to work with the Nuffield Unit of Tropical Animal Ecology in the Queen Elizabeth National Park. During the latter part of his stay in Uganda he taught at Makerere University. In 1970 he became a lecturer in botany at the University of Ghana, Legon, where for seven years his research interests were divided between the taxonomy of African Zingiberaceae and aspects of plant–animal relationships, including orchid pollination. After leaving Ghana he taught for a year at the University of Lancaster and then spent six months working in the herbarium of the Royal Botanic Gardens, Kew. Since 1980 he has worked as an independent botanical consultant, mainly in the Sudan but also in Nigeria and Venezuela.

JOHN K. MORTON spent many years in university posts in West Africa, first in Ghana, then, after a brief stay at Birkbeck College, London, in Sierra Leone where, as Professor of Botany at Fourah Bay College, he was able to make a study of the floras of the Loma Mountains and the Tingi Hills. He is the author of *West African Lilies and Orchids* and over 80 scientific papers. In 1968 he became Professor of Botany at the University of Waterloo, Canada, and Chairman of the Biology Department from 1974 to 1980. Professor Morton was President of the Canadian Botanical Association during its 1974–75 session.

WILLIAM W. SANFORD was born in the United States and still remains the citizen of that country. He has spent 20 years in Nigeria as Professor of Botany at the University of Ife. Before that, he carried out research in plant physiology at the Boyce Thompson Institute in America which is now located at Cornell University, Ithaca. He also spent some time as Fullbright teacher in biology in Greece. His major interests are in the structure of savanna and its stability and management.

M. ADEBISI SOWUNMI is Senior Lecturer in Environmental Archaeology, Palaeobotany, and Palynology and she has been an academic member of staff at the University of Ibadan since 1967. Dr Sowunmi's interests include vegetational and climatic history of West Africa during the Quaternary period, pollen morphology of extant Nigerian plants with special emphasis on climatic indicator species and food crops, the beginning of agriculture in Nigeria as indicated by botanical evidence, the ecological status of West African savannas, and the study of Nigerian honey pollen content.

MICHAEL D. SWAINE graduated from the University College of North Wales, and then worked first as Lecturer then Senior Lecturer in Botany at the University of Ghana from 1971 to 1979. Since then he has been a lecturer in tropical botany and a member of the Institute of South-east Asian Biology at the University of Aberdeen. Dr Swaine is maintaining research studies in Ghana, and has begun other studies on tropical forest in Papua New Guinea.

Contents

Foreword

Paul Richards
Emeritus Professor of Botany, University of Wales

The plant life of tropical Africa, though broadly similar to that of other tropical regions, has a very marked character of its own. Not only are there very many plants and animals peculiar to Africa, but also, curiously, there are some groups of plants, such as palms, which are very numerous in other parts of the tropics, but poorly represented in Africa. As I once wrote, Africa is in many respects the 'odd man out', the flora of tropical America and tropical Asia being more alike than either is to Africa. Yet, in spite of its great interest, African plant ecology has in recent years received much less attention than that of other parts of the tropics. For this reason, a book like this one, in which acknowledged experts give up-to-date accounts of some of the major aspects of West African plant ecology, is much to be welcomed.

A special feature of this volume is three chapters on aquatic, coastal and marine vegetation, which in earlier accounts of West African vegetation have usually received no more than cursory treatment. This is primarily a book by scientists for scientists, but there is much in it to interest and instruct the less expert. Some chapters, notably Professor Gillet's on the desert and Sahel, Dr Hedberg's on land-use and conservation, and the editor's on ecology and development, are on topics of urgent practical importance.

At a time when West African plant life is undergoing rapid and destructive changes, it is of great importance to record as much as possible about the vegetation, particularly the forest, before it is too late. The forest, which a hundred years ago occupied a considerable fraction of the area, seems doomed to disappear in the near future, except for very small conserved areas; and its hundred of plant and animal species, about most of which very little is known, will also be lost. The savannas and other types of vegetation are also suffering from exploitation and development, but perhaps less obviously.

These great changes are of course mainly due to the growth of human populations and technology during the past hundred years. They follow long-term cycles of climatic change which, as Dr Sowunmi shows in Chapter 11, have affected Africa over many millions of years. The difference between the recent man-made changes and those in geological time is that the former may be irreversible. During arid periods in the distant past the forest shrank in area, but some survived in refuges and when the climate became moister the forest was able to expand again. If nearly all the forest now disappears, it seems doubtful whether small conservation areas will be large enough to act as 'Noah's arks' from which the forest could recover lost ground, if and when that became possible.

I am glad to recommend this book to anyone with an interest in West African ecology.

Paul Richards
May 1986

Editor's Preface

The intention in compiling this work has been to bring together for the very first time within one volume authoritative and up-to-date accounts of the principal types of vegetation in West Africa and to outline the main processes that occur within them. For those not already familiar with West African vegetation a brief overall survey is given in Chapter 1. This, it is hoped, will provide the perspective for the much more detailed treatments given in later chapters. Experienced ecologists may safely omit this chapter and go straight to the more specialist accounts.

Put at the simplest there are two main kinds of vegetation in the region, namely forest and savanna. With the imminent total destruction of the former in prospect, its importance for mankind must clearly be on the wane and our chapter on it must of necessity sound something of a valedictory note — a last post for the last pole! Such sentiment is well expressed by Hall and Swaine in a book published recently (Hall and Swaine, 1981). In this they state:

> It is clear from the foregoing that very little of the forest as we know it today is likely to survive the next century. Our present book will be a valuable record for the 21st century botanist of the distributions which forest plants used to have, but it is unlikely to be of much direct help in his attempts to find them.

Nevertheless, it is important to set down now as much as possible of what is known of forest — not just as a record for posterity but as a perspective in which future exploitation of the forest zone can be planned and assessed.

The savanna, on the other hand, covers a very much greater area than the forest. In the past, though the northern savanna regions have been very intensively exploited for agriculture there have been large areas of the southern and middle savanna regions that have remained relatively untouched. With increasing human population pressure and the availability of better agricultural techniques it is inevitable that a great deal of attention will need to be paid to savanna in the future. So again it is hoped a basic understanding of the ecosystem will provide a framework in which rational decisions may be made. The foundation for such an understanding is given by Professor Sanford and Dr Isichei in Chapter 5.

The Sahel zone has often been in the news in recent years especially with regard to the terrible droughts that have afflicted it and its human and animal populations. There has also been much alarmist talk about the southward advancement of the Sahara Desert. Professor Gillet has made a very clear exposé of these topics in Chapter 6. He explains in detail the real changes that are happening. By indicating the underlying causes of these changes he suggests ways by which they may be combated and the situation brought under control.

In West Africa, as in other parts of the world, there is a very high concentration of human population near the coast. Thus the coastal belt, though narrow and of limited area, becomes disproportionately important in any discussion of vegetation and vegetation processes. One chapter (Chapter 8) deals with the ecology of this coastal strip; another reviews and summarizes an aspect of the subject that is not often included in books on plant ecology, namely the vegetation of the actual littoral zone and what lies beyond and deeper, the sublittoral region (Chapter 9).

Aquatic vegetation of another type is dealt with in Chapter 7. This is the vegetation that is found in freshwater conditions in lakes and ponds, rivers and streams. As well as dealing with the macrophytic vegetation of plants that occur floating on or in water bodies, or grow totally or partly submerged around them, it was originally intended to include also the microscopic vegetation of algae that occur as free-floating planktonic forms or as forms growing on the rock, sand, and mud, or on the larger hydrophytic plants in such aquatic environments. However, this latter is a large subject in itself and for reasons of space it had to be omitted.

In considering the dynamics of vegetational changes it is always good to have a long view to be able to put the problems in perspective. This is exactly what Dr Sowunmi does in Chapter 11. Here for the very first time is a synthesis of the evidence taken from a wide variety of sources leading to a view of the palaeoecology of plant life in the West African region. Such work not only gives insight into the past but illuminates the present, for it is only in the light of history, and in this case prehistory, that the present, often puzzling, distribution patterns of vegetation can be fully understood and appreciated.

A type of vegetation that is relatively minor in extent, though of considerable intrinsic interest is dealt with in Chapter 10. This is the vegetation that occurs on the mountains of this region. As with freshwater aquatic vegetation it is azonal in character, being dependent much more on altitude than latitude. Professor Morton considers in detail both the vegetation itself and the factors that are responsible for its existence.

Another aspect of plants that needs to be taken into account for a complete understanding of vegetation is the relationship they bear to animals and their influences. It is not simply the fact that plants ultimately provide the food for all animals: there are many other ways the two interact. Most fascinating of these, perhaps, are the ways by which co-evolution has occurred, for instance in pollination where a particular insect and a particular plant have had their attributes gradually sifted by evolution to produce a pair of organisms inseparately linked in structure and function to their mutual benefit. Such plant–animal interrelationsips are dealt with by Dr Lock in Chapter 2. The theme of pollination is also taken up by Professor Baker in Chapter 3 on the breeding systems of plants in West Africa.

Plant–animal relationships of the type mentioned in the preceding paragraph must have taken thousands upon thousands of years to develop their delicately tuned perfection. But as we know these products of natural evolution can be wiped out very rapidly in the modern world by the practice and mechanisms of economic exploitation. In former days when only a few species of trees were singled out for the use of man the ecosystems from which they were extracted had some chance of maintaining themselves even with the losses that they sustained. But what chance is there for the continuity of such ecosystems now that there are factories for the production of softboard, for example, where every possible type of woody fibre is grist for their mills?

It is such ruthless exploitation that has led some to the concept of conservation of natural renewable resources and to the idea of maintaining at least parts of the original ecosystems as intact as possible so that they may remain as areas for the preservation of genetic material, for study, and for recreation. It is heartening to know that many areas for plant and animal conservation have already been set aside in West Africa, though much still needs to be done to preserve them for posterity. Dr Hedberg deals with this difficult subject in Chapter 12.

It was the original intention to include chapters on West African climate and soils, but for various reasons this has not proved possible. However, excellent accounts exist for both of these topics elsewhere; for climate the reader is referred to Ojo (1977) and Griffiths (1972) and for soils to the comprehensive account of West African soils by Ahn (Ahn, 1970).

The chapters in this book have been written independently of each other by authors with different backgrounds and differing points of view. They would not necessarily agree with everything that has been written here, though in some cases what they say may overlap to some extent. I make no apology for such overlaps. The human mind has a great propensity to let slip things that are said just once and some of the important things stated in this volume bear repeated iteration, especially so when they are put forward from different points of view. Nor do I apologize for any lack of unanimity the reader may discern. Different selection of facts and the varying opinions that are thus derived only verify that the subject is a living and developing one that has not yet ossified into doctrine. It is left to the reader to read and compare the various accounts and by measuring them against his own experience to arrive at a synthesis that he will be proud to call his own!

Flowering plant names follow the *Flora of West Tropical Africa* (Hutchinson, Dalziel, Keay and Hepper, 1954 – 69) and marine algal names follow Lawson and John (1982) except where otherwise stated.

References

Ahn, P. M. (1970). *West African Soils*, Oxford University Press, London.

Griffiths, J. F., 1972. *Climates of Africa*, Vol. 10 of H. S. Landsbey (ed. in chief), World Survey of Climatology. Amsterdam.

Hall, J. B. and Swaine, M. D. (1981). *Distribution and Ecology of Vascular Plants in a Tropical Rain Forest: Forest Vegetation of Ghana*. W. Junk, The Hague.

Hutchinson, J., and Dalziel, J. M., revised by R. W. J. Keay and F. N. Hepper (1954 – 69). *Flora of West Tropical Africa*. Crown Agents, London.

Lawson, G. W., and John, D. M. (1982). The Marine Algae and Coastal Environment of Tropical West Africa (*Beih, 70 zur Nova Hedwigia*). J. Cramer, Vaduz.

Ojo, O., 1977. *The Climates of West Africa*. University Press, Ibadan.

Plant Ecology in West Africa
Edited by G. W. Lawson
© 1986 John Wiley & Sons Ltd

CHAPTER *1*

Vegetation and environment in West Africa

G. W. Lawson
Department of Biological Sciences Bayero
University, Kano, Nigeria

I. Introduction

The vegetation of West Africa presents in broad outline a relatively simple picture compared with other parts of tropical Africa (Fig 1.1, see front end paper). The terrain is generally low-lying, so that the vegetation falls into natural latitudinal zones determined by climate. There are few of the complicated patterns associated with great differences in altitude found, for example, in parts of East Africa. Mountains are a conspicuous feature in only a few places; such as the Loma–Nimba massif of Sierra Leone and Guinea, the Jos Plateau and the Obudu Plateau in central and eastern Nigeria respectively, and the Bamenda Highlands and Cameroun Mountain of Cameroun. It is true that less well-marked geological features such as the Gambaga, Mpraeso, and Akwapim escarpments, and the Atewa range of Ghana influence plant life to some extent but hardly to alter the overall pattern of major vegetation types. In addition one finds isolated inselbergs in many parts of West Africa, but here again, though they are well worth exploring to yield the occasional rare or unusual plant, their effect on the general pattern of vegetation is minimal. Thus for very considerable areas West Africa consists of what might be called monotonous stretches of country well below 1000 m in altitude. In the drier savanna region of the north this takes the form of gently undulating plains. In the southern forested region, where rainfall is higher, the situation is similar except that the valleys have been etched out more sharply by centuries of erosion.

The underlying geological features, of course, vary widely from place to place throughout the regions, but it is an observed fact as far as soils and the vegetation they bear are concerned that the major determining factor is the climate. Thus different types of underlying rocks, provided they are worked on for a sufficient period of time by climatic factors and provided they are at the

1

same altitude, eventually produce a generally similar type of soil with its associated vegetation type. This is not to say of course that all soils of West Africa have reached such maturity: until they have attained a state of equilibrium with the climate and vegetation soils may exert a disproportionate effect on the vegetation. Thus, for example, a sandy soil with low water-holding capacity will not have the same capacity to bear vegetation as a loamy soil in a region of equivalent rainfall.

So in West Africa with its relatively low-lying terrain the zones of vegetation have come to largely reflect the basic climatic zones. This has resulted in a series of vegetation zones that run in roughly parallel bands starting at the southern Guinea coast, where rainfall is high and well distributed throughout the year, through zones of increasingly drier vegetation until the Sahara Desert is reached in the north.

II. The forest zone

The forest zone characterized by tall trees extends inwards from the coast for up to roughly 300 km and corresponds to the climatic zone where there is a two-peak or double maxima annual rainfall distribution and where the total amount of rainfall in one year is not much below about 1250 mm, though Richards (1952) and Longman and Jeník (1974) put the minimum figure for genuine rain forest at about 2000 mm. This forest zone is not by any means homogeneous, however. In areas where soils are brackish and waterlogged vast tracts are covered by mangrove forest. This applies especially to the Niger delta area. Silt brought down by the rivers and deposited at their mouths is colonized by species of mangrove plants which stabilize the surface even though it is daily flooded by tides or seasonally flooded by fresh or brackish water (Keay, 1959). Mangrove trees, though superficially of similar form, have evolved from a number of unrelated families but are alike in their ability to exploit conditions that would be completely unsuitable for most other kinds of plants. The stabilization which they initiate, however, eventually leads to their own downfall as they moderate the originally harsh environment sufficiently for other less well-adapted plants to compete successfully with them and take over the territory they had previously won. Given sufficient time for the soil level to build up from accumulated silt and organic debris, mangrove swamp turns firstly into freshwater swamp forest and eventually into ordinary lowlands forest indistinguishable from that which occurs over solid earth further inland.

The mangrove swamps of the Niger delta grow in an area where they always maintain an open connection through the rivers with the sea, but there are other mangrove areas which surround lagoons that become completely cut off from the sea for greater or lesser periods of the year. This tends to occur in coastal areas with lower rainfall which would not be able to support forest and where, as a result, the change with time would be from mangrove to freshwater marsh rather than freshwater swamp forest and would eventually lead to a type of coastal savanna (Boughey, 1957).

Anomalous low-rainfall areas exist along parts of the West African coast. Though these areas lie within the region normally regarded as forest zone the small amount of water available is not enough to support forest (but see Lieberman, 1982). In these places it is replaced by a type of coastal savanna or thicket vegetation. One of the best known of these areas is the Accra plains in Ghana which has an annual rainfall even less than that of the Guinea savanna zone but better distributed as it still lies within the double maxima rainfall zone (Lawson, 1966; see also Kunkel, 1964).

The forest zone may best be regarded as being comprised of a number of subzones (Taylor, 1960; Hall and Swaine, 1976). Towards the south and in the higher rainfall areas where the dry season is short and does not exert much effect, the vegetation can be described as evergreen forest. Such forest where the soils are heavily leached and the trees can grow continuously without much interruption from climatic restraints may be regarded as 'true' rain forest and can be identified by a characteristic assemblage of species. But in West Africa this rain forest occupies a relatively small part of the forest zone. The much larger area to the north of this forest experiences a marked dry season to which many of its trees respond by shedding their leaves. For this reason this drier type of forest which has a less leached and therefore richer type of soil is often referred to as moist semi-deciduous forest or by some such similar term. It may in turn be subdivided into a number of subzones lying between the evergreen forest and the savanna, depending on rainfall and its distribution.

III. Savanna

North of the forest zone the vegetation has traditionally been described as savanna, that is, a type of vegetation which includes trees of a smaller size than the forest trees, and sufficiently widely spaced to allow a grass cover to develop beneath them. Such a definition is wide enough to include all the kinds of vegetation falling within the 2000 km or so of territory that lies between the forest and the desert. It is possible to divide the savanna into a number of zones, but because they tend to intergrade in each other, their boundaries are much less distinct than the boundary, say, between the forest and savanna. For this reason the maps that have been produced of the savanna zones are many and varied. Broadly speaking, however, three main zones have traditionally been accepted in West Africa. These are, from south to north the Guinea, Sudan, and Sahel zones (Chevalier, 1900). There are many other ways of classifying savanna: it can be subdivided on the basis of the physiognomy of the vegetation (see Hopkins, 1965) or on other criteria (see Chapter 5), but from an ecological point of view there is much to be said for following the traditional classification.

To begin with, then, the Guinea zone lying north of the forest will be considered. This lies in a wide band where the average annual rainfall ranges from about 900 mm to 1250 mm and where it falls largely in a single peak when plotted graphically or as a histogram with the months of the year as the horizontal axis and the amount precipitated each month as the vertical axis. There is thus a fairly long dry season running from about October to May, i.e. about eight months, and this is the limiting factor as far as the vegetation is concerned. Only those trees that can survive this drought period can grow in such savanna and the ones that maintain themselves there are variously adapted to do so. The major adaptation is that almost all of them become deciduous during that period. Loss of leaves cuts down transpiration and therefore conserves water but it also, of course, cuts out photosynthesis and therefore reduces growth to a minimum during the dry season. Such intermittent growth means that these trees also have to be adapted to build up a store of sufficient energy in the wet season to enable them to metabolize, albeit at a much reduced rate, during the dry season and also to produce a new flush of leaves at the beginning of the next wet season before photosynthesis can recommence.

The more stringent conditions of life in the Guinea zone compared with those of the forest mean that the trees are smaller and more widely spaced. This in turn means that the canopy, even during the best growth period, is never a continuous one and the light it allows to pass provides the conditions for the development of what is the main difference between forest and savanna, namely a more or less continuous layer of grass below the trees. These Guinea savanna grasses are tussocky and grow rapidly, producing tall stems and leaves during the wet season and finally flowers and fruits at its close. It is not until the erect parts have died and become dehydrated during the dry season, however, that they exert their maximum effect on the rest of the vegetation. In this condition they form a mass of readily ignitable material. The bush fires caused by hunters and farmers represent one of the most important events in the annual cycle of savanna, for it is this feature that ensures that only those plants that have fire-resistant properties can survive. In the case of the trees this usually means thick bark and a considerable development of underground parts from which new suckers can arise in the event of the destruction of these parts above ground. The survival of the herbaceous vegetation also depends on well-developed underground organs and the geophytic habit is therefore widespread in this region.

That part of the savanna that lies immediately north of the forest is sometimes spoken of as derived savanna, the supposition being that the area was originally forest. It is believed that the opening up of the forest canopy for farming allowed grasses to colonize the soil surface, following which fires occurred destroying more of the forest species and making way for the invasion of savanna trees to replace the forest ones. In fact this area consists of a patchwork of plots of forest and savanna standing side by side and is sometimes referred to as forest–savanna mosaic.

One curious fact about the Guinea zone is that compared to both the forest lying to the south and the Sudan zone to the north it is much less populated by human beings and therefore less disturbed by agriculture and the other activities of man. Why this should be so is not immediately clear. Various suggestions have been made. One is that this is the zone of tsetse flies which interfere with agriculture at least as far as cattle-raising is concerned. Another is that the soils are poor (Hopkins, 1965). Again it has been pointed out that the population may have been much reduced during the slave-trading era and that the area has never recovered from this. It has also been suggested that the north was developed by its contacts, trading and otherwise, from across the Sahara and the south was developed by its contacts with sea traders from Europe leaving the central zone underdeveloped. On the face of it one would expect agriculture to flourish more in the Guinea zone with its higher rainfall than in the drier northern regions. On the other hand, of course, these better conditions for plant growth could have made agriculture more rather than less difficult. As already mentioned, the Guinea zone is characterized by a profusion of growth by small woody plants many of which possess extensive underground systems with remarkable powers of regeneration. The control of such growth and the battle against weeds may have been too much for farmers with simple implements to cope with. Perhaps the relatively low population of the Guinea zone is due to a combination of several such factors.

The Sudan zone, then, tends to be much more disturbed by human activities. Though the trees in undisturbed Sudan savanna are more widely spaced than in Guinea savanna to start with, their numbers are further reduced by farming until they appear to be very scattered indeed. In addition their destruction is not random but selective. Trees that have some economic value are often left to grow while the less useful are destroyed. Thus such species as *Parkia clappertoniania*, the locust bean tree, whose fruits may be eaten, *Butryrospermum paradoxum*, the shea-butter tree, *Borassus aethiopicum*, and *Acacia albida*, which is notable for retaining its leaves during the dry season and therefore is a valuable source of green matter for cattle during that period, are left to grow into large trees and grace the otherwise denuded landscape. Other useful trees such as the baobab, *Adansonia digitata* and the date-palm, *Phoenix dactylifera* may be planted. Especially around the larger towns the exploitation of woody vegetation to provide firewood is intense. Around Zaria, in northern Nigeria, for instance, the silk-cotton trees (*Ceiba pentandra*) have been lopped so systematically that they have acquired a striking and grotesque growth form quite unlike that of untouched trees of the same species. In parts of the Sudan savanna the characteristic doum-palm *Hyphene thebaica* is common and is easily recognized by its branched habit. The grasses are shorter than those of Guinea savanna and do not attain the luxuriant growth of those in that zone.

The annual average rainfall in the Sudan zone is about 600–900 mm which is an appreciable amount, but it is the long dry season of six to eight and a half months that is critical for the vegetation, and the plants that grow in this region have to be well adapted to such conditions to be able to survive.

Even more rigorous conditions obtain in the Sahel zone where the annual average rainfall is about 250–600 mm, but where the dry season may extend from eight and a half to as much as eleven months. Whereas the trees of the Sudan savanna consisted of a mixture of broad-leaved types, such as predominate in the Guinea savanna, and fine-leaved types, it is the latter that form the bulk of Sahel vegetation, small species of *Acacia* are especially abundant and, as these plants usually have stipules modified into sharp thorns, Sahel is often given the name thorn scrub. The grass cover is low and sparse so that even when dried out completely it does not have enough mass and proximity to provide fuel for bush fires. Clearly, then,Sahel species do not have the same need to be adapted to fire resistance as those found further south.

Within all types of savanna are areas that may, because of specially favourable soil conditions, support a type of vegetation that is not characteristic of the zone. This is especially evident along the banks of watercourses in savanna where riparian or riverain forest may occur due to the exceptionally good soil-moisture conditions (Fig. 1.2). The map shown in Fig. 1.1. is from the most recent and authoritative map of African vegetation, namely the *UNESCO/AETFAT/UNSO Vegetation Map of West Africa* (White, 1983). The classification of savanna vegetation given is not quite the traditional one (Chevalier, 1900). In particular, whereas the zone labelled undifferentiated woodland Sudanian (No. 29) corresponds to the traditional Sudan zone, some authors have formerly described the area labelled Sudanian woodland with abundant *Isoberlinia* (No. 27) as northern Guinea savanna. Again the zone labelled mosaic of lowland rain forest and secondary grassland — Guinea – Congolian (No. 11) corresponds to their southern Guinea savanna including perhaps the derived savanna. Thus the Sudan zone is regarded as a wider zone and the Guinea zone as a somewhat narrower zone than in the traditional view as followed here.

IV.   Coastal vegetation

Though, as has been indicated above, the savanna zones lie north of the forest there are some areas along the coast and south of the forest that have been sometimes loosely described as 'savanna' because they consist of a mixture of grass and woody species. One such well-known area is the Accra plains of Ghana where the vegetation consists of short grass and clumps of thicket

Fig 1.2   Riparian forest in Guinea savanna along the banks of the River Volta in Ghana.

associated with old termite mounds. It has been described as 'steppe' by some authors. The main reason why this area is not forest is because it has an exceptionally low rainfall with an annual average not greater than that of the Sudan zone though better distributed as it is of the two-peak type. The balance between thicket and grass appears to be maintained by the bush fires that sweep the area annually during the dry season. In the West this Accra plains vegetation grades off into continuous thicket, in the north to dry forest, and in the east it joins the Guinea savanna where it breaks through the Volta gap.

Other types of coastal grassland are found in areas of much higher rainfall as in western Ghana and Ivory Coast. This type of grassland is interspersed by patches of forest often dominated by oil-palm (Fig. 1.3). Here edaphic factors appear to be in control and the absence of continuous forest is because of periodic waterlogging of the soil.

Whatever the main type of vegetation along the coast, forest, savanna, thicket and grassland, or mangrove swamp, there is usually a narrow strip of characteristic maritime vegetation immediately adjacent to the sea. This is in the area where there are strong coastal winds from the sea often bearing destructive salt, and where the soil is so sandy that it lacks stability and any water-holding capacity. Such stressful conditions mean that only a handful of highly adapted plants are able to endure in such an environment. Similar conditions are found on shores throughout the tropics, and as the plants that can grow in them are often distriubted in sea-water currents there is a tendency for them to be very widespread and for the vegetation they comprise to be pan-tropical in distribution. The most extreme conditions, of course, are found at the very outermost parts of the shore. Even a few tens of metres inland there is more shelter, less salt deposition, and a better and more stable soil so that a wider variety of less tolerant plants is found, including some shrubby species.

Between and below tide marks in West Africa in areas where mangroves do not occur only marine algal vegetation is found, and that is restricted to rocky areas that provide a good substratum for algal attachment.

Fig. 1.3   Patches of forest with abundant oil palms in edaphic grassland near Axim in south-western Ghana.

V. Desert

Below about 250 mm of rainfall per annum even the dry type of savanna vegetation described earlier cannot maintain a foothold and the result is semi-desert or desert. In such areas the plant life is very restricted indeed and reaches its best development only in the dried beds of rivers or wadis where there is the possibility of underground water. However, the sparsity of vegetation on the surface is somewhat deceptive as many of the persistent drought-resistant plants have very extensive root systems which may be in competition with each other below the surface.

As well as the specially adapted plants that endure the almost unending drought conditions of the desert there are others that evade the dry conditions completely and exploit the very short wet season to the full by completing their life cycles entirely during that period. After rain commences such plants germinate, grow to maturity, and produce flowers and seeds within a few weeks. They have no special adaptations to resist desiccation, relying on their very fast growth rates to beat the time limit of the wet season. On the return of dry conditions they persist as seeds until the next rains. Such short-lived annuals as these may also succeed as weeds in other areas much further south, an example being the small star-grass *Dactyloctenium aegyptium*.

VI. Freshwater vegetation

Brief mention must also be made here of the vegetation found in bodies of fresh water such as lakes and rivers. Microscopic phytoplankton are found in all such bodies, but larger forms of plant life reach their best development in those with relatively static conditions. All lakes and ponds, however large, are temporary phenomena tending to fill up gradually with silt, debris, and organic matter until they are converted into dry land. Plants play an important role in this process. In deep water only floating plants such as water lettuce can exist, but if enough light can penetrate, submerged aquatic plants rooted on the bottom will also grow. Such plants slow the water currents and cause them to drop their carried particles which, together with the organic matter produced by the plants themselves, help to raise the level of the bottom. In the shallower areas thus produced, plants rooted under water but with aerial parts that break surface are common. The zones formed by such plants as those mentioned above round the margins of lakes represent a temporal succession in space, for as the lake gradually fills in they move towards the centre. Eventually woody plants are able to establish in the waterlogged soil, resulting, in areas of high rainfall, in swamp forest, and ultimately rain forest. In areas of lower rainfall the characteristic type of vegetation cover appropriate to the area will be produced in time on the land thus reclaimed.

VII.   Montane vegetation

By far the greater part of West Africa is covered by lowland vegetation, but there are some areas of high land where the lowland vegetation gives way to a characteristic montane type of vegetation. It is only above about 1000 m that the changed climatic conditions are sufficient to modify the vegetation. In forested areas the zone between about 1000 and 2000 m is occupied by a type of forest similar to the lowland forest, but with some modifications and may therefore be conveniently referred to as submontane rain forest. Because of the heavy mists and consequent high humidity that envelop this forest for long periods it is called 'mist forest' by some authors such as Keay (1959) for example. Such conditions allow for the development of a rich epiphytic flora and the trees are thickly covered by mosses and also by flowering plants such as orchids and begonias. Lianes are also common. The canopy of tall trees is often somewhat discontinuous, and in such places where it is broken, and especially near streams, tree ferns may be locally abundant. Good examples of such forest may be seen, for example, on the Obudu plateau on the extreme eastern border of Nigeria and adjacent to the Bamenda Highlands of Came-roun. On the the Obudu Plateau such forest tends to cling to the valleys and gives way abruptly to grassland on the more open but gently rolling hills on the summit of the plateau. A sharp line between forest and grassland has sometimes been attributed to the fire factor but at such altitudes, wind is also probably important. High wind may cause extreme desiccation and the trees that are vulnerable to this factor are restricted to sheltered localities provided by the cuttings etched by streams where the improved soil-moisture conditions are additionally, of course, in their favour.

The highest land in West Africa is represented by the volcanic Cameroun Mountain, and Keay (1959) has pointed out that above the submontane forest mentioned above, and at heights varying between about 2600 and 3400 m, varying with aspect and slope, there is a zone of true montane forest quite distinct from the lowland forest. In structure this forest is more stunted and the canopy is represented by two storeys rather than three. The drier nature of this forest, whose characteristic trees include *Syzygium staudtii* and species of *Schefflera*, is indicated by the fact that though epiphytes are still common they consist mainly of drought-tolerant lichens rather than mosses. Such forest is much more likely to be swept by fire, and at higher levels where there are tussocky grasses such as *Andropogon distachyus* between the trees, condi-tions are much more akin to savanna than forest.

Above about 3000 m, the height varying with local conditions, the trees disappear completely and only a few small shrubs such as *Senecio* represent the woody vegetation. Here the grass is shorter and forms more compact tufts with narrower rolled leaves. The genera include *Festuca* and *Deschampsia* which are well known in temperate countries, emphasizing the fact that the zonation on tropical mountains parallels in a miniature way the vegetation zones from the equator to the poles. The summit of Cameroun Mountain at about 4056 m presents conditions near the limit for any plant life, as evidenced by the fact that much of the volcanic ash is completely bare of vegetation and only patches of mosses and lichens together with a short grass *Pentaschistis mannii* retain a precarious foothold.

The overall view presented in this chapter has, it is hoped, served to put West African vegetation in perspective: the chapters that follow will deal with each individual type of vegetation in the much greater detail that they deserve and with the ecological factors that have been briefly mentioned.

References

Boughey, A. S. (1957). Ecological studies of tropical coastlines. 1. The Gold Coast, West Africa. *J. Ecol.*, **45**, 665–687.

Chevalier, A. (1900). Les zones et ces provinces botanique de C.A.O.F. *C.R. Acad. Sci. Paris*, **130**, 1205–1208.

Hall, J.B., and Swaine, M. D. (1976). Classification and ecology of closed-canopy forest in Ghana. *J. Ecol.*, **64**, 913–951.

Hopkins, B. (1965). *Forest and Savanna*. Ibadan.

Keay, R. W. J., (1959). *An Outline of Nigerian Vegetation*. Federal Government Printer, Lagos.

Kunkel, G. (1964). The artificial coastal savanna of Liberia. *Geographische Zeitschrift*, **52**, 324–328.

Lawson, G. W. (1966). *Plant Life in West Africa*. Oxford University Press, London.

Lieberman, D. (1982). Seasonality and phenology in a dry tropical forest in Ghana. *J. Ecol.*, **70**, 791–806.

Longman, K. A., and Jénik, J. (1974). *Tropical Forest and its Environment*. Longmans, London.

Richards, P. W. (1952). *The Tropical Rain Forest*. Cambridge University Press, Cambridge.

Taylor, C. J., (1960). *Synecology and Silviculture in Ghana*. Thomas Nelson, Edinburgh.

White, F. (1983). *UNESCO/AETFAT/UNSO Vegetation Map of Africa*. Descriptive memoir and map. UNESCO, Paris.

Plant Ecology in West Africa
Edited by G. W. Lawson
© 1986 John Wiley & Sons Ltd

CHAPTER *2*

Plant-animal interactions

J.M. Lock
Cambridge, England

I. Introduction

The simplest kind of interaction is the eating of plants by animals. It is possible to make a distinction here between large herbivores (such as cows and elephants) which tend to eat a large number of different plant species, and completely avoid only a few, and plant-feeding (phytophagous) insects, most of which feed on one plant species, or a small range of species, and avoid all others. These two types of plant feeding will be discussed below. There are also more subtle interactions, such as pollination and seed dispersal, in which both the plant and the animal benefit; these will be discussed, as will the related question of seed predation by animals and insects. The feeding of animals on plants has effects on plant individuals, and also on plant communities. Finally, there are many more indirect effects of vegetation on the animal habitat, especially forest, where the considerable vertical extension of the plants interferes with incoming radiation (sunlight), wind flow, and rainfall to such an extent that a very wide range of animal habitats can exist in a very small area.

II. Plants as a food source

1. Feeding by large mammals

Large herbivores are limited in their ability to use plants as food by their lack of an enzyme which can break down the cellulose cell walls of plants and release their contents. Most of them depend upon symbiotic micro-organisms to break down the cellulose walls, and have a digestive system which retains the food for the relatively long period needed for the micro-organisms to work. Exceptions are those animals which feed on very soft and easily digested foods such as fruits, and the elephant, which merely eats a great deal of food and passes it through a very long gut.

It is generally known that most grazing animals do not feed randomly on all plants, but select particular species, or parts of species, from the available herbage. While this might seem a simple fact to demonstrate, it is not at all easy to assess accurately either the total amount of food consumed or its composition in terms of plant species and parts. The ideal method, which can as a rule only be used with domestic animals, is to fit an oesophageal fistula. This is a tube in the animal's neck, leading into the oesophagus, which is inserted by a relatively simple operation. This tube is so arranged that all food passing down the oesophagus can either be diverted into a collecting bag, or continue uninterrupted into the digestive system. The collected material can then be removed and sorted into species and parts of species, and analysed if appropriate. Comparison of the composition of the material eaten with the composition of the pasture allows assessment of selection by the animal. Unfortunately this method requires relatively sophisticated facilities and management, and can only be applied to domestic animals. An alternative to this is direct observation. An observer accompanies the grazing animal, and records both the species eaten and also the species which are present but which are not eaten (Field, 1970). While this may seem a simple task, few people can consistently recognize every leaf in a grass sward, particularly if a number of different species are interwoven. In West African savannas, most of the grasses grow as individual tussocks, so that assessment by this method is easier than in temperate grasslands. However, one still has to be very close to the animal to distinguish which grasses are being eaten and avoided, and even the tamest domestic animals will not tolerate an observer within a metre or so without special training. The problems are much greater with wild animals.

Another method of assessing the diet of herbivores is to identify the residual plant fragments in the faeces (Stewart, 1965). The identifiable fragments are of the outermost layer of the leaf, the cuticle, which is very resistant to digestion. Cuticle fragments can be extracted from faeces by solution of the other material with nitric acid followed by bleaching. They can then be identified under the microscope by comparison with a reference collection of slides prepared from named specimens. The method can be extremely useful. Drawbacks are many, however. In West African savannas there are usually quite large numbers of different grass species in a small area; 20–25 in 1 ha would not be unusual. This makes identification difficult because the number of possibilities is large. Allowance must also be made for the different sizes of fragments; it is not satisfactory if a single fragment is given the same weight as two of half the size. Likewise, not all grass species have cuticles which are equally resistant to digestion; this should at least be borne in mind even if correction for it is difficult. Finally, as a general rule it is not possible to distinguish different species of dicotyledons, which can form an important proportion of the diet, from cuticular characters. As in any other method, a detailed knowledge of the species in the feeding area and their relative abundance is necessary if feeding preferences are to be assessed.

2. Food preferences in large herbivores

A number of studies have shown that certain species are preferred over others by grazing animals. There are, however, large differences both between species and between different populations of the same species. Thus at Shika, Nigeria, cattle consumed readily 9 woody plant species; they occasionally consumed a further 19 species. Sheep in the same area readily consumed 8 species; however only 4 species were readily consumed by both species. The sheep occasionally consumed a further 11 species; of these 5 were also consumed by cattle either regularly or occasionally. Cattle at Katsina regularly fed on 11 woody plant species; only 4 species were regularly eaten at both Katsina and Shika, although there is no record of whether all species were available at both sites (de Leeuw, 1979).

What determines the selection of particular species by a grazing animal? This question has been discussed extensively by Arnold and Hill (1972). They show that taste and smell are clearly important, and colour may also be used. A further aspect is the 'bulkiness' of the material. Cattle and other ruminants require substantial volumes of material, and even if their requirement for protein, the usual limiting factor in diet in tropical savannas, can be satisfied by relatively small quantities of protein-rich young grass, they may seek out old and fibrous herbage to make up the necessry intake volume. Preferences between species, however, are probably caused by different smells and tastes. There are five main taste characteristics detectable by man — saltiness, sweetness, sourness, bitterness, and astringency. The last requires some explanation; it is the 'rough' feeling in the mouth given by some unripe fruits, tannin-rich wines, etc. and is caused by the precipitation of proteins by tannins in the material. Do animals taste in the same way as man? A few experiments suggest that they do, and that they detect the same five major categories of taste. Tests on the palatability of the eggs of various wild birds suggested that rats, hedgehogs, and man all found the eggs of various birds equally palatable or unpalatable, and that they were united in disliking eggs which tasted bitter to man (Bate-Smith, 1972). Again, cows appear to be able to detect a range of substances dissolved in their drinking water, these substances covering the range of the five major taste sensations.

It seems that cows, at least, can use their sense of taste to select particular grasses. Normally cows avoid the lush green grass that appears around cow-dung piles. If such grass is sprayed with sucrose, they will eat it; the attraction of the sweetness overcomes the unpleasantness of the dung patch (Marten and Donker, 1964). Tannins in high concentration can inhibit feeding;

cattle feeding on the legume *Lespedeza cuneata* in America fed on it much less when its tannin content rose from 4.8 to 12 per cent of the dry weight (Wilkins *et al.*, 1953). Various West African grasses contain aromatic compounds (terpenes) whose smell may deter cattle from feeding on them. A common introduced species in southern Ghana, *Bothriochloa bladhii*, is strongly scented and seems to be avoided by cattle, as is *Cymbopogon afronardus* in Uganda (Harrington and Thornton, 1969); all *Cymbopogon* species are aromatic, and non-African members of the genus such as *C. citratus* (lemon grass) and *C. nardus* (citronella grass) yield essential oils which are used in perfumery and elsewhere. Various poisonous plants are avoided by animals, presumably because they taste bitter (many alkaloids and cardiac glycosides are bitter), or because the animal tastes a small quantity of the plant, suffers, and avoids the plant subsequently.

The results of food preferences shown by large herbivores can now be examined. Clearly, if animals graze selectively on a few species out of the total flora of an area, then those species will tend to decline and species which are consistently avoided will tend to increase. Such changes are most likely to happen where the grazing animals belong to only a few species; in a well-developed savanna ecosystem, such as is still found in a few places in East Africa, there will be a sufficiently large number of different herbivores, with different food preferences, to ensure that a roughly equal grazing load is placed on all the plant species in the area (Lamprey, 1963). However, such a situation is now very much the exception. A much commoner situation is to have only one or two species, such as cattle and goats. The effects of preferential grazing can arise in two ways. Preferred species can be physically removed — which will obviously lead to their decline — or they may be prevented from reproducing themselves. The speed with which this causes a change in the composition of the vegetation will obviously depend upon the life span of the plants concerned; annuals may decline in a year or two unless there are large seed stocks in the soil, while trees may not decline for many years although a lack of young individuals may soon become obvious.

These changes are often obvious in West African grasslands but have not often been described. On the Obudu Plateau, south-eastern Nigeria, grazing of the natural *Hyparrhenia–Andropogon* grassland leads to the development of a grassland dominated by *Sporobolus africanus*, a densely tufted grass which is relatively unpalatable (Hall and Medlar, 1975). Close relatives of *S. africanus*, particularly *S. pyramidalis*, very often come to dominate heavily grazed grassland in many parks of Africa; in western Uganda, where hippopotamuses are the major grazers, this has been attributed to the strong roots of *S. pyramidalis* which make it difficult to uproot (Lock, 1972). Other grass species are often completely uprooted by the grazing of the hippopotamus, which detaches grass by pulling with the lips, rather than by cutting with the teeth. Poisonous species, such as *Calotropis procera* and woody species of *Solanum*, are often relatively common around villages in the Sudan and Sahel zones where grazing by sheep and goats can be very intense.

Effects on tree populations are little described from West Africa; in East Africa elephants have been frequently implicated in the destruction of woodlands in various national parts (Laws, 1970). This destruction can take place in two ways. Some trees, particularly *Acacia* spp., are pushed over bodily; the elephant then feeds on the foliage and smaller branches. At the same time the elephants tend to feed on young trees, and thus produce a very rapid vegetation change because they both remove the adult trees and also prevent regeneration. However, it would seem that regeneration is not completely prevented; rather, the young trees persist among the grass in a suppressed form, and when elephant-grazing pressure is reduced, as it has been recently in Uganda by the killing of most of the elephants, then regeneration is very rapid (Spence and Angus, 1970). Other trees, particularly species of *Terminalia*, are damaged by the elephants removing the bark. This exposes the cambium, which is normally protected by the thick bark, to the heat of fires and leads to canopy thinning. This in turn allows more grass growth and provides more fuel for fires, which become hotter and do more damage to the bark of the trees. The activities of the elephants thus initiate a chain reaction which leads to the destruction of large areas of woodland, leaving dead trees standing in the grassland (Buechner and Dawkins, 1961).

Such changes, occurring as they do in national parks, cause considerable problems. One school of thought would have us leave everything alone and allow nature to take its course, with a destruction of trees followed later by a decline in elephant numbers which might eventually allow an increase in tree numbers again. The other school of thought would crop the elephants and keep their numbers at a level where they do not affect the vegetation significantly. Essentially the course to be taken depends on the size of the area available, and the importance attached to the preservation of vegetation as well as animals. Large areas, perhaps with discrete populations of elephants in different areas, might be allowed to take their course; smaller, more specialized reserves cannot afford this luxury as they may be damaged irreparably in the process.

3. Feeding by insects

In contrast to large herbivores, insects usually feed on a very restricted range of plant species; either a single species, a number of closely related species in the same genus or family, or, less commonly, a few apparently unrelated species. For instance, the caterpillars of the citrus swallowtail butterfly (*Papilio demodocus*) feed only on plants in the family Rutaceae, which includes the cultivated species of *Citrus* such as oranges and lemons. The very beautiful blue butterflies of the genus *Iolaus* feed only on species of the genus *Tapinanthus*, common semi-woody parasites on the branches of cocoa and other trees (Carcasson, 1982). This specificity and its causes were unexplained until 1959, when Fraenkel made the suggestion that feeding specificity of insects is controlled by the so-called 'secondary plant substances'. These are compounds which may occur in plants in considerable quantities, but which do not appear to have any clear metabolic or other role. It had been considered that such substances were anomalies of metabolism, or waste products; such an explanation, however, seems unlikely, as many of these compounds are complex and require a good deal of energy for their synthesis. In general, natural selection does not permit the expenditure of energy on processes which do not have a useful end-product. Fraenkel suggested that these compounds act both as a stimulant to the feeding of a few insects, which, it is presumed, have developed the enzyme systems to deal with them, and as a deterrent to the vast majority of insects (see p. 31).

III.   Secondary plant substances and their significance

Secondary plant substances exist in enormous variety; perhaps 30 000 compounds have been characterized, including at least 6000 alkaloids, and this large number derives only from the rather small proportion of the world's flora that has been thoroughly studied. It is worth looking at some of the classes of compound which are commonly encountered. Alkaloids are perhaps the best known, because many of them are used as drugs or poisons. They are a chemically diverse group; all are basic, and all contain at least one nitrogen atom, and many are toxic or have physiological effects in man. Many have an extremely bitter taste; quinine (**1**) (from *Cinchona*) is one of the bitterest substances known. Some alkaloids, such as quinine, strychnine (**2**) (from species of *Strychnos*, particularly *S. nux-vomica*), and morphine and codeine (**3**) (from the opium poppy, *Papaver somniferum*), are important drugs when used in very small quantities; others, such as nicotine (**4**) (from tobacco, *Nicotiana tabacum*) and solanine (**5**) (from the potato, *Solanum tuberosum*) are familiar compounds which are not, however, used medicinally, although nicotine is used as an insecticide.

   The cyanogenic glycosides are an important group of secondary compounds. In their normal form they are not poisonous, but when the tissue in which they occur is damaged, an enzyme is released which splits the glycoside molecule, releasing cyanide, which is a powerful poison of cellular respiration, inhibiting terminal oxidation in the cytochrome system. The most familiar example of a plant containing a cyanogenic glycoside is cassava (*Manihot esculenta*) (Dalziel, 1937). The glycoside linamarin (**6**) occurs throughout the plant, and is particularly abundant in the young leaves, and in the tubers. In West Africa, most of the cyanide is removed during the processing of the tubers into *garri*, but from time to time imperfect processing leads to serious poisoning. Some cassava varieties, called 'sweet', have the glycoside in the tubers largely confined to the outer layers, so that most of it is removed by peeling; others, the 'bitter' varieties, have higher concentrations throughout the tuber. Such varieties are often grown where rats, pigs and porcupines are common, as they are very much less palatable than the sweet varieties.

An intriguing group of secondary plant substances are the non-protein amino acids. They exert their poisonous effect by mimicking the protein-building amino acids and interfering with protein synthesis. The seeds of *Canavalia ensiformis* (jack-bean) and *Dioclea megacarpa* contain large quantities of canavanine (**7**), an analogue of arginine (**8**) and another arginine analogue, indigospicine (**9**), is found in *Indigofera spicata*. The seeds of the common West African leguminous shrub, *Griffonia simplicifolia*, contain 6–10 per cent of their dry weight of 5-hydroxy-L-tryptophan (**10**). The abundant introduced leguminous shrub *Leucaena leucocephala* contains large quantities of mimosine (**11**) which prevents its use in large quantities as animal fodder, for which it would otherwise be very suitable. Its main effect is to make most of the hair on the animal fall out. It may interfere with the metabolism of the essential amino acids tyrosine (**12**) and phenylalanine (**13**), but this is still somewhat uncertain. These non-protein amino acids may have a dual function: as well as acting as chemical defences, they may also be valuable storage materials.

There are a few toxic plant proteins, two of which come from familiar West African plant species. The attractive black-and-red seeds of *Abrus precatorius* (crab's eyes) contain the toxic protein abrin; a single seed can be suffcient to cause death in humans. The castor oil plant, *Ricinus communis*, contains the extremely toxic protein ricin.

So far all the toxins mentioned have been nitrogen-containing, but there are also many nitrogen-free secondary plant substances. Examples are the cardiac glycosides, such as ouabain (**14**), which is extracted from species of *Acokanthera* and *Strophanthus* and used as an arrow poison in parts of East Africa. Similar substances are also contained in the leaves of members of the milkweed family (Asclepiadaceae); it will be shown later how these are used by the caterpillars which feed on these plants. Other nitrogen-free toxins are the saponins (**15**) which obtain their name from their ability to form a lather with water; their action in mammals is on the blood; they cause bursting of the red blood cells (haemolysis). *Securidaca longipedunculata* and *Balanites aegyptiaca* are savanna trees both of which contain saponins in the leaves and bark (Dalziel, 1937); both are used as soap substitutes, and both are very toxic to fish and are therefore used as fish poisons. Another fish poison of rather different structure (a flavonoid) is rotenone (**16**); it is extracted from the root of *Derris elliptica*, a forest climber of South-east Asia, and is also a powerful insecticide. A related substance, tephrosin, also a powerful fish poison, occurs in the West African legumes *Tephrosia densiflora* and *T. vogeli*; both are extremely effective fish poisons (Dalziel, 1937). Another powerful natural

Alkaloids

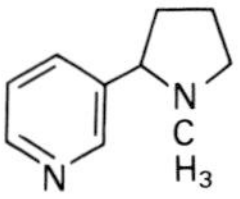

Nicotine
(**4**)

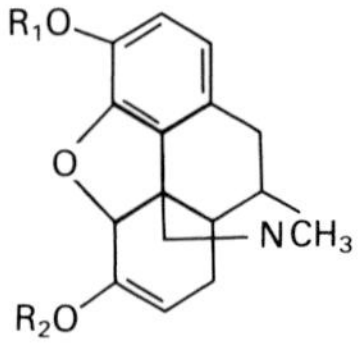

Morphine $R_1 = R_2 = H$
Codeine $R_1 = CH_3$, $R_2 = H$

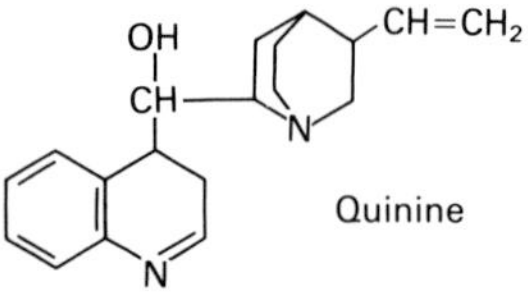

(**3**)

Quinine
(**1**)

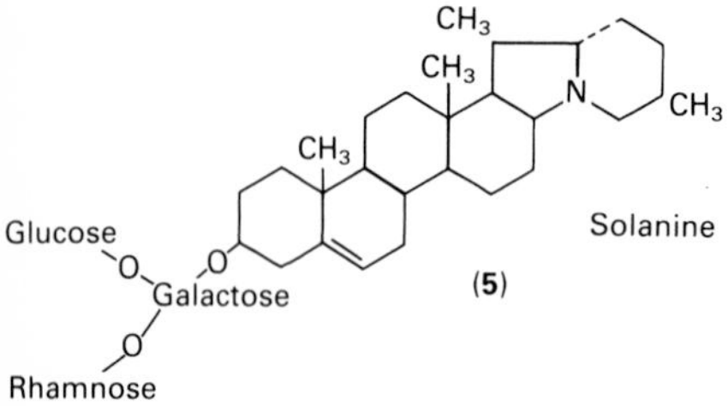

Solanine
(**5**)

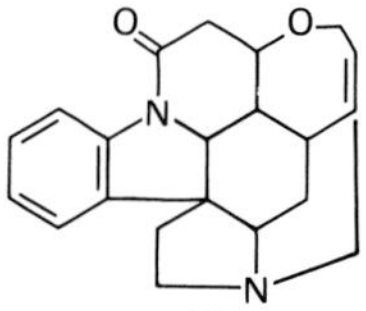

Strychnine
(**2**)

Cyanogenic glycoside

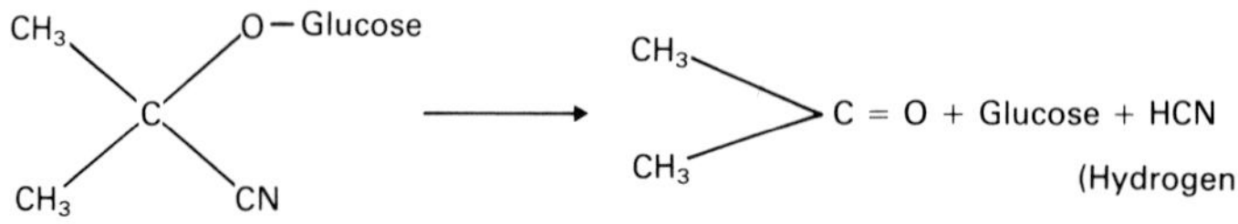

Linamarin

$$C = O + Glucose + HCN$$

(Hydrogen cyanide)

(**6**)

Non protein amino acids

$NH_2C' = NH.NHO(CH_2)_2CHNH_2CO_2H$

Canavanine
(**7**)

$NH_2C = NH.(CH_2)_4CHNH_2CO_2H$

Indigospicine
(**9**)

Protein-building amino acids

$NH_2C = NH.NH(CH_2)_3CHNH_2CO_2H$

Arginine
(**8**)

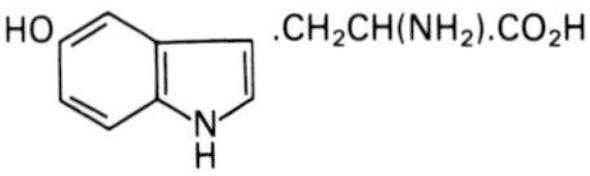

$.CH_2CH(NH_2).CO_2H$

5-hydroxy-L-tryptophan
(**10**)

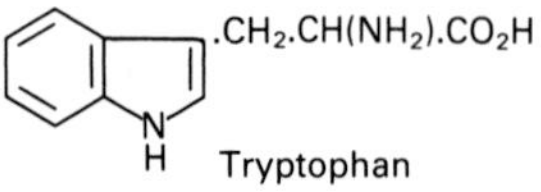

$.CH_2.CH(NH_2).CO_2H$

Tryptophan

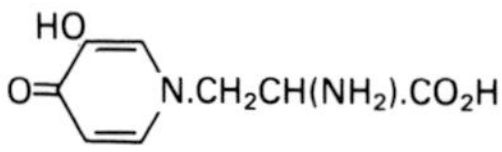

$N.CH_2CH(NH_2).CO_2H$

Mimosine
(**11**)

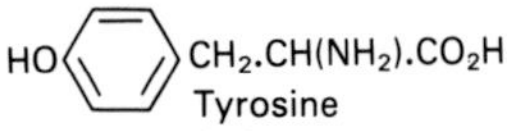

$CH_2.CH(NH_2).CO_2H$

Tyrosine
(**12**)

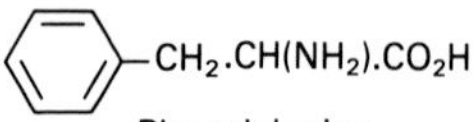

$-CH_2.CH(NH_2).CO_2H$

Phenylalanine
(**13**)

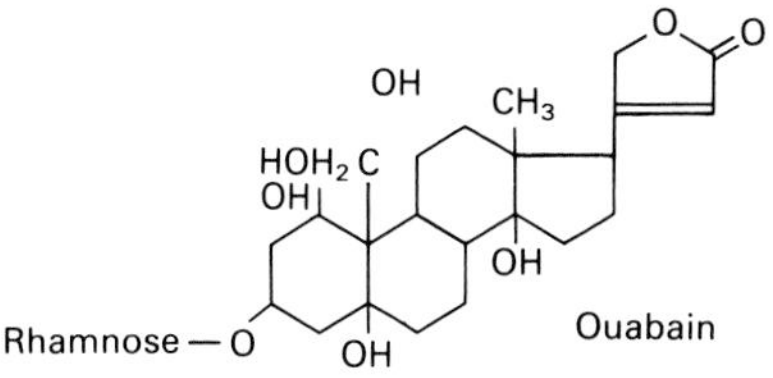

Ouabain
(14)

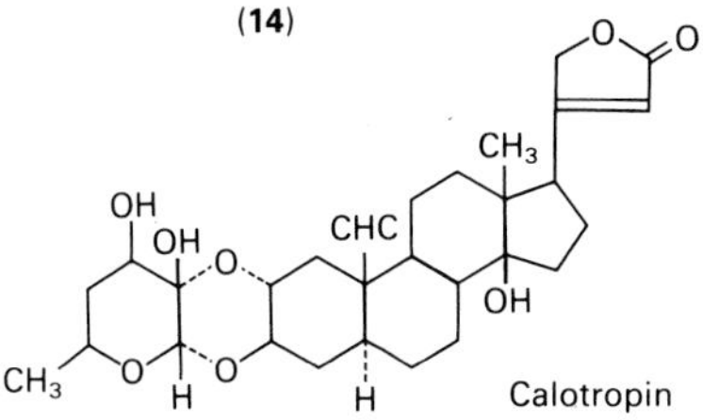

Calotropin
(18)

Cardiac glycosides

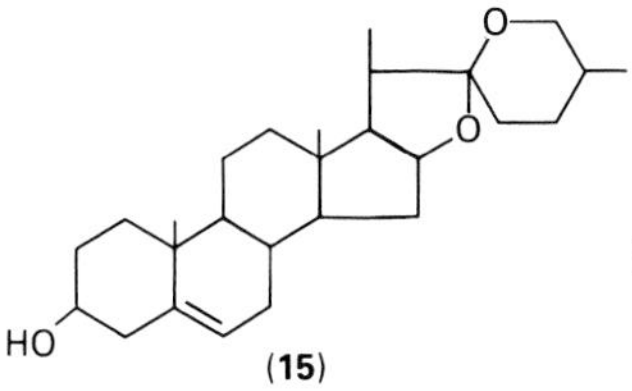

Diosgenin (a saponin)
(15)

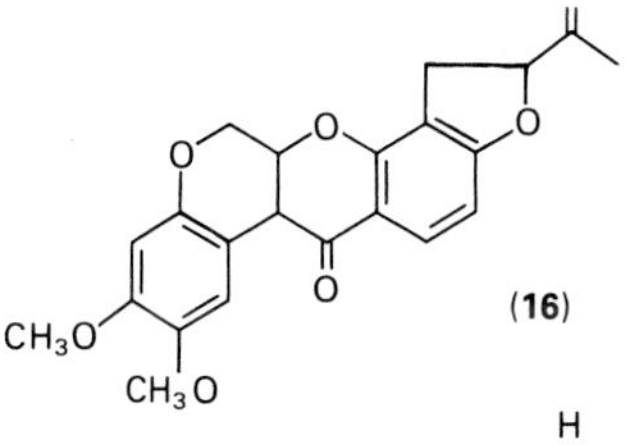

Rotenone
(16)

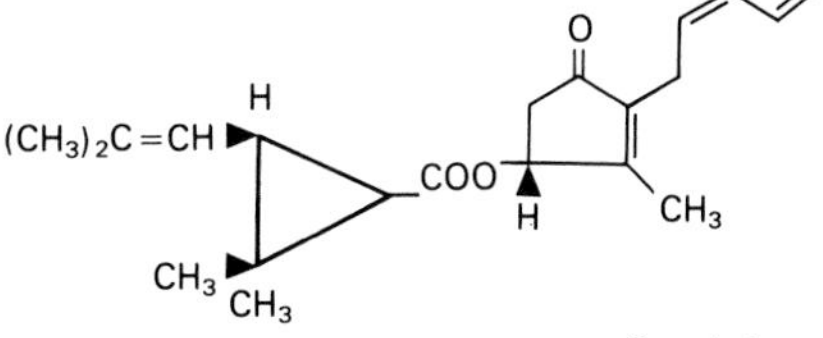

Pyrethrin
(17)

insecticide is pyrethrin (**17**), obtained from the flowers of _Chrysanthemum cinerariaefolium_ which is grown as a commercial crop in the highlands of Kenya (Purseglove, 1968).

All these plants which contain powerful poisons have their own insect pests. Tobacco is often damaged by leaf-eating caterpillars; pyrethrum is attacked by thrips (e.g. _Thrips tabaci_), and caterpillars attack both the leaves and capsules of _Ricinus_. Presumably these insects have developed enzyme systems which denature the toxins and render them innocuous. In most cases the mechanism of detoxification is unknown, but in animals which can feed on cyanogenic plants there is an enzyme called rhodanese. This detoxifies cyanide by adding a sulphur atom to give thiocyanate; the sulphur comes from β-mercaptopyruvic acid ($HSCH_2COCO_2H$) which is in turn converted to pyruvate (Jones, 1972). Habitual feeders on cyanogenic plants contain rhodanese; other species, such as man and sheep, develop higher levels of rhodanese as a result of exposure to frequent small doses of cyanide.

While some insects detoxify poisonous secondary plant substances, a few behave in a very different way, and store the poisonous compounds so that the adult insect is toxic and distasteful to predators as a result of the feeding habits of the larva. In West Africa, the larvae of the danaiid butterfly _Danaus chrysippus_ feed on various members of the family Asclepiadaceae (Edmunds, 1974), including _Calotropis procera_ which contains very toxic cardiac glycosides such as calotropin (**18**). These substances are very toxic to higher animals and also very bitter. It has been shown that birds quickly learn to avoid the adult butterflies because of their toxin content; various non-toxic butterfly species such as _Hypolimnias misippus_ are virtually identical in colour and pattern to _Danaus_ and are also believed to be avoided.

It seems likely that it is the presence of these so-called secondary plant substances which determines feeding specificity in insects. This is determined in two ways. First, the adult female insect is extremely selective about which plants are used for oviposition. She flies around the plant, repeatedly settling and palpating the surface with her proboscis; during this process she is presumably tasting and/or smelling the plant to determine its suitability. The hatched caterpillar is also very particular about its food and will usually refuse to feed on other plant species. If, by subterfuge, it is made to feed on an abnormal food plant, it will usually die, presumably because it is unable to deal with the toxins contained in that plant.

It seems likely that plants have an additional line of defence against insects which is behavioural. Many forest shrubs and trees do not grow and produce new leaves continuously, but in sudden and often synchronous flushes — periods of growth in which most of the buds on the tree grow very rapidly and produce several leaves and their supporting stem within a few weeks (Hopkins, 1970). Cocoa (_Theobroma cacao_) is a typical and familiar example, and also shows the red colours of the young leaves which may be correlated with high concentrations of phenolic compounds which act as a deterrent to leaf-eating insects. It has been suggested that if tropical forest trees grew continuously in the year-long favourable conditions that they enjoy, insect pests would be able to build up to levels at which few leaves would survive. Flushing allows the tree to complete its growth before insect populations have time to increase and take advantage of the large quantity of food (Janzen, 1975).

IV.   Special relationships between plants and ants

It has been noted how the toxins in some plants are stored by particular insects for their defence. The presence of these poisons in the plant is thus of benefit both to the plant and to the insect, but the presence of the insects cannot be considered as being of any benefit to the plant. There are, however, some cases in which insects, particularly ants, form associations with plants in which the ants receive food and shelter, while the plant receives protection from other organisms because of the aggressiveness of the ants which live on it. The best-known West African example of such an association is that between the small forest tree *Barteria fistulosa* (Passifloraceae) and the ant *Pachysima aethiops*. This has been known for a long time but was not studied in detail until 1972, by Janzen. *Barteria* is a small tree of the rain forest zone of southern Nigeria, and is found mainly in disturbed places. Young trees develop horizontal, hollow, somewhat thickened branches which are colonized by queens of *Pachysima*. The colony lives in the hollow branches, and feeds on the exudations from scale insects which the ants place in depressions on the inner surface of the hollow branches. The worker ants wander about on the *Barteria* plant, and on the ground near it. Janzen observed that they remove leaf fragments and other rubbish which falls on to the *Barteria* leaves, and that the leaves of occupied trees were relatively free of epiphyllae and rubbish compared to unoccupied trees. The ants also bite off the tips of climbers and other plants in an area 2–3 m in diameter around the stem of the *Barteria* tree; Janzen found no climbers on occupied trees and an average of 1.9 climbers on each unoccupied tree. The ants also keep the *Barteria* relatively free from insect attack, and as a result of this occupied trees are larger and leafier than unoccupied trees of the same age (Table 2.1).

At the present time insects are the major potential predators on *Barteria*, although man, who destroys forest for agriculture, is also in a sense a major predator. Janzen suggests that the major protection afforded by *Pachysima* is against browsing mammals. The ants have a very unpleasant sting whose effects are both painful and long-lasting. The ants are slow-moving and do not find insects quickly if they settle on *Barteria*, but because they are present both on the tree and on the ground around, they nearly always come quickly into contact with a large herbivore. The trees are avoided by farmers when they are clearing the bush; they are well aware of the unpleasant effects of the sting of

Table 2.1. Comparison of *Barteria* trees with and without *Pachysima* ants

	Average height (m)	Average number of leaves	Average number of branches
Heavily occupied trees	2.8	60	6.2
Unoccupied trees	2.3	13	2.2

Pachysima and it is reported that unfaithful wives were formerly punished by being tied to an occupied *Barteria* tree. It seems that the leaves are palatable, as a monkey has been seen to feed intensively on an unoccupied tree but to snatch only a quick handful from an occupied one (McKey, 1974). Although the activities of elephants in forest can produce a very favourable habitat for *Barteria*, they do not browse the tree. This association between *Barteria* and *Pachysima* thus appears to be genuinely of mutual advantage. The tree is protected from insects, large herbivores, and from competition from other plants. The ant receives food and a nesting site.

A rather similar association is found in several *Acacia* species in the savannas of East Africa (Hocking, 1970). These small trees have, like most species in the genus, paired stipular spines at the base of the leaves. In the species in question, however, the base of the spine pair is fused and greatly swollen to give a subspherical structure up to 3 cm in diameter, called an ant-gall. These are hollow, and are inhabited by ants of the genus *Crematogaster*. When the ant-gall-bearing tree is disturbed by a browsing animal, the ants rush out, attacking the intruder and producing an unpleasant smell. It seems likely that this association is also a mutually beneficial one, although it is not totally effective as giraffes can often by seen feeding on *Acacia drepanolobium*, the best-known gall *Acacia*.

Only a few plants have such highly developed associations with ants as *Acacia* and *Barteria*. However, many tropical plants have small glands on their leaves, usually on the petiole but sometimes elsewhere on the lamina, or on the stem close to the nodes. These glands secrete a sugary solution, and are called extra-floral nectaries. Ants visit them and feed on the sugar. It has been suggested that these organs, by attracting ants to the plant, increase the protection that is normally afforded to a plant by the visits of predatory ants which eat herbivorous insects. It has also been suggested that the extra-floral nectaries divert the attention of the ants from the flowers, where they are often nectar thieves, contributing nothing to pollination.

Some particularly aggressive ants, such as the tailor-ant *Oecophylla*, appear to exclude other insects from the trees where they have their colony; this has been suggested as a possible means of biological control (Majer, 1976). By encouraging *Oecophylla*, harmful insects such as capsids can be reduced in numbers, although there may be problems if the ants encourage harmful scale insects.

V.   Pollination

Although many plants are pollinated by wind, and a few by water, it is true to say that the great majority of plant species rely on some other living organism to transfer pollen from one individual to another. Pollinating organisms are usually insects, although in the tropics there are many plants which are pollinated by birds or bats. Pollination relationships are mutually beneficial; the plant gains the advantages associated with genetic interchange, while the pollinator is rewarded with food. Pollination has an extensive literature; Proctor and Yeo (1973) and Faegri and Van der Pijl (1979) provide excellent general accounts on which this brief summary draws extensively.

Bird- and bat-pollination (ornithophily and cheiropterophily) are frequent in the tropics. Only here is the climate sufficiently constant throughout the year to allow enough food, in the form of nectar, to always be available. Even so, some nectar-feeding birds undertake considerable migrations in order to be sure of a continual food supply.

In Africa, the main nectar-feeding birds are the sunbirds (Nectarinidae), although other species such as white-eyes (Zosteropidae) and some starlings (Sturnidae) are also flower visitors; sometimes, of course, it is difficult to know if a bird visits a flowering tree for the nectar, or to feed on visiting insects. Unlike the New World humming-birds (Trochilidae), sunbirds cannot hover and must perch on the flowers at which they are feeding. Birds in general have good colour vision; although this conclusion is based mainly on work with humming-birds, there is no reason to assume that it does not apply to the sunbirds as well. Their sense of smell, however, is poorly developed. Flowers regularly visited by sunbirds in West Africa are species of *Bombax* (Bombacaceae), *Erythrina* spp. (Papilionaceae), and many species of Loranthaceae. All of these have red flowers, little scent, and fairly robust flowers, or, in the cases of *Erythrina* and Loranthaceae, rather small flowers in sessile clusters which can easily be reached by a bird perched on the supporting stem. They also have copious nectar.

The role of bats as pollinators was only recognized quite recently. Much of the credit for this is due to Harris and Baker (1959) who made observations on several bat-pollinated tree species while at the University of Ghana. Bats fly at night. They have well-developed vision, and also a well-developed sense of smell. They are unable to hover, and tend to alight on the flower in order to feed. Bat flowers open at night, or, at least, produce nectar at that time. They are often pale in colour, and usually strongly scented, the smell being often rather unpleasant or overpowering to the human nose. The individual flowers are either large and strong, or grouped into firm masses which can support the weight of a bat, and resist the tearing of sharp claws. Often the flowers hang down on long stalks, or are held in dense masses clear of the main mass of the foliage; both of these arrangements make the flowers easily accessible to bats. Among species pollinated by bats in West Africa are the baobab, *Adansonia digitata* (Bombacaceae), the silk-cotton, *Ceiba pentandra* (Bombacaceae), and the species of *Parkia* (Mimosaceae). *Adansonia* and *Parkia* have pendant inflorescences, bearing a single flower in *Adansonia* but made up of thousands of tiny flowers in *Parkia*. Both *Adansonia* and *Ceiba* have white flowers; all produce copious nectar. Other bat-pollinated flowers include the introduced balsa tree *Ochroma lagopus*) and the savanna tree *Maranthes polyandra*, formerly *Parinari polyandra* (Lock and Marshall, 1976; Lack, 1978). The pollinating bats are mostly small species, particularly *Nanonycteris veldkampi* and *Micropteropus pusillus*, but the large fruit-bat *Eidolon helvum* has been recorded as visiting *Ceiba*.

The vast majority of flowers pollinated by another living organism are pollinated by insects. In order to understand the form and colour of insect-pollinated flowers, it is important first to consider the sensory abilities of insects. Human colour vision covers wavelengths ranging from about 400 nm, seen as violet-blue, to about 700 nm, seen as deep red. Not all insects have colour vision; of those which can see colours, most of those investigated show different responses to those of man. Bees have been investigated more thoroughly than any other insect; their vision extends from about 300 nm, in the ultraviolet (UV) and invisible to man, to about 650 nm, the boundary between yellow and orange. Red light is not visible to them and appears as black (unless, of course, UV light is also reflected, when it will be perceived as UV). A number of other insects, including flies and butterflies, appear to have similar ranges of colour vision. Thus, instead of the three main ranges of colour perceived by man—red, green, and blue—insects perceive yellow, blue, and UV as their main colours.

The majority of insect-pollinated flowers appear blue or yellow to us; some also appear white and these, in at least some cases, appear blue-green to bees because they cannot detect the red light in the mixture of red, blue, and green which is reflected and appears white to the human eye. We know little of the apparent colours of tropical flowers, particularly red ones, some of which are regularly visited by butterflies, particularly swallowtails (*Papilio* spp.). A flower which appears red to the human eye can appear black to an insect, if it reflects no other wavelength, or UV, if it reflects both red and UV light.

In addition to their general colour, many flowers have markings which appear to direct attention to the site where nectar is stored. These markings are usually in a contrasting colour; blue on yellow or yellow on blue. Recently, by photographing flowers with UV illumination, using UV sensitive films and lenses transparent to UV, it has been possible to show that many flowers, which are to us plain-coloured, do in fact bear markings which would be visible to a UV-sensitive insect eye (Eisner *et. al.*, 1969). Such markings have been shown to be caused by the patchy distribution of pigments, just as patterns visible to the human eye are caused.

Many flowers are scented, and there is plenty of evidence to suggest that insects are well able to detect these scents and to find flowers even if they are concealed. Scents are of many types, but most are pleasant to the human nose. Exceptions are those flowers which rely on flies for pollination; many of these produce scents which mimic those of rotting meat, or faeces. Night-flowering plants often have particularly strong scents; presumably this helps to overcome the difficulty of using sight after dark.

Thus flowers provide a range of clues which can be used by insects seeking them. Coloured flowers stand out from their green surroundings, and are often arranged in dense inflorescences which accentuate the visibility even further. Scent may attract from distances even greater than those at which a flower can be seen by an insect. When the insect reaches the flower, markings on the petals lead it to the reward, and the shape of the flower may ensure that the insect enters in a particular way.

The reward offered by a flower to its pollinator is usually nectar. This is a sugary solution secreted by glands in the flower; these may be prominent (nectaries) or they may be diffuse and difficult to see. Some flowers have a tube or sac (the spur) in which the nectar accumulates. Nectar is mostly a solution of sugars in water. The commonest nectar sugars are sucrose, fructose, and glucose. Different flower species have different mixtures; there may also be some variation between individuals and with time of day. The quantity of nectar secreted often follows a daily rhythm, although fluctuations in humidity, which affects the water status of the plant and also nectar evaporation, can make interpretation difficult. Until rather recently it was assumed that nectar only contained sugar, but Baker and Baker (1975) have shown that the nectar of many insect-pollinated flowers does in fact contain significant quantities of amino acids. These are present in greater quantities in flowers which are pollinated exclusively by butterflies and humming-birds which suck up the nectar. Flowers pollinated by bees, which can obtain nitrogen from pollen, tend to contain less amino acid. The only other substances which have been detected in nectars are lipids, which are not abundant and do not occur in all species.

Insect pollinators are of many kinds, showing differing degrees of specialization to nectar feeding. Perhaps the most specialized are the hawk-moths (*Sphingidae*). These have very long tongues — up to 15 cm in the convolvulus hawk-moth (*Herse convolvuli*), a common West African species whose caterpillars feed on *Ipomoea*. These species fly at night. Flowers on which they feed open at night, have white flowers, strong sweet scents, and nectar concealed at the base of a long tube. The insect can hover in front of a flower, feeding without alighting merely by extending the proboscis. Typical West

African hawk-moth flowers are various lilies (*Crinum* spp.), *Sansevieria, Nicotiana*, and the introduced bignoniaceous tree *Millingtonia*.

Butterfly-pollinated flowers are also common. Most have tubular flowers which are open by day, nectar concealed in a fairly short tube, and blue, yellow, or white flowers. Most are symmetrical, and are often held in a more or less vertical position so that the butterfly can easily alight on the flower or inflorescence and probe into the flowers. Typical butterfly-flowers include the species of *Mussaenda* (Rubiaceae), and many Compositae. In these, as in hawk-moth-flowers, pollen is deposited on the proboscis of the insect.

Bee-flowers are of many kinds, as there are many kinds of bees with a considerable size range and also a wide range of tongue lengths. Most bees alight on flowers to feed, and both suck up nectar and collect pollen. Bee-flowers are often asymmetrical, and held in a horizontal attitude so that the lower petal forms a landing-place for the bee. The flower is often tubular, and some force is often needed to obtain the nectar by displacing one or more of the floral parts. Such displacements often lead to the style and/or the stamens moving to come into contact with a particular part of the bee. Because of their economic importance, a good deal is known about the feeding habits of the honey-bees. Individuals are able to communicate with each other, and so many individuals from one colony can all feed at one source. Furthermore, individuals have a time sense which is well enough developed to allow them to visit the same food source at the same time on successive days. They are thus potentially very important pollinators. However, the vast majority of bees do not form colonies but are solitary. They are also important pollinators; it has been shown that in the tropics they fly very considerable distances and also follow similar routes each day, which may be important in tropical forests where individuals of a single plant species may be widely scattered (Janzen, 1971b). Many common West African flowers are visited and probably pollinated by bees, including *Crotalaria* (Leguminosae), various Labiatae, and *Aframomum* (Zingiberaceae). Some bees, in search of pollen, may visit plants which are normally wind-pollinated; thus it is not unusual to see bees visiting the male flowers of maize (*Zea mays*). While the activities of these bees may be purely destructive, it is likely that their activities lead to pollen being scattered into the air, so that it can then be carried to the stigmas.

Finally, mention should be made of two classes of flower which rely on deception for pollination. The first class of these produce unpleasant smells like those of dung or rotting meat, and are visited by flies. In some cases the flies merely visit the open flower and carry away the pollen; such is the case in *Stapelia nobilis*, a frequently cultivated succulent in West Africa. This flower is dull red in colour, like meat; flies are deceived to such an extent that they may lay eggs on the flower. A second type of flower with an unpleasant smell is found in several families: Asclepiadaceae (*Ceropegia*), Aristolochiaceae (*Aristolochia, Pararistolochia*), and Araceae (*Amorphophallus*). Here the flowers are funnel-shaped, with a narrow entrance — in Araceae a spike of flowers is surrounded by a funnel-shaped bract which serves a similar purpose. The flower produces an unpleasant smell which attracts insects. These enter the trap-like flower and have great difficulty in escaping. While inside the flower they deposit any pollen that they may be carrying, on the stigma, and pick up pollen from the anthers which usually dehisce after the stigma has ceased to be receptive. Little is known of the West African plants which have this kind of pollination mechanism; in other parts of the world it has been shown that a particular plant species is often visited by only one species of insect or a few closely related species. Is this so in West Africa?

The second class of flower which is pollinated as a result of deception are certain orchid flowers. It should be said at once that no orchids with this mechanism are known from West Africa, but the pollination mechanisms of orchids are frequently peculiar and would repay investigation, although anyone tackling such a problem should be very patient as insect visitors are often few and far between. Certain orchids of southern Europe, and another in Australia, are pollinated by male wasps which mistake the flowers for a female of the species and attempt to copulate with it. In the process they detach the pollen masses (the pollen of orchids is aggregated into two lumps, called pollinia), and transfer them to the next flower that they visit. The orchid flower achieves this apparently difficult deception by imitating the general appearance of the female insect, by having a similar, if not identical, smell, and by having floral parts which are of a particular shape and texture to stimulate the male. The deception probably only works because the male insects emerge in the spring before the females and are therefore inexperienced when deceived by the orchid flower. It has, however, been found that some species of orchid are visited by male bees in preference to females of their own species! Instances such as this, where a plant species is pollinated by one sex of a single insect species, represent the height of co-evolution in pollination relationships.

VI. Dispersal of fruits and seeds by animals

The pollination of a flower leads to the formation of seeds, and animals are often involved in the dispersal of these. Often, seeds with hooks or other adhesive mechanisms are thought of as being the only ones which are animal-dispersed, but this is a very false picture. In Ghanaian forests it was found, in most sites, that between 60 and 75 per cent of all species present had fleshy, presumably animal-dispersed, fruits (Hall and Swaine, 1981). The vast majority of seeds which are dispersed by animals are contained within fleshy fruits, or have fleshy outer seed coats or arils. These are eaten by animals, and the seeds then either pass through the animal and are deposited with the dung, or are regurgitated. Often the seeds are undamaged by this process, and their germination may even be improved, either by the removal of soluble inhibitors from the seed coat, or by the softening of hard outer coats which would otherwise prevent the entry of water into the seed.

The size of seeds that can be dispersed in this way varies from *Borassus*, with its three seeds in each fruit, each weighing about 250 g, to the tiny seeds of figs. The seeds of *Borassus* are dispersed by elephants, which eat the whole fruits and pass the seeds in their droppings, where they may often be seen germinating. Other seeds which have been observed to be dispersed by elephants include the savanna tree *Detarium senegalense*, and the forest trees *Balanites wilsoniana*, *Panda oleosa*, and *Desplatsia* spp. There are undoubtedly many more, but with the decline in elephant populations in West Africa research on this topic is now difficult, and it seems likely that trees dependent on this animal for dispersal may now depend on chance carriage of their seeds by fruit-eating birds and monkeys (Hall and Swaine, 1981). Seeds dispersed by elephants have hard coats which resist digestion in the long intestine of the elephant. Fruit-eating birds, bats and monkeys consume large quantities of the smaller size classes of fruits, and birds and bats, particularly, can be responsible for long-distance dispersal as they migrate, or make daily flights to roosts or breeding sites.

In extreme cases it has been suggested that dispersal by animals may be responsible for the pattern of vegetation in an area. Such a case is described for an island in Lake Victoria, Uganda, where large populations of vervet monkeys (*Cercopithecus aethiops*) feed on fruits (Jackson and Gartlan, 1965). The seeds are deposited in their dung, and the monkeys tend to defecate on prominent places, such as rocks and termite mounds. Small thickets are almost invariably centred on a feature of this kind, and consist of species which are eaten by the monkeys and dispersed by them. Once the thicket clumps are established, they are held in check by the fires which occur in the surrounding grassland and a balance is struck between grassland and thicket vegetation. Thus the monkeys are responsible for the establishment of new thickets, but not for the continuing vegetational structure. Grassland with thicket clumps is a vegetation type which is widespread in Africa in regions close to the equator with a low bimodal rainfall pattern. Examples are found in western Kenya, Uganda, and in the coastal regions of Ghana. Although monkeys have not been implicated in the spread of thickets in these other areas, it is true to say that virtually all the common species in thicket clumps in Uganda and Ghana have fleshy fruits which are potentially bird- or animal-dispersed.

Fruits contain considerable amounts of protein and other nutrients; however, if dispersal is to be effective, then the fruits must not be eaten before the seeds are ripe. Not surprisingly, there are very dramatic biochemical changes in fruits as they ripen. In the flesh, starches are converted to sugars, leading to a marked change in taste. Chlorophyll breaks down, and other coloured compounds come to predominate, giving the familiar yellows, reds, and blacks of ripe fruits instead of the green of unripe fruit. The astringent taste of unripe fruit, often caused by tannins and other phenolic compounds, together with the acidic component which is often dominant, gives way to sweet tastes as sugars increase. Volatile compounds which can be smelt often increase, but we know relatively little of these because they are often present in very small quantities, and sense organs are often very sensitive even to such tiny quantities. The compound giving the distinctive smell of cucumber, nona-2,6-dienal ($CH_3CH_2CH=CHCH_2CH_2CH=CHCHO$), can be detected at a concentration of 0.0001 ppm (Harborne, 1977). In some cases the compounds contributing to a distinctive flavour are present in such minute quantities and in such complex mixtures that it has not yet proved possible to identify them. This is true of both coffee and chocolate.

VII.   Seed predation by insects

While little work on the topic has been carried out in West Africa, the question of seed predation by insects should be mentioned here. Unlike mammals and birds, insects do not disperse seeds, they eat and destroy them; they are predators not dispersers. Their activities, however, are related to dispersal in the following way. Various aspects of their activities make the neighbourhood of an adult plant a dangerous place for its seeds and seedlings. In South America studies of the seedling establishment of the large climber *Dioclea megacarpa* showed that seedlings which arose close to the parent tree were much less likely to survive than those which germinated further away (Janzen, 1971a). This was attributable to effects of caterpillars, which fed on the adult and frequently fell to the ground from its crown. Once on the ground the caterpillars look for an alternative food source; if they encounter a seedling of *Dioclea* they eat the young shoot tip. This is of course fatal to the seedling, although it would not be so to the adult plant which has large numbers of shoot tips and can afford to lose a few to caterpillars.

   Extensive dispersal also means that the seeds are less likely to be consumed by seed predators. If all the seeds of a plant are deposited in a small area near the parent, the predator has no problem in moving from one to another; if they are widely dispersed, the predator has to search for each one. An extreme case of this is provided by plant introductions. A plant may be introduced but its predators are unlikely to be introduced with it. Thus *Leucaena leucocephala*, a leguminous shrub, is a native of Central America, where 95 per cent of its seed crop is destroyed by beetles (Janzen, 1971a). In West Africa, where it has been introduced as a possible forage crop (but see p. 19), it has no seed predators and produces very abundant seedlings.

   It was mentioned earlier that the production of leaves in flushes can be regarded as a way of avoiding excessive insect predation. Likewise the very irregular flowering and fruiting, often at long intervals, displayed by many forest plants, can also be regarded as a means of avoiding excessive predation on seeds. The synchronous nature of this flowering and fruiting, often over large areas, can likewise be regarded as a means of evading seed predators. The extreme case of this is shown by the bamboos, most of which flower at very long intervals of as much as scores of years (Janzen, 1976). At flowering times, all the members of a population flower and die, even members which have been transplanted to botanic gardens on the other side of the world, implying the existence of some kind of very accurate internal biological clock. It has been pointed out that in cases like this, the flowering and fruiting pattern will be sharpened by seed predation, because the seeds of a plant which flowers asynchronously are extremely unlikely to survive. Only seeds from the mass flowering, when the predators are swamped with food, have much chance of producing an adult plant.

References and Bibliography

Arnold, G. W., and Hill, J. L. (1972). Chemical factors affecting selection of food plants by ruminants. In J. B. Harborne (Ed.) *Phytochemical Ecology*. Academic Press, London and New York.

Baker, H. G. and Baker, I. (1975). Studies of nectar-constitution and pollinator-plant coevolution. In L. E. Gilbert and P. H. Raven (Eds.), *Coevolution of Animals and Plants*. University of Texas Press, Austin and London.

Bate-Smith, E. C. (1972). Attractants and repellants in higher plants. In J. B. Harborne, (Ed.), *Phytochemical Ecology*. Academic Press, London and New York.

Buechner, H. K., and Dawkins, H. C. (1961). Vegetation changes induced by elephants and fire in Murchison Falls National Park, Uganda. *Ecology*, **42**, 752–766.

Carcasson, R. H. (1982). *The Butterflies of Africa*. Collins, London.

Dalziel, J. M. (1937). *The Useful Plants of West Tropical Africa*. Crown Agents, London.

Edmunds, M. (1974). *Defence in Animals*. Longman, London

Eisner, T., Silbergeld, R. E., Aueshansky, D., Carrel, J. E., and Howland, H. C. (1969). Ultra-violet video-viewing: the television camera as an insect eye. *Science, N.Y.*, **166**, 1172–1174.

Faegri, K., and Van der Pijl, L. (1979). *The Principles of Pollination Ecology*, 3rd ed. Pergamon Press, Oxford.

Field, C. R. (1970). Observations on the food habits of tame warthog and antelope in Uganda. *East African Wildlife Journal* **8**, 1–17.

Fraenkel, G. (1959). The *raison d'être* of secondary plant substances. *Science, N.Y.*, **129**, 1466–1470.

Gilbert, L. E., and Raven, P. H., Eds. (1975). *Co-evolution of Animals and Plants*. Texas University Press, Austin and London.

Hall, J. B., and Medlar, J. A. (1975). Highland vegetation in south-eastern Nigeria and its affinities. *Vegetation*, **29**, 191–198.

Hall, J. B.. and Swaine, M. D. (1981). *Distribution and Ecology of Vascular Plants in a Tropical Rain Forest : Forest Vegetation in Ghana*. W. Junk, The Hague.

Harborne, J. B. (1972). *Phytochemical Ecology*. Academic Press, London and New York.

Harborne, J. B. (1973), *Phytochemical Methods*. Chapman & Hall, London.

Harborne, J. B. (1977). *Introduction to Ecological Biochemistry*. Academic Press, London and New York.

Harrington, G. N., and Thornton, D. D. (1969). A comparison of controlled grazing and manual hoeing as a means of reducing the incidence of *Cymbopogon afronardus* Stapf in Ankole pastures. *East African Agric. For. J.*, **35**, 154–159.

Harris, B. J., and Baker, H. G. (1959). Pollination of flowers by bats in Ghana. *Nigerian Field*, **24**, 151–59.

Hocking, B. (1970). Insect associations with the swollen-thorn acacias. *Trans. Royal Entomological Soc. London*, **122**, 211–255.

Hopkins, B. (1970). Vegetation of the Olokemeji Forest Reserve, Nigeria. VI. The plants on the forest site with special reference to their seasonal growth. *J. Ecol.*, **58**, 765–793.

Jackson, G., and Gartlan, J. S. (1965). The flora and fauna of Lolui Island, Lake Victoria. A study of vegetation, men and monkeys. *J. Ecol.* **53**, 573–597.

Janzen, D. H. (1971a). Escape of juvenile *Dioclea megacarpa* (Leguminosae) vines from predators in a deciduous tropical forest. *Amer. Nat.*, **105**, 97–112.

Janzen, D. H. (1971b). Euglossine bees as long-distance pollinators of tropical plants. *Science, N.Y.*, **171**, 203–205.

Janzen, D. H. (1972). Protection of *Barteria* (Passifloraceae) by *Pachysima* ants (Pseudomyrmecinae) in a Nigerian rain forest. *Ecology*, **53**, 885–892.

Janzen, D. H. (1975). *Ecology of Plants in the Tropics*. E. Arnold, London.

Janzen, D. H. (1976). Why bamboos wait so long to flower. *Annual Review Ecol. and Systematics*, **7**, 347–391.

Jones, D. A. (1972). Cyanogenic glycosides and their function. In J. B. Harborne (Ed.), *Phytochemical Ecology*. Academic Press, London and New York.

Lack, A. (1978). The ecology of the flowers of the savanna tree *Maranthes polyandra* and their visitors, with particular reference to bats. *J. Ecol.* **66**, 287–295.

Lamprey, H. F. (1963). Ecological separation of the large mammal species in the Tarangire Game Reserve, Tanganyika. *East African Wildlife J.*, **1**, 63–92.

Laws, R. M. (1970). Elephants as agents of habitat and landscape change in East Africa. *Oikos*, **21**, 1–15.

de Leeuw, P. N. (1979). Species preferences of domestic ruminants grazing Nigerian savanna. In S. S. Ajay and L. B. Halstead (Eds.), *Wildlife Management in Savanna Woodland*. Taylor & Francis, London.

Lock, J. M. (1972). The effects of hippopotamus grazing on grasslands. *J. Ecol.*, **60**, 445–467.

Lock, J. M., and Marshall, A. G. (1976). Probable pollination of *Parinari* polyandra by bats. *Nigerian Field*, **41**, 89–92.

McKey, D. (1974). Ant-plants: selective eating of an unoccupied *Barteria* by a colobus monkey. *Biotropica*, **6**, 269–270.

Majer, J. D. (1976). The influence of ants and ant manipulation on the cocoa farm fauna. *J. Appl. Ecol.*, **13**, 157–175.

Marten, G. C., and Donker J. D. (1964). Selective grazing induced by animal excreta. I. Evidence of occurrence and superficial remedy. *J. Dairy Sci.*, **47**, 773–776.

Proctor, M. C. F., and Yeo, P. F. (1973). *The Pollination of Flowers*. Collins, London.

Purseglove, J. W. (1968). *Tropical Crops: Diocotyledons*. Longman, London.

Richards, P. W. (1952). *The Tropical Rain Forest: An Ecological Study*. Cambridge University Press.

Rosenthal, G. A., and Janzen, D. H., Eds. (1979). *Herbivores — Their Interaction with Secondary Plant Metabolites*. Academic Press, London and New York.

Spence, D. H. N., and Angus, A. (1970). African grassland management — burning and grazing in Murchison Falls National Park, Uganda. In E. Duffey and A. S. Watt (Eds.) *The Scientific Management of Animal and Plant Communities for Conservation*. Blackwell Scientific Publications, Oxford.

Stewart, D. R. M. (1965). The epidermal characters of grasses, with special reference to East African plains species. *Bot. Jahrb. fur Systematik, Pflanzengeschichte und Pflanzengeographie*, **84**, 63–174.

Verdcourt, B., and Trump, E. C. (1969). *Common Poisonous Plants of East Africa*. Collins, London.

Watt, J. M., and Breyer-Brandwijk, M. G. (1962). *Medicinal and Poisonous Plants of Southern and Eastern Africa*. E. and S. Livingstone, Edinburgh and London.

Wilkins, H. L., Bates, R. P., Hewson, P. R., Lindahl, I. L., and Davis, R. E., (1953). Tannin and palatability in sericea lespedeza, *Lespedeza cuneata. Agronomy.* **45**, 335–336.

Plant Ecology in West Africa
Edited by G. W. Lawson
© 1986 John Wiley & Sons Ltd

CHAPTER *3*

Breeding systems of plants in relation to West African ecology

Herbert G. Baker
Botany Department, University of California,
Berkeley, California 94720, USA

I. Introduction

Although pollen grains may be carried from flower to flower by wind or water or by animals (insects, birds, bats, and non-flying mammals) (Kevan and Baker, 1983), it is important genetically and ecologically whether that pollen comes from the same plant as it lands on, or from a different plant (which will possibly be different in genetical constitution). Not only is it possible to have a wide range of sexual breeding systems from obligatory self-pollination to total cross-pollination, but there can also be the phenomenon of apomixis (the substitution of asexual reproductive processes for the usual sexual processes) which reproduces the parental genotype exactly. Consequently, the breeding systems of seed plants are very important ecologically. Continuous reassortment of the genes by outcrossing can provide adaptation to a wider range of prevailing habitat conditions and the ability to tolerate environmental changes as time passes.

Sexual reproduction by plants was recognized in West Africa before it was generally accepted in Europe. Thus, in 1700, a Dutch traveller named Willem Bosman wrote to a friend in Holland an account of his experiences in West Africa:

> There grow multitudes of papay-trees [*Carica papaya*] all along the Coast, and these are of two sorts, viz. the male and female, or at least they are here so called, on account that those named males bear no fruit but are continually full of blossoms, consisting of long white flowers; the female also bears the same blossom, though not so long, nor so numerous.

Some have observed that the females produce their fruits in greater abundance when the males grow near them; you may, Sir, believe what you please: but if you do not, I shall not charge you with heresy (Pinkerton, 1814).

The papaya is a fruit tree, probably native in Central America (Purseglove, 1968), although it is not definitely known from the wild at the present day, but it illustrates one of the outcrossing mechanisms that is found in tropical plants that are truly wild — dioecism (Baker 1976).

II. Dioecism

Dioecism (separate male and female plants) is less common as an outcrossing mechanism than self-incompatibility (the failure of seed-setting from self-pollination of a flower with both stamens and carpels), but recent work (e.g. Bawa and Opler, 1975; Bawa, 1980) has indicated that it is more common in tropical forest trees than in their temperate-zone counterparts. Lists compiled by E. W. Jones (1955) in Nigeria (at the Okomu Forest Reserve) show 40 per cent of dioecious species in the tree flora. A similar situation appears to prevail elsewhere in the tropics (e.g. Central and South America and South-east Asia).

Examples of dioecious forest tree species include members of the Ebenaceae and *Sphenocentrum jollyanum* (Menispermaceae). In the understorey of West African forests *Mallotus oppositifolius* (Euphorbiaceae) is dioecious. *Antiaris africana* (Moraceae) is variously described as dioecious or monoecious (see below).

Dioecism is also relatively frequent among wind-pollinated species, such as the date-palm, *Phoenix dactylifera*, various cycads, and *Chlorophora excelsa* (Moraceae).

Functional dioecism occurs in species that have morphologically very similar male and female flowers, so that their dioecism was not appreciated until recently. The similarity is related to the need for the same insect to be attracted to both kinds of flower (Baker, 1976).

Weeds, which are opportunists, and will often form a seed-reproducing population following the introduction of a single seed into a disturbed area, are seldom dioecious for that would require the simultaneous establishment of a male (staminate) and a female (pistillate) plant in close proximity (Baker, 1974).

Some recent studies have suggested that often dioecism has another function besides the insistence upon seed-setting by cross-pollination (Willson, 1979; Freeman et al., 1980; Baker, 1983). If the male and female trees have slightly different ecological requirements (usually greater performance in moist valley bottoms for the females, and an ability to grow in more stressful sites by the male trees), there can be a more intensive exploitation of an area by the tree species in question. It has also been suggested by K. Bawa (1980) and T. Givnish (1980) that female, as opposed to hermaphrodite trees, may produce a larger crop of fruits because no effort is put into making and dispersing pollen. Most of the trees concerned have fleshy fruits and it is postulated that animal seed-dispersers (birds, bats, and climbing mammals) will be more successfully attracted by the extra display of fruits.

III. Monoecism

Monoecism, the production of separate male and female flowers on the same plant, is less obviously an outcrossing mechanism, but it is probably true that the separate male and female flowers do lead to lessened self-pollination. Many members of the Moraceae, including the fig genus *Ficus* and the breadfruit genus *Artocarpus*, are monoecious, as are most members of the Euphorbiaceae. Maize (*Zea mays*) is an example, as are species of the sedge genus *Carex* (Cyperaceae). Once again, there is a special frequency of monoecious species in wind-pollinated plants, but this is not exclusive, for the Cucurbitaceae (e.g. *Momordica charantia*) are insect-pollinated but character-istically monoecious. In the wind-pollinated species the staminate and pistil-late flowers tend to be morphologically very distinct in their adaptations to dispersing and catching pollen, respectively (Proctor and Yeo, 1973). Howev-er, the same tendency that was noted with regard to dioecism, in which the staminate and pistillate flowers of tropical forest trees are remarkably similar in morphology, has the same explanation — that both must attract the same visiting insects. The abundantly represented family, the Meliaceae, shows this clearly (Styles, 1972).

IV. Self-incompatibility

Probably a majority of the forest trees and many of the undergrowth species have hermaphrodite flowers (containing both stamens and carpels) but this does not usually mean that they can be self-fertilized, because many of them are self-incompatible (Baker *et al.*, 1983).

There are several kinds of self-incompatibility mechanism but only a general account need be given here.

One of the first kinds of self-incompatibility to be described (by Charles Darwin in 1862) is *heterostyly*. Here there are two or three kinds of self-incompatible, cross-compatible plants in a population (Darwin, 1877).

Where there are two kinds, the population is said to show distyly. The two kinds are, respectively, long-styled and short-styled (Fig. 3.1a). Each kind of plant has a whorl of anthers at the level in the corolla tube corresponding to the stigma height in the other type. Only cross-pollinations between anthers and stigmas at the same height are fully effective in producing seeds.

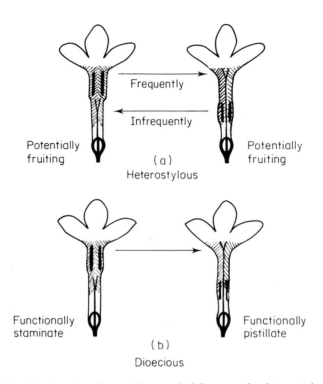

Fig. 3.1   (a). Sections through long-styled flowers of a heterostylous species of *Mussaenda* (only two stamens shown in each case for increased clarity) with extreme development of scaly hairs in the corolla tube. (b). Similar sections through functionally pistallate and staminate flowers in dioecious *Mussaenda* probably derived from a heterostylous ancestor. (From Baker, 1959).

   In the much rarer cases of tristyly there are three kinds of flowers, each containing two whorls of anthers and one of styles of a complementary length. Once again, only cross-pollination between anthers and stigmas at the same height produce a full setting of seed.
   Heterostylous species are particularly common in the Rubiaceae, where the genus *Psychotria* provides many examples and *Mussaenda* has several forest and savanna species that are distylous (Baker, 1958). Also distylous are shrubby species of *Byrsocarpus* (Connaraceae) (Baker, 1962). Some species of *Oxalis*, growing as grassland herbs, are tristylous, as is *Pemphis acidula* of the Lythraceae.
   Much more common is so-called homostylous self-incompatibility. Here there is no morphological difference between the self-incompatible members

of a population, but the physiological reaction of their pollen and styles are such that pollen tubes from cross-pollination usually grow quickly down the style to the ovules, whereas those from pollen grains from the same plant as provides the style are inhibited in their growth.

In the cocoa genus, *Theobroma* (Sterculiaceae) it has been found that incompatible pollen tubes may penetrate the style as far as the ovules before being inhibited (Cope, 1962). This delayed reaction may be unusually common in tropical forest trees according to K. Bawa (1980), who has worked mostly with forest trees in Costa Rica.

V. Herkogamy

Dioecism and self-incompatibility enforce outcrossing; monoecism encourages it. Another morphological feature of flowers may also encourage outcrossing in plants that are not dioecious or self-incompatible. This is the placing of anthers and stigmas in such relative positions that pollen is not easily conveyed from the anthers to the stigmas of the same flower. This is obviously the case in the African Glory Vine, *Gloriosa superba* (Liliaceae), where the stamens stand out widely from the corolla, as does the style. Physical contact between anthers and stigmas is out of the question. The flowers of the baobab, *Adansonia digitata* and the various species of *Hibiscus*, where the style is exserted from a staminal column, are other good examples of herkogamy. Herkogamy, however, does nothing to prevent 'cross-pollination' between separate flowers on the same plant (which is genetically equivalent to self-pollination).

VI. Dichogamy

Self-pollination may also be avoided by separation in time of the maturation of the stamens and carpels (dichogamy). If the anthers release the pollen before the stigma is receptive to pollen grains the condition is one of *protandry*, while *protogyny* is the converse situation, the stigmas being receptive before the pollen is released. Protandry is by far the more common of the two.

Strong protandry is shown, for example, by members of the Compositae (where the styles actively push out pollen from a ring of anthers and 'present' it to pollinators before becoming receptive to pollen from another source). Protandry is also to be seen in the many-flowered inflorescences of the dawa-dawa, *Parkia clappertoniana*, Mimosoideae/Leguminosae) where the ball of flowers is in a staminate condition for one night, to be followed by expansion of receptive styles on a succeeding night (Baker and Harris, 1957).

Protogyny is considerably rarer in occurrence, but it may be seen in the flowers of *Aristolochia* vines in the Aristolochiaceae. When flies are first attracted to these flowers the stigmas are receptive to the pollen that they may bring with them. For a period of time the insects are trapped in the flowers, then the stamens release their pollen coincidentally with the freeing of the insects (Faegri and van der Pijl, 1978). A similar phenomenon is to be seen in the inflorescences of various aroids (Araceae) where the separate pistillate and staminate flowers mature in sequence. Some grasses, such as *Pennisetum clandestinum* are strongly protogynous (Baker, 1978).

Both protandry and protogyny may be incomplete in the sense that there may be some overlap in the shedding of the pollen and the receptivity of the stigma. This may provide an insurance of seed-setting should cross-pollination fail to occur (provided, of course, that the plants are self-compatible).

VII. Gynodioecism

Another breeding system that may encourage the production of cross-fertilized seed while allowing at least some seed-setting by self-pollination in some plants in the population, is gynodioecism (Darwin, 1877). In this system, populations are mixtures of plants with hermaphrodite flowers along with other plants that have sterilized stamens, so that they are functionally female. The latter flowers can only set seed by cross-pollination while, provided that the hermaphrodites are self-compatible, some seed may be set by them even if because of unfavourable conditions there is a failure of cross-pollination.

Gynodioecism is to be seen in many grass genera among wind-pollinated plants and in some dicotyledonous herbs and shrubs in the families Labiatae, Verbenaceae, and Rubiaceae. It is less common among trees.

Various other combinations of hermaphrodite and unisexual flowers occur. *Solanum* species tend to show andromonoecism (hermaphrodite and male flowers on the same plant) and there are also cases of androdioecism (hermaphrodite plants and male plants) and gynomonoecism (hermaphrodite and female flowers on the same plant), but the functional significance of these arrangements is obscure (Anderson, 1979).

VIII. Autogamy

In some ecological circumstances there is selection for reliability of seed-setting rather than the production of cross-fertilized seed. This may be the case where there is a shortage of potential pollinators (rare in the tropics) or in circumstances where the rapid multiplication of plants with similar genetic constitutions (genotypes) is called for, as is the opportunistic occupation of territory by weeds (Baker, 1974).

Autogamy is the term given to the regular self-fertilization of seed plants. In autogamous species, the flowers are usually small and relatively inconspicuous; anthers and stigmas are in close proximity and the pollen that falls on the stigma is able to cause fertilization of the egg-cells in the ovules. There is unlikely to be any significant reward to flower visitors and these plants may remain unvisited. Their seeds may be small and produced in large numbers. They are capable of starting a new population (or re-creating an old one) from a single seed and, because they have had a history of self-pollination for many generations, there is no 'inbreeding depression' when such colonization takes place. Indeed, the genetical uniformity of the progeny of an autogamous weed may hasten the occupation of the site because the great majority of the plants that are produced will be as well adapted to life in that habitat as was the parental plant (Baker, 1955, 1974). The disadvantage of autogamy may come to light when environmental conditions change and new genotypes are needed.

The small flowers of the weedy *Hibiscus micranthus* contrast strikingly with the large showy flowers of such species as *H. rosa-sinensis* and *H. schizopetalus* which are self-incomptabile. *Ageratum conyzoides, Eupatorium microstemon*, and *Tridax procumbens* are weeds in the Compositae that are autogamous in West Africa. Also, some *Oldenlandia* species (Rubiaceae) and some annual *Euphorbia* species (Euphorbiaceae) are other autogamous weeds that have become widespread in the tropics (Adams and Baker, 1962).

A generalization that can be made is that seed-setting by self-pollination is least frequent in long-lived trees and becomes progessively more frequent as one moves through shrubs and perennial herbs to annuals and 'ephemerals'.

An extreme form of autogamy was first investigated scientifically by Charles Darwin (1877). This is *cleistogamy,* in which pollination takes place in a flower that never opens. The same plant may produce chasmogamous (normally opening) flowers and also cleistogamous flowers. Once again, this may be looked upon as a system in which the benefits of outcrossing may be attained by some seeds but the certainty of seed production by self-pollination also may be assured.

Cleistogamous flowers may be seen in *Portulaca oleracea* (Portulacaceae) among the weeds, *Arachis hypogaea* (Leguminosae), the peanut, among cultivated plants, and some mistletoe species of the genus *Tapinanthus* (Loranthaceae) which are parasites.

IX. Apomixis

Certainty of reproducing a well-adapted genotype can be assured if the genetical recombination that is a feature of sexual reproduction is avoided. This is apomixis, which has been defined as 'the regular substitution of asexual reproduction for sexual reproduction' (Winkler, 1908).

Some plants are apomictic because they rely almost exclusively on vegetative reproduction by bulbs, rhizomes, and stolons. Examples of this are provided by such weeds as *Oxalis corymbosa* which, although it flowers, never produces seeds. It is propagated exclusively by bulbils (small bulbs) produced on the rootstock. But in many *Allium* species the bulbils replace many or all of the flowers in the inflorescence.

Apomixis is more frequently achieved by the setting of seed without the usual process of gamete-fusion — so-called 'agamospermy' (Nygren, 1967). There are several types of agamospermy in which diploid egg-cells are produced that develop parthenogenetically (i.e. without fusion of the egg-cell with a male gamete). Many perennial tropical grasses, such as the Guinea grass, *Panicum maximum*, and other members of the tribes Panicoideae, Paniceae, and Andropogoneae are apomictic.

However, it seems that the commonest form of agamospermy in the tropics is 'adventitious embryony' (Baker, 1960). In this process, the normal sequence of sexual stages may be gone through, but soon after the embryo begins to form in the embryo-sac extra embryos, formed by division of the cells of the nucellus, are budded into the embryo-sac, so that there are several embryos competing for dominance in the developing seed. Usually the winner is one of the extra embryos which reproduces exactly the characters of the parent plant.

Polyembryony of this sort gives a useful clue to whether a plant is apomictic (by adventitious embryony) or not. For example, the mango tree (*Mangifera indica*) has varieties that reproduce sexually and those that are apomictic; the former have seeds with only one embryo, the latter have polyembryony (Purseglove, 1968).

Other examples of apomictic species are provided by the genera *Citrus* (Rutaceae) and *Bombacopsis* (Bombacaceae) (Baker, 1960).

X.   Vegetative reproduction

Reliance only upon vegetative reproduction is a form of apomixis (see above), but many perennial plants rely upon vegetative reproduction as an adjunct to reproduction by seeds.

Vegetative reproduction is important in closed communities where space for seedling establishment does not occur frequently, such as in 'closed' grassland or in marshes. But vegetative reproduction may also be a significant adaptation in unstable habitats such as strand and dune situations and in aquatic habitats. Also, in the depths of the forest there may be little light reaching the forest floor and this may militate against seedling establishment and give success to those plants that can reproduce vegetatively through drawing on a better source of nourishment than a seed. *Geophila* spp. (Rubiaceae) and *Costus* spp. (Costaceae) are forest herbs reproducing by creeping stolons. Generally the forest canopy trees do not show vegetative reproduction (Baker *et al.*, 1983).

Plants that show exaggerated vegetative reproduction are usually outcrossers and their vegetative reproduction can be looked at as an insurance that their existing genotypes will continue to exist even though they experiment with new genotypes in the seeds that they produce and disperse.

There are dangers associated with vegetative reproduction if the plants are self-incompatible or dioecious in that a particular clone may dominate an area and bear flowers that cannot be cross-pollinated with successful seed-setting. This is amply demonstrated by the forest root-parasite *Thonningia sanguinea* (Balanophoraceae) which is dioecious and exists in unisexual aggregations produced by underground vegetative reproduction. Seed-setting in this species is very rare.

XI. Evolution of breeding systems

The breeding systems of all plants provide them with adaptation to their environments. When the environment changes there may be a change in the breeding system. Thus, we believe that self-incompatibility systems break down to give self-compatibility, even autogamy, in circumstances where the rapid production of large numbers of genetically similar seeds is called for (Baker, 1955, 1958, 1965, 1974). The ancestors of the common weed *Ageratum conyzoides* were self-incompatible species in the American tropics; autogamy has given the weed the opportunity to establish populations opportunistically. Similarly, many apomictic weeds have their ancestry in outcrossing sexual plants.

On a different line, heterostylous self-incompatibility can evolve into dioecism as shown by the genus *Mussaenda* (Rubiaceae) (Baker, 1958, 1959), probably as a result of the more frequent transfer of pollen from high anthers in the short-styled form to high stigmas in the long-styled form (rather than the opposite transfer from low anthers to the short-style). This has probably come about because of the choking of the corolla tube with long hairs (Fig. 3.1b). Mutations sterilizing the low anthers in the long-styled plants and the low styles in the short-styled plants will not be selected against and will tend to accumulate in time.

XII. Conclusion

The breeding system of a species, whether it be sexual or asexual, cross-pollinating, or self-pollinating, is a part of the adaptation of that species and as such is a matter of autoecological importance; but the next task to be carried out by ecologists is to see how these breeding systems fit together in an ecosystem context.

References

Adams, C. D., and Baker, H. G. (1962). Weeds of cultivation and grazing lands. In J. B. Wills (Ed.), *Agriculture and Land Use in Ghana*, pp. 402–415. Oxford University Press, London.

Anderson, G. J. (1979). Dioecious species of hermaphrodite origin is an example of a broad convergence. *Nature*, **28**, 836–838.

Baker, H. G. (1955). Self-compatability and establishment after 'long-distance' dispersal. *Evolution*, **9**, 347–348.

Baker, H. G. (1958). Studies in the reproductive biology of West African Rubiaceae. *Jl. West African Sci. Assoc.*, **4**, 9–24.

Baker, H. G. (1959). Reproductive methods in speciation in flowering plants. *Cold Spring Harbor Symposia in Quantitative Biology*, **24**, 177–191.

Baker, H. G. (1960). Apomixis and polyembryony in *Pachira oleaginea* Decne. (Bombacaceae). *Amer. Jl. Bot.*, **47**, 296–302.

Baker, H. G. (1962). Heterostyly in the Connaraceae, with special reference to *Byrsocarpus coccineus* Schum. et Thonn. *Bot. Gazette*, **123**, 206–211.

Baker, H. G. (1965). Characteristics and modes of origin of weeds. In H. G. Baker and G. L. Stebbins, (Eds.) *The Genetics of Colonizing Species*, pp. 147–172. Academic Press, New York.

Baker, H. G. (1974). The evolution of weeds. *Ann. Rev. Ecol. and Systematics*, **5**, 1–24.

Baker, H. G. (1976). Mistake pollination as a reproductive system with special reference to the Caricaceae. In J. Burley and B. T. Styles (Eds.), *Variation: Breeding and Conservation of Tropical Forest Trees*, pp. 161–169. Academic Press, London.

Baker, H. G. (1978). Invasion and replacement in Californian and neotropical grasslands. In J. R. Wilson (Ed.), *Plant Relations in Pastures*, Ch. 24, pp. 367–384. CSIRO, East Melbourne.

Baker, H. G. (1983). Comments on the functions of dioecy in seed plants. *Amer. Naturalist* (in press).

Baker, H. G. , Bawa, K. S., Frankie, G. W., and Opler, P. A. (1983). Reproductive biology of plants in tropical forests. In F. B. Golley and H. L. Lieth (Eds.), Ecosystems of the World, Vol. 14A, *Tropical Forest Ecosystems*, Ch. 12, pp. 182–218. Elsevier, Amsterdam.

Baker, H. G., and Harris, B. J. (1957). The pollination of *Parkia* by bats and its attendant evolutionary problems. *Evolution*, **11**, 449–460.

Bawa, K. S. (1980). Evolution of dioecy in flowering plants. *Ann. Rev. Ecol. and Systematics*, **11**, 15–40.

Bawa, K. S., and Opler, P. A. (1975). Dioecism in tropical forest trees. *Evolution*, **29**, 167–179.

Cope, F. W. (1962). The mechanism of pollen incompatibility in *Theobroma cacao* L. *Heredity*, **17**, 183–195.

Darwin, C. R. (1877). *The Different Forms of Flowers on Plants of the Same Species*. John Murray, London.

Faegri, K., and van der Pijl, L. (1978). *The Principles of Pollination Ecology*, 3rd ed. Pergamon Press, Oxford.

Freeman, D. C., Harper, K. T., and Ostler, W. K. (1980). Ecology of plant dioecy in the intermountain region of western North America and California. *Oecologia*, **44**, 410–417.

Givnish, T. J. (1980). Ecological constraints on the evolution of breeding systems in seed plants: dioecy and dispersal in gymnosperms. *Evolution*, **34**, 959–972.

Jones, E. W. (1955). Ecological studies on the rain forest of southern Nigeria. IV. The plateau forest of Okomu Forest Reserve. *Jl. Ecol.*, **43**, 564–594.

Kevan, P. G., and Baker, H.G. (1983). Insects as flower visitors and pollinators. *Ann. Rev. Entomology*, **28**, 407–453.

Nygren, A. (1967). Apomixis in the angiosperms. *Encyclopedia of Plant Physiology*, **18**, 551–596. Springer-Verlag, Berlin.

Pinkerton, J. (1814). '*Bosman's Guinea.*' *A General Collection of the Best and Most Interesting Voyages and Travels in All Parts of the World*, Vol. 16, pp. 337–547. Longman, London.

Proctor, M., and Yeo, P. (1973). *The Pollination of Flowers*. Collins, London.

Purseglove, J. W. (1968). *Tropical Crops: Dicotyledons*, 2 vols. Longman, London.

Styles, B. T. (1972). The flower biology of the Meliaceae and its bearing on tree breeding. *Sylvae Genetica*, **21**, 175–182.

Willson, M. F. (1979). Sexual selection and dioecy in angiosperms. *Amer. Naturalist*, **119**, 579–583.

Winkler, H. (1908). Über Parthenogenesis und Apogamie in Pflanzenreich. *Prog. nei. Bot.*, **2**, 293–454.

Plant Ecology in West Africa
Edited by G. W. Lawson
© 1986 John Wiley & Sons Ltd

CHAPTER *4*

Forest structure and dynamics

M. D. Swaine and J. B. Hall*
Department of Plant Science, The University,
Aberdeen AB9 2UD, Scotland

I. Forest structure

1. Introduction

Forest and savanna form two easily distinguished vegetation types in West Africa. The environmental conditions prevailing in the two formations are so different that few species occur in both types (Swaine, et al., 1976). The boundary between forest and savanna is very abrupt in most places so that the distribution of forest in West Africa is clear, even on images obtained from satellites 917 km above the earth. An example of such a view of West African vegetation is shown in Fig. 4.1, which shows part of South-east Guinea. There is a clear distinction between the lighter-toned savanna vegetation to the north and the dark tones of the forest zone further south. The image was produced from information gathered by sensors on the satellite which measured reflected light in the red wavelength — darker areas on the figure representing low reflection. The scene was taken during the dry season when savanna vegetation is generally lacking in chlorophyll which absorbs strongly in red wavelengths. Thus forest, mostly composed of evergreen trees, appears dark compared with savanna. Within the forest zone, other detail is apparent; roads and towns appear white, as well as clouds; farms and farm bush are less dark than closed-canopy forest. At the lower right of the image, a fine pattern of forest fringing the rivers can be seen extending into the savanna zone.

The forest zone of West Africa was originally covered entirely by forest much like that found in forest reserves today. As the human populations in the area have grown, so the influence of man on the vegetation has increased, and today the forest zone is made up of a mosaic of different land uses, each

---

* Deceased

Fig. 4.1   LANDSAT image of part of South-east Guinea taken in the dry season in the early morning of 18 December 1973. The forest zone appears dark, savanna lighter. Closed-canopy forest (lower centre) is shown by the darkest tones. The roads on the lower left, appearing white, meet at Voinjama (v) just inside Liberia. Other towns visible as white patches are Macenta (m), Kerouane (k), Beyla (b), and Famorodougou (f), all in Guinea. Other clear features are topographic lineaments, and in the savanna (lower right), a fine dendritic pattern of fringing forest may be seen (see Chapter 5, p. 106).

representing different degrees of conversion from the original natural forest. At one end of the scale are the forest reserves, many of which must still be much as they were before agriculture became extensive. Various stages of shifting agriculture presently occupy the largest part of the forest zone. During the fallow period the forest tends to re-establish, though it does not usually recover fully before being reused in the cycle of cultivation. Areas of permanent agriculture, such as oil-palm plantations, commercial food farms, and other agricultural projects have few species in common with natural forest. Finally, roads and towns represent the extreme form of disturbance. This chapter will be concerned almost exclusively with closed-canopy forest, the kind nowadays mostly confined to forest reserves.

Moving closer from the satellite view to an aerial photograph of part of a forest reserve, it is possible to see something of the structure of the forest itself (Fig. 4.2). Individual tree crowns can be recognized, and patches of shorter vegetation are evident where the forest is regrowing in gaps caused by the death of one or more of the bigger trees. The crowns of the big trees differ in size, shape, and tone because there are many different species which can grow large enough to form part of the canopy. The crowns of some trees may be as much as 50 m across, and the trees may thus be almost as wide as they are tall.

2. Life-forms

From ground level (Fig. 4.3), the forest may seem impossibly complex — every other plant encountered seems to be a different species, and much of the foliage is inaccessible in the canopy. The undergrowth may be very dense near rivers and other places where the canopy is broken, so that it is difficult to obtain an overall view of the vegetation. As a first step towards understanding this complex assemblage of species, plants may be grouped according to their life-forms.

The predominant life-form is, of course, the tree, but it should be realized that not all the species of tree present are capable of growing big enough to form part of the upper canopy. These differences in the potential stature of woody plants were recognized by Raunkiaer (1934) in his classification of phanerophytes in four classes based on the maximum height which a species could attain. Thus megaphanerophytes exceed 30 m at maturity, and many exceed 50 m in West Africa (e.g. *Ceiba pentandra*, the silk-cotton tree); mesophanerophytes are between 8 and 30 m at maturity (e.g. *Funtumia africana* — Apocynaceae); microphanerophytes between 2 and 8 m (e.g. *Baphia pubescens* — Papilionaceae); and nanophanerophytes less than 2 m (e.g. *Pycnocoma macrophylla* — Euphorbiaceae). In this last class, the distinction between shrubs and trees is not easy, but shrubs branch at ground level. True trees which do not exceed 2 m at maturity are sometimes called pygmy trees. In any area of fully developed forest, the smaller trees will be a mixture of all phanerophyte classes at different stages of growth, and without knowing the specific identity of each tree, it will be impossible to distinguish the classes.

Another group of species, some of which can grow large enough to reach the upper canopy, are the climbers. Their stems are relatively slender (even the largest rarely exceed 10 cm in diameter) but may be very long. Specimens over 200 m long have been measured. Their stem anatomy is often quite different from that of trees, with phloem tissue penetrating the lignified xylem cylinder

Fig. 4.2. Aerial photograph of closed-canopy forest.

Fig. 4.3. Interior of moist semi-deciduous forest at Kade, Ghana; site for the work published by Lawson *et al.* (1970).

(Obaton, 1960). This gives the stem considerable flexibility and resistance to breaking and must help the plant to survive when its support falls. Tangled loops of climbers are common in the understorey of West African forest, and it is evident that a climber which falls from the canopy often grows back — hence the great stem lengths sometimes encountered.

The largest climbers are often called lianes to distinguish them from other climbing species which do not grow very tall. Many of the smaller species are typical of disturbed forest and farm regrowth, where the canopy is generally rather low, but there is a distinctive class of slender climbers that are found only in the understorey, including some which may be called root or bole climbers because they are typically found climbing the bases of forest trees, to which they are attached by their adventitious roots.

Most forest-floor herbs are small, but some members of the Zingiberaceae and Marantaceae may exceed 2 m in height (e.g. *Aframomum, Megaphrynium*). Many are typical of gaps in the forest where there is more light, but others, such as *Costus engleranus* (Zingiberaceae) and *Geophila obvallata* (Rubiaceae) are found in deep shade, and may be damaged by exposure to full sunlight. Forest vegetation has few grasses and these are of species which are never found in savanna. They typically have much broader leaves than savanna species, presumably as an adaptation to growth in low light intensity (e.g. *Leptaspis cochleata, Olyra latifolia*).

The most abundant and widespread epiphytes, occurring in all types of forest, are bryophytes and lichens. Vascular epiphytes (Johannson, 1974) include herbaceous and woody species. Ferns and orchids are the commonest herbs, and are most diverse in the wetter forest types (Sanford, 1968, 1969). Most woody epiphytes are restricted to forests of constantly high humidity, but epiphytic species of *Ficus*, of which about 50 species occur in West Africa, are exceptional in occurring in all kinds of forest and even savanna.

A few species are parasitic on other plants. *Thonningia sanguinea* (Balanophoraceae) is a widespread holoparasite which lacks chlorophyll and lives on the roots of a variety of woody species. It is visible above ground only when it flowers. Hemiparasites, which do have chlorophyll, are represented by a variety of species of Loranthaceae which live on the branches of established trees. Saprophytes obtain their nutrients solely from decaying organic matter and also lack chlorophyll. Flowering plant saprophytes are rather uncommon in West African forest, growing as inconspicuous plants on the forest floor, especially in wetter regions. Fungal saprophytes, however, are abundant in the litter and soil, and are very important in the decomposition process.

3.   Tree populations

Although the life-forms described above are to be found in any area of forest, it is the trees which largely determine the structure of the forest as a whole.

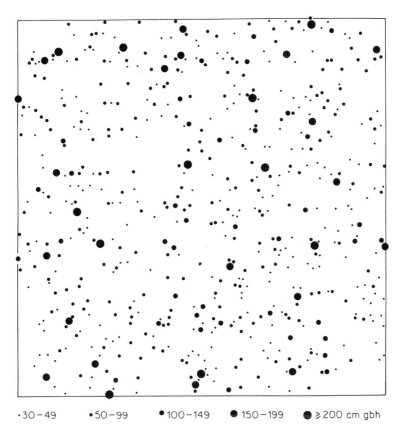

Fig. 4.4. Map of the trees >30 cm gbh (girth at breast height) on 1 ha of forest at the same site as Fig. 4.3. The dot sizes represent trees of different girth classes.

Figure 4.4 is a map of the trees with stems 30 cm or more in girth in 1 ha of forest at Kade, Ghana. The different sizes of dots represent different classes of stem size. There are relatively few big trees (megaphanerophytes) which form the upper canopy, and the intervening spaces are filled by the smaller-size classes. The distribution of the trees over the plot is more or less random, there being no evidence for aggregation of trees into clusters. This is to be expected in a stable forest where all the available space is occupied. The number of trees in each girth class in this and another nearby hectare is shown in Fig. 4.5; the exponential decline in numbers with increasing size is again typical of natural forest and represents a stable age/size distribution. It may be observed that few microphanerophytes exceed 100 cm in girth, and few mesophanerophytes exceed 200 cm. The biggest tree in the 2 ha sample measured 680 cm in girth.

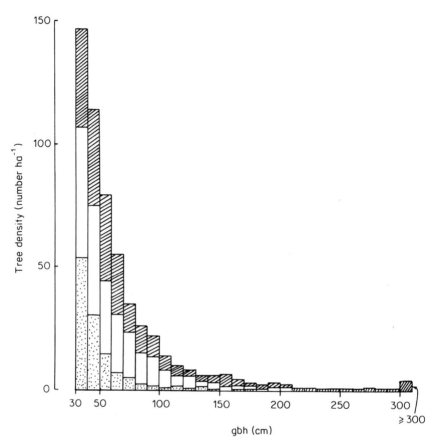

Fig. 4.5. Girth-class distribution of trees >30 cm gbh in a 2 ha sample of forest at Kade, Ghana. The columns for total trees in each class are divided up according to the contribution of micro- (stippled), meso- (open) and megaphanerophytes (hatched).

Figure 4.6 is a vertical section (profile diagram) of a small area (8 × 60 m) of forest in the Bia National Park, Ghana. The crowns of the trees are shaded according to the phanerophyte class to which the species belongs. Some authors (e.g. Richards, 1952) claim that tree crowns occupy distinct layers or strata in the canopy, whereas others (e.g. Rollet, 1974) deny the existence of stratification. Clearly, any tree must have passed through all the smaller sizes during growth to its present size, and we may expect future canopy trees to be drawn from the existing population of smaller meso- and megaphanerophytes. Stratification could therefore only be found if trees were to increase in height quickly between strata, pausing, as it were, within each stratum. Strata may be useful abstractions, but it seems unlikely that they are discontinuous.

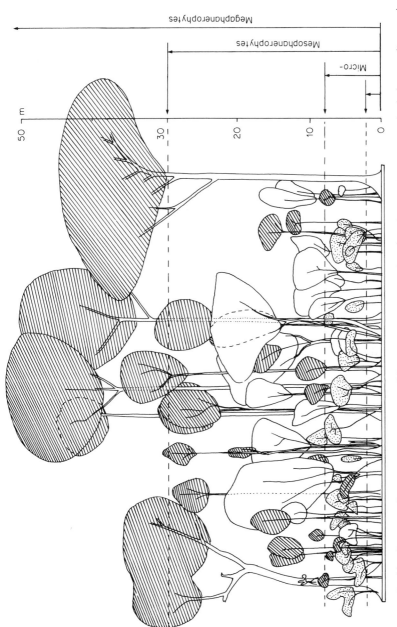

Fig. 4.6. Profile diagram of a 60 m × 8 m strip of forest in Bia National Park, Ghana. Crowns are shaded according to the phanerophyte class of the species (key as in Fig. 4.5). Only trees >3 m in height are included, so that no nanophanerophytes are shown. Note that the megaphanerophytes occur at all levels, but meso- and microphanerophytes are absent from heights above about 25 and 10 m respectively.

4. Variation within natural forest

Although the broad features of forest structure and composition as outlined above may be found in virtually all West African forest, there are variations from place to place at varying scales, especially in floristic composition. In the following sections this variation will be examined starting from the smallest scale. Microvariation comprises differences which occur over a few metres; local variation relates to topography and canopy gaps at scales of tens or hundreds of metres; zonal variation is caused by broad differences in climate and geology; with geographic variation we consider differences between remote blocks of forest under similar environment; such variation must depend largely on historical factors.

(a) Microvariation

If the species growing in adjacent small patches of tropical forest are compared many differences will be found. In Table 4.1 species lists of all vascular plants are compared for two adjacent 25 m² plots. There are relatively few species common to both samples — the plots seem to have fewer resemblances than differences. Such wide discrepancy between local patches of forest under essentially the same environmental conditions arises partly because of the natural floristic richness of the vegetation. In a hectare of forest there may be between 200 and 500 species of vascular plant (Fig. 4.7) any of which could inhabit a particular patch. Clearly, not all these species can be accommodated on a small 25 m² area, so that the actual composition is made up of a limited number of species taken from the complete list of potential occupants.

Such diversity, and the supposed fortuitous nature of its determination, is sometimes taken as evidence that the distribution of plants in tropical forest is largely determined by chance. Although chance must play a part in determining floristic composition, closer study shows that the occurrence of individual

Table 4.1 Comparison of species composition in an adjacent pair of 25 m² forest samples at Kade, Ghana. All vascular plants were recorded

	Sample	
	A	B
Number of species	46	64
Number of species confined to one sample (% sample total)	24 (52)	42 (65)
Number of species common to both samples (% total for both samples)	22 (25)	
Total number of species	88	

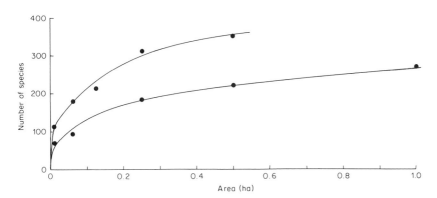

Fig. 4.7. Species–area curves for all vascular plants at two sites in Ghana: moist semi-deciduous forest at Kade, and wet evergreen forest in Neung Forest Reserve. Note the much greater diversity under higher rainfall.

plants is more often determined by environmental conditions and by the interaction between species or individuals.

The growth form of many a species imposes a pattern on its distribution in the forest. *Costus engleranus*, for example, is a small fleshy herb of the Zingiberaceae which grows in deep shade in the forest litter. It tends to form patches of 4–8 m in diameter determined by the limited growth of its rhizomes. Some understorey trees form large morphologically determined patches by the layering of their drooping branches as in *Sloetiopsis usambarensis* and *Scaphopetalum amoenum* (Jeník, 1969). These patches are sometimes so dense that many potential occupants of the area are excluded, creating local patches of very low diversity (Hall and Swaine, 1976, p. 929).

Other examples of species which influence the distribution of their neighbours include *Okoubaka aubrevillei* (Santalaceae) and *Barteria fistulosa* (Passifloraceae). *Okoubaka* is regarded throughout its range as sacred (Aubréville, 1959) because it is believed to kill surrounding trees. Our (unpublished) observations in Ghana have shown that *Okoubaka* is parasitic on the roots of other woody plants, and this may explain its ability to keep its competitors at a distance. In Nigeria and Cameroun, *Barteria fistulosa* provides a home for fierce ants which not only deter potential predators of the tree but also bite through the stems of young climbers competing with it. *Barteria* trees are thus commonly found in small openings in the forest (Janzen, 1972).

The means by which plants disperse their seeds can also introduce pattern into the distribution of individuals. The seedlings of wind-dispersed trees such as *Khaya ivorensis* (Fig. 4.8) are often asymmetrically distributed around the parent as a result of the predominance of certain wind directions. *Cynometra*

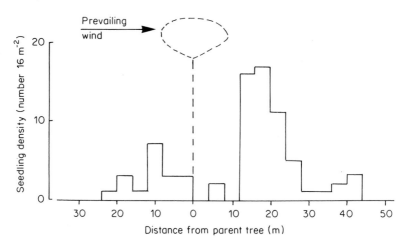

Fig. 4.8. Density of wind-dispersed *Khaya ivorensis* regeneration around a parent tree, along a transect aligned with the prevailing wind. The distribution is skewed to the lee side of the parent. Neung Forest Reserve, Ghana.

ananta and *Bussea occidentalis*, for example, have explosively dispersed seeds, which are unlikely to be dispersed more than about 50 m from the parent tree (Swaine and Beer, 1977; Swaine and Hall, 1983). Whatever the means of dispersal, many seeds will fall to the ground close to the parent trees, and it is not unusual to find large populations of a species associated with a single large parent. Further away from the parent, the species may become very rare or entirely absent. Janzen (1970), however, has proposed as a general rule that predation may be greater in the close vicinity of the parent, leading to better survival at some distance from it. The observations of Alexandre (1977) on *Turreanthus africanus* in Ivory Coast support Janzen's hypothesis.

(b) Local variation

One of the most important causes of pattern in rain forest is the distribution of light in the lower parts of the canopy. This light environment is extremely complex, and therefore difficult to describe. Light intensity varies dramatically both horizontally (Fig. 4.9) and vertically; a patch of forest floor which is brightly illuminated may be in the deepest shade a few minutes later as the sun moves across the sky. These rapid changes mean that the light conditions at any particular spot must be assessed by integrating incoming radiation over long periods. In addition to temporal and spatial variation in light intensity, the quality of light (spectral composition) is also very variable, and is likely to have profound influences on the responses of forest plants.

Despite the difficulty of accurately defining the light environment, it is clear that certain areas in the forest regularly receive more light than other areas.

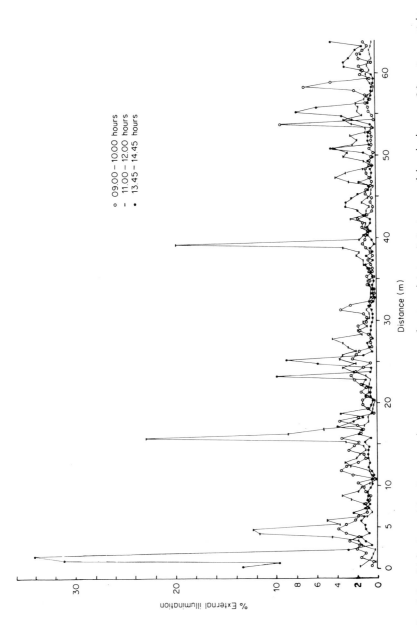

Fig. 4.9. Light intensities (expressed as a percentage of external intensity) at ground level along a 64-m transect in closed-canopy forest at Kade, Ghana. Measurements were made every 0.5 m of the light reflected 20 cm above a 25 × 25 cm white card, three times during a single day. Notice the high variation over short distances and between different times of the same day. The overall average for the three periods was 1.64 per cent of external illumination.

Even quite small breaks in the canopy are marked by patches of light-demanding ferns and marantaceous herbs. Canopy gaps are created when a tree, or part of it, dies. The increase in light in the forest below may be slight if the tree does not fall over, or if the lower canopy remains intact. If the tree falls, for example by windthrow, which is a common cause of death among large trees, it may knock over other trees and will in any case create a much larger gap. Emergent trees are commonly over 50 m in height and may have crown diameters over 30 m. Such gaps often expose considerable areas of forest floor to full sunlight and to the diurnal cycle of temperature and humidity which are so pronounced outside the protection of the forest canopy. In such gaps, seeds lying dormant in the soil may be stimulated to germinate and a patch of pioneer species may establish (see below, p. 88). Forest thus naturally consists of a mosaic of patches at different stages in the development from a gap (Fig. 4.10).

Breaks in the canopy, as we have seen, are very variable, sometimes only a single pioneer becomes established, sometimes a dense grove. Often, when a gap is formed by tree fall, a neighbouring large canopy tree may remain standing in the gap so formed, and it is not uncommon to find old emergents growing immediately adjacent to young pioneer species such as *Musanga cecropioides*. For this reason it becomes very difficult to define the limits of a gap, even immediately after it has been formed.

The frequency of secondary tree species in natural forest depends on the proportion of its area which has been disturbed by tree death within the life-span of the pioneer tree species. An abundance of secondary species in forest cannot be taken as evidence for a recent history of human clearance as is sometimes claimed, but merely of a history of frequent tree death.

Local variation in forest vegetation also arises as a result of the topography. Most forest soils in West Africa are very ancient, and over the thousands of years of their development rainfall has tended to wash minerals and clay particles down-slope towards the streams and rivers. As a result, fairly clear differences develop between soils at different elevations on the undulating topography, even on gentle slopes. These differences repeat themselves over successive valleys and hills — the system is then called a catena (Ahn, 1970).

Such changes in soil conditions cause parallel differences in the forest growing on them as is clearly demonstrated by the distinctive appearance of swamp forest, found on the lower parts of the catena. Chief among the causes for the difference between swamp forest and forest higher on the catena must be the poor aeration of the seasonally or permanently waterlogged soils. Because of the accumulation of minerals and clay particles washed down from above, these soils may, however, be relatively fertile. Swamp forest normally has few tall trees, except for stands of the characteristic swamp genus, *Mitragyna* (Rubiaceae), and the canopy is thus rather sparse, allowing a dense undergrowth to develop. Perhaps the most widespread and typical species of West African swamp forest is the palm, *Raphia hookeri*, often in association

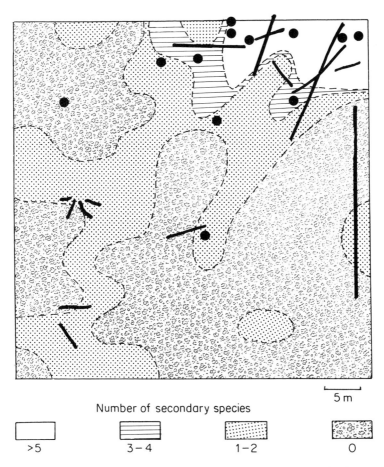

Number of secondary species

>5	3 – 4	1 – 2	0

Fig. 4.10. The pattern of tree and branch fall (black bars) on a 0.25 ha forest plot in relation to the regrowth of secondary (pioneer) species. Dots represent pioneer trees >30 cm gbh which are clearly associated with areas of higher frequency of pioneer species.

with climbing palms in the genera *Calamus* and *Ancistrophyllum*. Even when tall trees are present, as in some places where waterlogging is infrequent, the diversity of tree species is low.

On the better-drained parts of the topography, catena differences in the vegetation are more subtle. Structurally, there seems to be little to distinguish the upper parts from the lower, but individual species sometimes show clear differences in abundance along the topographic gradient (Fig. 4.11).

The variaton in forest on the freely draining parts of the catena were the subject of a detailed study by Lawson *et al.* (1970). Table 4.2 is an abstract

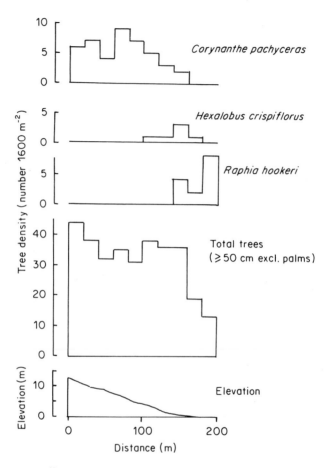

Fig. 4.11 Catena effects in Ghanaian forest: density of trees on a 200 m transect in Bia National Park, Ghana, showing species apparently confined to upper, lower and bottom positions on the catena.

from the same study when two 1 ha plots were subdivided into four blocks, and all trees of 30 cm gbh and over were recorded. One of the plots was located some 200 m lower down the slope, but still on well-drained soil. The commonest tree in the area, the upper canopy tree, *Celtis mildbraedii*, shows a clear difference in density between the two plots; *Irvingia gabonensis*, and *Carapa procera*, although not very common, are absent from the upper sample. Similar trends may be seen in the detailed study of a single valley in Ijaiye Forest Reserve, Nigeria (Chachu, 1976).

The reasons for the differences in species abundance along catenas remains to be determined, but Lawson *et al.* (1970) could find no cause in microclimate, and it is likely that soil conditions provide the controlling influence.

Table 4.2.   Density of selected tree species on eight 0.25 ha plots at different positions on a catena in forest at Kade, Ghana

Species	Catena position							
	Upper				Lower			
	1	2	3	4	5	6	7	8
Celtis mildbraedii	31	25	28	31	11	10	16	12
Carapa procera	0	0	0	0	6	1	4	8
Irvingia gabonensis	0	0	0	0	1	2	1	2
Trichilia prieuriana	8	10	5	10	2	5	4	2

(c) Zonal variation

Over distances of hundreds of kilometres, differences in climate begin to have a significant influence on forest vegetation. The climatic factor which seems to be of paramount importance is rainfall. This exerts its influence not just by varying the amount of available water but by indirect effects on the forest through the soil. Soils under high rainfall, particularly such ancient soils as those typical of West Africa, are usually strongly leached: small particles of clay minerals and organic matter, together with ions dissolved in the soil solution, are washed down through the soil profile in much the same way as they are moved across the catena. Leached soils are poor in minerals such as phosphorus, calcium, and potassium, and have rather low pH. Acid soil conditions inhibit uptake by plants of the already scarce mineral salts. It is not surprising, therfore, to find that forests on leached soils show relatively low productivity and are often of reduced stature compared with forests under moderate rainfall.

It is sometimes assumed that the climate of the wetter forest regions is less seasonal than lower rainfall climates in that there are fewer 'dry' months in the year. In fact all African forest, with the possible exception of the central Congo basin, is subject to a dry season, but higher rainfall during the rainy season means that the soil will be wetted more thoroughly and the carry-over into the dry season may thus be extended.

Variation in the underlying geology of forest land may have an important influence on the vegetation through the soil which has developed over it. In Nigeria and parts of the Ivory Coast, for example, relatively young (Tertiary) sediments form the sandy parent material for large areas of forest soils. In the Niger delta rainfall is high, but the soils, being young, are not so heavily desaturated as older soils on the Basement Complex. The gradient of forest with rainfall in this part of Nigeria is confounded by soil differences of geological origin, and a distinctive forest community can be recognized (Hall, 1977).

Table 4.3. Summary of variation in forest vegetation and environment in Ghanaian forest types. Figures are means (or ranges) of several samples. Adapted from Hall and Swaine (1976, 1981)

	Forest type[a]					
	WE	ME	MS	DS	SM[b]	SO[c]
Annual rainfall (mm)	>1750	1500–1750	1500–1750	1250–1500	1000–1250	<1000
Soil pH	4.2	4.7	5.4	5.9	5.8	6.1
Soil exchangeable bases (m-equiv. 100 g^{-1})	1	5	8	12	12	15
Tree density (number ha^{-1}>30 cm gbh)[d]	445	505	497	489	469	456
Leaf litter fall (g m^{-2} yr^{-1})	630	—	730	—	—	380
Evergreen species (% canopy trees)	81	78	65	42	52	—
Number of species (on 625 m^2)	138	120	105	75	47	25

[a] Forest types: WE, wet evergreen; ME, moist evergreen; MS, moist semi-deciduous; DS, dry semi-deciduous; SM, southern marginal; SO, south-east outlier. Forests very like the first four types given here occur widely in West Africa.
[b] An unusual forest type scarcely reported outside Ghana.
[c] Small outlying patches of forest in the savanna zone.
[d] The differences are not significant.

Forest classifications relating principally to climatic differences have been devised for Ivory Coast (Guillaumet and Adjanohoun, 1971), Nigeria (Keay, 1959), and Ghana (Table 4.3 shows trends in forest types along a rainfall gradient in Ghana). The names applied to the major forest types or formations differ, but the general features are common to all. With increasing annual rainfall there is a decline in soil fertility and an increase in acidity; an increase in canopy height and litter production, followed by a decline under the higher rainfall regimes; an increase in the proportion of evergreen species in the canopy and a steady increase in species diversity.

The productivity of forest in West Africa (as reflected by canopy height and litter production) is probably controlled by two principal factors: availability of moisture where annual rainfall is less than about 1000 mm, and soil nutrients where rainfall exceeds about 1750 mm annually.

The relative rarity of deciduous species in the wetter forests has been attributed to the supposed superior ability of evergreen species to conserve mineral nutrients. Evergreen species normally retain their leaves for longer periods, and shed them more uniformly through the year than deciduous species, and thus, it is argued, have greater control over the nutrient cycle (Monk, 1966). Unfortunately, the comparative data on leaf longevity and nutrient cycling necessary to test this hypothesis are lacking.

As yet there is no convincing hypothesis to explain why species diversity continues to increase with rainfall when soil fertility is declining.

Another form of variation in forest vegetation which arises principally from climatic differences is that found on mountains. Although high mountains are few in West Africa, the effects of increasing altitude which are so clearly demonstrated on Mount Cameroun (Hall, 1973; Richards, 1963), are also evident to a lesser extent on lower hills. Montane or submontane forest can be found in Sierra Leone (Loma), and Liberia (Nimba — Adam, 1970; Schnell, 1952), Ivory Coast (near Man), Nigeria (Obudu — Hall and Medler, 1975), and even on hills as low as 500 m (Atewa range, Ghana — Swaine and Hall, 1977).

With increasing altitude the air temperature falls (approximately 1°C for each 100 m rise in altitude), giving rise to increased mistiness, reduced sunshine, higher rainfall (with the corresponding decline in soil fertility), and reduced decomposition rates. The combination of these factors gives rise to a distinctive forest, generally of lower stature than the surrounding lowland forest, with a rich epiphyte flora and the occurrence of temperate genera such as *Rubus* as well as a notable collection of endemic species, probably because areas with similar conditions are so isolated.

In other sites the effects of increased altitude are offset by the rocky nature of the topography, leading to poor water retention in the soil. Such hills may support only savanna vegetation despite quite high rainfall. In the Volta region of Ghana the higher hills are also exposed to strong, highly desiccating harmattan winds in the dry season (Jeník and Hall, 1966).

Forest near the boundary with savanna is occasionally encroached on by fires started in adjacent savanna patches. Forest is normally regarded as relatively incombustible, but evidence from Ivory Coast (Spichiger and Pamard, 1973) and Ghana (Hall and Swaine, 1976, 1981; Swaine *et al.*, 1976) suggest that these infrequent ground fires may have a significant influence in determining the nature of such forests.

Ground fire, fuelled by leaf litter accumulated during the dry season, kills the plants in the herb layer, and most of the saplings and poles of the tree species. Larger trees, however, are rarely killed (Fig. 4.12), but the opening of the lower canopy permits many seeds in the soil seed bank to germinate and allows the invasion of other plants. Such forests commonly have a large number of pioneer or weedy plants in their species complement, and may be mistaken for old farm regrowth. New seedlings of canopy species will also benefit from the increased light levels and if they can grow large enough in time to survive the next fire, the forest will maintain itself.

These forests are also characterized by tree species known to be somewhat fire resistant (*Chlorophora excelsa*, *Afzelia africana*, *Anogeissus leiocarpus*, *Elaeis guineensis*) and by the absence of some species (e.g. *Hymenostegia*

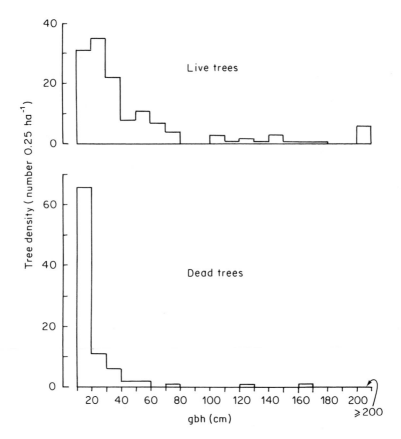

Fig. 4.12. Girth-class distributions of trees on 0.25 ha in forest recovering from a ground fire which occurred three months previously. Notice the paucity of small stems in the surviving trees (cf. Fig. 4.5), and the concentration of deaths in the smaller-size classes. Nsemre Forest Reserve, Ghana.

afzelii) which might be expected in view of the suitable climate, but whose thin bark presumably makes them unable to survive fire. *Elaeis guineensis*, the oil-palm, is regarded by Zeven (1967, in Moore, 1973) to be native to this kind of forest, and it is of interest to note that its seeds show enhanced germination following a heat treatment (Purseglove, 1972).

(d) Geographic variation

It has been pointed out that forest in West Africa varies in relation to rainfall: from Sierra Leone to Nigeria wet, or 'evergreen' forest can be distinguished from the drier 'semi-deciduous' forest. The forests under the wetter climates are, however, separated by intervening areas of lower rainfall, and it is reasonable to ask, for example, if wet forest in Nigeria is the same as wet forest in Ghana, or Ivory Coast. Climatic and soil conditions may be very similar, but do they have the same floristic composition?

The great distances between these areas of similar environment, and the intervening zones of drier forest and savanna, provide a considerable barrier against dispersal and the interchange of genetic material (see Hall and Swaine, 1981, pp. 34–37), so that it is not surprising to find that some species are found only in one of the three main blocks of wet forest. Figure 4.13(a) shows the recorded distribution of four closely related species of *Cola*, of which three are confined to one of the three blocks indicated by the 2000 mm annual rainfall isohyet (Papadakis, 1966). In contrast, *Cola chlamydantha* is found in all three blocks, but is absent from drier forest. The causes for the different distribution types exemplified in Fig. 4.13(a) are not fully understood, but past changes in the climatic zones of the region (see Chapter 11) must be considered as well as present-day environmental differences. If the total forest flora for any West African country is examined, it is found that only a small fraction (e.g. about 1 per cent in Ghana) is endemic (i.e. found only in that country), and most of these are from the more isolated wet forests with high diversity. It follows that the bulk of the flora of any country may also be found in other West African countries, so that there is much overlap in the composition of forests in different countries. Generally speaking, species of the drier forest types are more widespread than those of the wetter forests, so that comparisons between such forests are more easily made.

It would seem feasible, therefore, to provide a classification of the whole of the West African forest zone based on floristic data, but so far no one has attempted a survey on a scale large enough to make this possible. Hall and Swaine (1981) report comparisons between 155 samples in Ghanaian forest with a few similar samples in Ivory Coast and Nigeria. The great majority of the species in records from Ivory Coast and Nigeria were also recorded from Ghana, though the proportion was less in comparisons between wetter forest samples. From these and other comparisons, it is possible to identify (Swaine and Hall, 1976) in Ivory Coast, Togo, Benin, and Nigeria, convincing examples of the forest types originally defined in Ghana. Species lists acquired from the literature for Casamance in Senegal could be equated with dry semi-deciduous forest in Ghana. A tentative map of forest types based on these data for the central part of West Africa is presented in Fig. 4.13(b) (see also Waterman *et al.*, 1978). The chorology of the West African forest zone is thoroughly discussed by White (1979), and for the whole of Africa in White (1983).

II. Forest dynamics

1. *Introduction*

In the previous sections, forest structure and variation have been considered principally in spatial terms, generally avoiding detailed consideration of

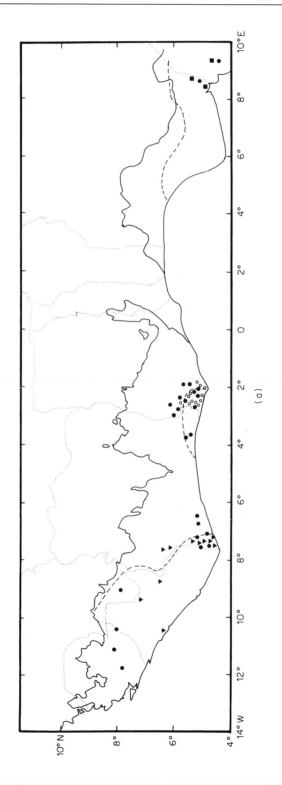

(a)

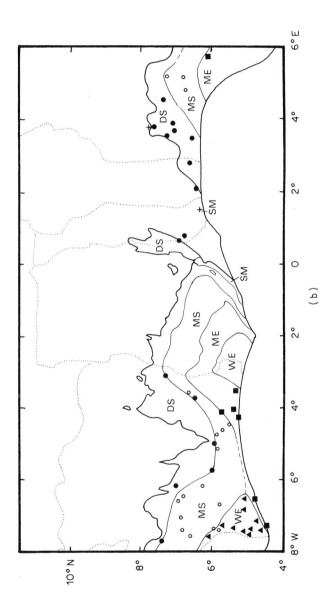

Fig. 4.13. Distribution of forest types (b), and four wet forest *Cola* species (a) in West Africa. The forest–savanna boundary, drawn from LANDSAT imagery, is shown by a heavy line. National boundaries are dotted. (a) Wetter areas in the forest zone are indicated by the 2000 mm rainfall isohyet (broken lines). The four *Cola* species are: (●) *Cola chlamydantha*, found in all three wet blocks; (▼) *Cola buntingii*, only in the western block; (○) *Cola umbratilis*, only in the central; and (■) *Cola argentea*, only in the east. (b) The boundaries of forest types defined in Ghana (Hall and Swaine, 1976, 1981) are shown by the thinner continuous lines. Symbols indicate the location of floristic samples outside Ghana which were identified with Ghanaian forest types by the method of Swaine and Hall (1976): (▲) wet evergreen; (■) moist evergreen; (○) moist semi-deciduous; (●) dry semi-deciduous; (+) southern marginal.

changes in time. Although the dynamic nature of the vegetation has been implicit in many of the features which determine pattern in forest structure and variation, the temporal aspects will now be examined in more detail.

Even on a first visit to a forest site, there is much evidence to be seen of the continual process of change. Individual plants may bear new, limp foliage, often pale in colour, or red. The crowns of some trees may be leafless and in places there may be freshly fallen flowers or fruit. These are all manifestations of the seasonal changes in forest.

On a subsequent visit to the same site, a few months later, the same signs of change may be seen, but involving different species. Further changes may now be apparent, however: paths newly blocked by fallen trees or branches, new trees may have grown in gaps, and existing trees will have grown larger. How many of these changes may be seen will depend on how closely the forest was observed on the first visit, but a number of questions arise, the answers to which will help understanding of how the forest maintains itself in a dynamic equilibrium: How often do the trees set seed? What happens to the seedlings? How fast do the trees grow? Do some trees grow faster than others? How long do they live? What causes the death of trees? Are the trees which die replaced by the same, or different species? What would happen if the forest were felled?

Part of the answer is known to many of these questions, but there are many points of detail which have yet to be understood, and which will be indicated in the following sections.

2.   Phenology and seasonality

Phenology is the study of leaf, flower, and fruit production and other events which seem to be related to climate. The control of these events is not fully understood — climatic change during the year, although of importance, may not be solely responsible. The level of herbivore damage, of disease, and the physiological status of the plant may also influence phenological events. A fuller account of the topic may be found in Longman and Jeník (1974).

(a) Leaf demography

In almost all plants, leaves are the centre of productivity: they fix atmospheric carbon and incorporate it into organic compounds which the plant needs; they influence water and mineral salt uptake by controlling the rate at which water is lost through the stomata; they are the organs which sense light quality and intensity, and moisture conditions, providing information on climate and general environmental conditions necessary for the control of various aspects of plant growth. A knowledge of how leaves are produced, when they die, and how they perform during their life is fundamental to an understanding of the whole plant's strategy for survival in the forest.

In closed-canopy forest most species produce leaves sporadically, rather than continuously. The reasons for this periodic flushing of new growth are not fully understood, but a number of factors seem to be involved. It is likely, for example, that leaf production is only possible when water is readily available. Although moisture stress undoubtedly influences leaf production in seasonally dry forests, periodic flushing is also seen in forests under constant-wet climates, when moisture is probably never limiting, at least in the understorey. The usual explanation offered to account for this is that periodic flushing minimizes herbivore damage. If leaves are produced continuously, the leaf predator population can build up to high levels because food is always available, and the plant will suffer accordingly. Periodic production, especially if it is unpredictable, will not allow predators to maintain high population densities, and new leaves may well escape damage (Fig. 4.14).

This theory assumes that herbivores prefer young leaves and that mature leaves are unattractive. There is increasing evidence that this is so. Figure 4.15 shows that most of the damage sustained by the leaves of *Drypetes parvifolia* occurs during the first month of life, generally before the leaves are fully expanded, and when they are rich in carbohydrate and mineral nutrients (Table 4.4). Following maturation of the leaf, further predation is almost eliminated. The resistance of mature leaves to attack can be attributed to a variety of factors: mature leaves are more lignified, and have thicker cuticles. The role of secondary compounds in leaves is not yet fully appreciated, but they may well have defensive functions in both young and mature leaves.

These arguments lead to the recognition of particular strategies for leaf production in forest species: leaves are expensive to the plant in proportion to how much energy is invested in their production, and thick, robust leaves with chemical defences must cost more than lightweight, relatively undefended leaves. It follows that 'expensive' leaves should be made to function for as long as possible, and we may expect them to be relatively long-lived. In contrast 'cheap' leaves can be produced more often and may be short-lived.

Information on leaf longevity is scarce, but two examples from a recent study in dry forest in Ghana provide an interesting contrast. *Grewia carpinifolia* is a shade-intolerant climber which produces leaves more or less continuously when moisture conditions permit, but they live for only a short time (one or two months). Its leaves are thin, and often heavily damaged by insects. In contrast, the shade-tolerant understorey tree *Drypetes parvifolia* produces leaves most erratically, without any obvious relation to climate. Although some flushes of leaves are heavily damaged by insects when young, those which survive beyond the first month normally live for three years or more.

Evergreen species such as *Drypetes parvifolia* retail their leaves through several dry seasons; while deciduous species will be leafless, at least for a short time, in the dry season each year (e.g. *Triplochiton scleroxylon*). Leaves of

Fig. 4.14. Leafy shoot of the small forest tree *Greenwayodendron (=Polyalthia) oliveri* (Annonaceae) showing three cohorts of leaves of different ages. The oldest leaves are heavily colonized by epiphyllous lichens and algae; the youngest are freshly expanded. All flushes appear to have escaped damage by herbivores.

deciduous trees, therefore, never live longer than about one year, and sometimes, as in *Grewia carpinifolia*, for much shorter periods. it was suggested earlier (p. 64) that the evergreen and deciduous habits may represent strategies related to nutrient conservation, but it is not easy to unravel

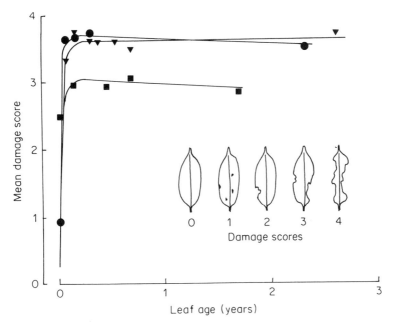

Fig. 4.15. Development of herbivore damage in three cohorts of the leaves of *Drypetes parvifolia* (Euphorbiaceae). Virtually all damage occurred in the first two months of life. Some slight improvement in the overall level of damage thereafter is due to the preferential loss of the more heavily damaged leaves. Damage scores were assessed by eye for each leaf on a five-point scale: (0) perfect; (1) blemished; (2) 1–10 per cent loss of leaf area; (3) 10–50 per cent loss; (4) >50 per cent loss.

the complex interactions between plant physiology, predators, climate, and their effect on leaf mortality. Much work remains to be done.

Whatever the causes of leaf fall, there is no doubt that over most of West Africa, leaf fall is concentrated in the dry seasons. This is particularly evident in dry forest (Fig. 4.16a), but is also clear in moister areas (John, 1973) and is still evident in high-rainfall areas such as the Nini-Suhien National Park in South-west Ghana (Fig. 4.16b).

(b) Flowering

Flowering is notoriously unpredictable in many forest plants, relatively few flowering regularly each year. A unique record of flowering and fruiting over six years in Malaysia was kept by Medway (1972), and provides a good demonstration of the variety of flowering patterns. In West Africa, many understorey trees are rarely seen in flower, as they flower infrequently and often only briefly.

Table 4.4 A comparison of chemical characteristics in young and old leaves of some African forest trees. After Waterman *et al.* (1980), Gartlan *et al.* (1980), and unpublished data of M. Royan. Values are % dry wt

Species	Leaf age	Potassium	Non-struct. carbohy-drate	Total phenolics[a]	Digestibility[b]
			Chemical character		
Anthonotha	Young	23	12	30.9	25
macrophylla[c]	Mature	19	4	7.6	4
Pancovia turbinata[c]	Young	39	14	9.1	69
	Mature	13	4	3.9	20
Lophira alata[c]	Young	38	nr	5.3	19
	Mature	31	3	6.3	9
Cassipourea	Young	27	nr	10.2	48
ruwenzoriensis[d]	Mature	10	6	5.3	24
Chaetacme aristata[d]	Young	25	7	5.3	59
	Mature	9	1	0.4	24

[a] Folin–Denis assay.
[b] Loss (%) in dry wt of leaves after 24 hr in rumen of fistulated sheep.
[c] Samples from Douala-Edea National Park, Cameroun.
[d] Samples from Kibale forest, Uganda.
nr Data not available.

Pollination has been observed with the necessary close scrutiny in only a few species, but pollination syndromes (Faegri and van der Pijl, 1971) can be recognized in many species. Small, scentless, and drab-coloured flowers, associated with wind-pollination are commoner in the species of the upper canopy, while the larger, fragrant, and more conspicuously coloured flowers associated with the attraction of animals are commoner in the understorey.

Flowering among forest plants occurs at all times of the year, but there is a clear peak in the number of species flowering in the latter half of the main dry season (January–March), and an equally clear minimum in the wettest months (June–July, Fig. 4.17). Enormous numbers of flowers are produced by some trees. One specimen of *Heritiera utilis* (Sterculiaceae), 37 m tall and with a 20 m diameter crown, produced an estimated 6.6×10^6 flowers in September 1978. Three months later, an estimated 9000 seeds were distributed on the forest floor; all of those sampled were dead, and no new seedlings could be found.

(c) Fruiting and dispersal
Fruit production depends of course on successful flowering and pollination, and is likewise very unpredictable in many species, with certain exceptions

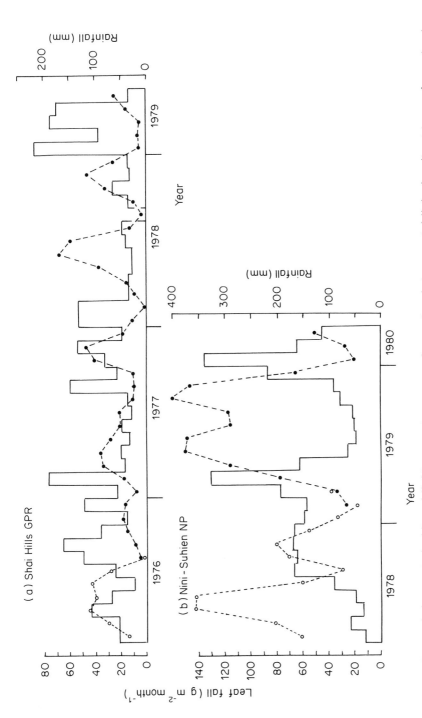

Fig. 4.16. Seasonal changes in leaf litter fall (histograms) over several years in relation to rainfall (broken lines) in two forest sites in Ghana: (a) Shai Hills Game Production Reserve; (b) Nini-Suhien National Park. Rainfall records are means of the two previous months; solid dots are actual measurements at the sites; open circles are based on mean annual rainfall data from nearby meteorological stations. Despite the large differences in annual rainfall between the two sites, leaf fall at both is clearly associated with dry periods.

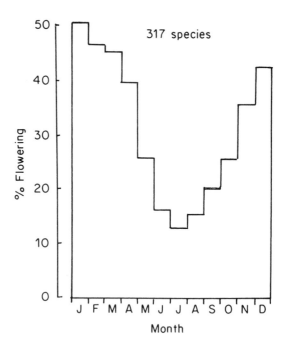

Fig. 4.17. Percentage of Ghanaian forest species in flower during the year in a sample of 317 vascular plants, showing a strong peak in the dry season. Species without clear seasonal flowering are omitted. Data from Hall and Swaine (1981).

such as *Ceiba pentandra*, which sets seed regularly every year in February and March (Fig. 4.18). As with the pollination syndromes of flowers, dispersal mechanisms may be suggested on the basis of fruit and seed morphology. Fleshy or arillate fruits will be dispersed by animals; winged or plumed fruits and seeds by wind; and others by explosively dehiscing fruits. These categories can be recognized more readily than the pollination syndromes, and some interesting correlations between habitat and climate can be drawn.

Wind-dispersal is common among the canopy species (Keay, 1957), while explosive dispersal is rather rare, and found especially among understorey plants. Fleshy fruits are found in the great majority of species, and are common at all levels in the canopy. We know very little about which animals are important for dispersal, but birds and bats must be involved, especially for small-seeded species, monkeys and other arboreal mammals for larger seeds, and ground rodents for plants which flower near ground level (e.g. *Chytranthus carneus*, Sapindaceae). Elephants are the only known effective dispersal agent

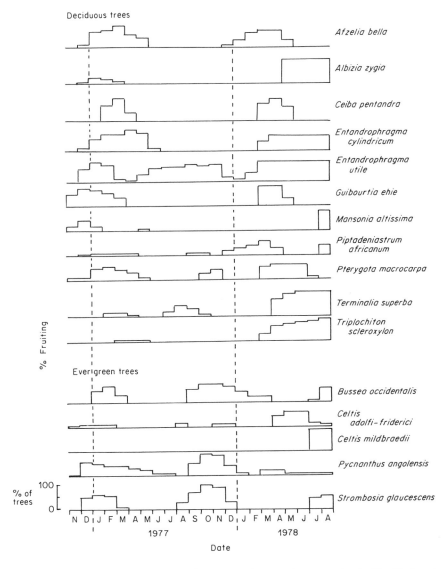

Fig. 4.18. Fruiting over 22 months of 16 megaphanerophyte species in Bia National Park, Ghana. Each species is represented by a sample of 10 mature trees. *Afzelia bella* and *Ceiba pentandra* show an annual pattern of fruiting, but none of the other species. Unpublished data of Claude Martin.

for some of the very large-seeded species such as *Panda oleosa* and *Balanites wilsoniana* (Alexandre, 1978).

Animal-dispersed species tend to fruit mostly in the latter half of the main dry season — the lean season for most animals; fleshy fruits are least abundant in the wettest months (Fig. 4.19). Wind-dispersed species show a strong peak of fruit production at the very end of the main dry season (Fig. 4.19) at a time when the first violent storms of the rainy season are beginning. Significantly, this is also the time when gaps are likely to be created in the forest. Fruiting in explosively dispersed species is closely associated with the months when the harmattan winds move into the forest zone: 80 per cent of explosively dispersed species in Ghanaian forest fruit in December or January (Fig. 4.19). These fruiting patterns are described for Ivory Coast by Alexandre (1980).

3. Tree Growth Rates

Trees are difficult to measure because they are so big, and because most of the processes of change take place in the canopy. At ground level, however, quite a good measure of tree size and growth can be obtained from measurements of trunk diameter or girth. As the tree grows, it lays down new wood in concentric rings around the stem. In temperate regions the annual flush of new growth is clearly marked by 'annual rings' which can be examined by taking a boring radially through the trunk. The spacing of these rings gives a measure of the growth rate, and a count of all the rings to the centre of the trunk will provide the age of the tree. Unfortunately in tropical trees, such clear growth rings are very rare; they may not be produced every year, or several may be produced in a single year depending on the species and on the particular growth conditions.

Measurement of growth in tropical trees must therefore be done by repeated careful measurement at the same point on the trunk over a period of years. Even this simple though slow method is beset with problems. All West African forest grows under seasonally variable rainfall, so that the trunks of the trees expand and contract depending on the water stress in the xylem. Trunks measured in the dry season will be thinner than in the previous wet season. The difference due to shrinkage may be quite as great as that due to growth in a whole year (Fig. 4.20). Other difficulties include the occurrence of buttresses on many of the larger forest trees. During growth, they increase in height and may rise up past the girthing point, confusing measures of growth rate. Damage to the cambium of the tree often causes swellings over the wound by the growth of callus which may cause further problems for the surveyor.

Despite these problems, certain points may be made about the progress of girth increase in forest trees.
1. Girth growth is typically greater among the larger trees, presumably reflecting their greater capacity for photosynthesis and nutrient uptake (Fig. 4.21).

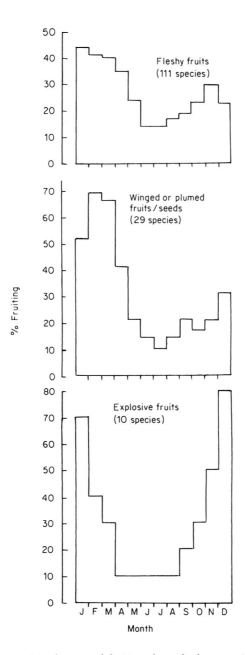

Fig. 4.19 Distribution of fruiting through the year in forest species with different fruit types. The histograms show the percentage of species in each group which are normally in fruit each month. Species without clear seasonal fruiting are omitted. Fruiting for all species is concentrated in the dry season, but clear differences exist between the three dispersal syndromes in the timing of peak fruiting. Data from Hall and Swaine (1981).

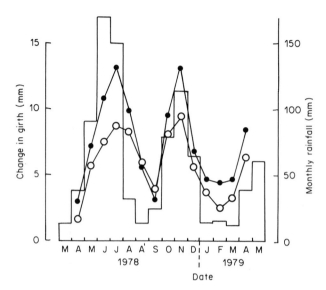

Fig. 4.20 Seasonal fluctuation in the girth of two trees of *Millettia thonningii* (Papilionaceae). The trees were measured with permanent dendrometer bands at four-week intervals. The initial gbh of each tree was (●) 580 mm, (○) 290 mm. Note the close relation between girth (lines) and rainfall (histogram — means of previous two months' rainfall), and that the permanent growth between April 1978 and April 1979 is considerably less than the annual variation.

2. Many trees, especially those in the understorey, show little or no increase in girth over many years. Suppressed trees are inhibited from further growth by competition with other trees for light and root space, and survive with minimal photosynthesis at or near compensation point. Foresters commonly assume that suppressed trees can be released by the removal of overshadowing trees.

3. Girth growth is extremely variable; different species have different capacities for increase; growth rates may vary between different soils, in different climatic regions, from season to season, and from year to year. Even individuals of the same species and size growing at the same site may show marked differences. In Fig. 4.21, the fastest-growing trees of *Celtis mildbraedii* achieved mean rates of the order of 3 cm yr^{-1} at girths over 100 cm, though others of similar size only grew 1 cm yr^{-1}. Since the biggest trees are few in number, and often difficult to measure accurately, the performance of trees late in life is not well defined, though it is often claimed that growth rates decline after the tree has reached maturity. In Fig. 4.21 there is no evidence for such a decline in *Celtis mildbraedii*, nor is there in the average growth rates for all species.

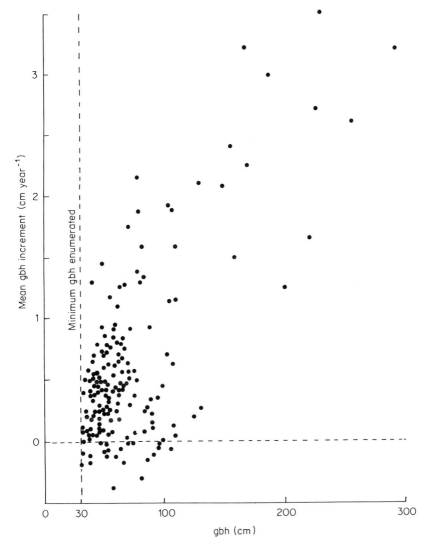

Fig. 4.21. Girth increments in trees >30 cm gbh as a function of tree girth in *Celtis mildbraedii* at Kade, Ghana. The plotted increments are means of three estimates spanning nine years. Larger trees show faster growth, but there is great variation.

The high variance which attaches to measures of growth rate in populations of forest trees means that the prediction of future performance, necessary for the calculation of the forester's sustainable yield (Baidoe, 1970), is very uncertain.

Evidently, girth is not a reliable guide to the ages of trees, though estimates have been made by calculating how long the average tree would take to pass through the various size-classes leading to the present size of the tree (e.g. Ogden, 1981; Nicholson, 1965). Since the slowest growth is usually found among the smallest trees, a large part of the age of many trees will be contributed by the time spent as seedlings and saplings. Few records incorporate measures of such small trees, but the limited data available confirm very low increments for the majority of the trees (Table 4.5).

Although most trees show relatively slow growth, it may be of more interest to consider the fastest-growing trees, as these are the ones most likely to reach maturity and form part of the canopy. Successful trees may be just those which have not suffered suppression early in life. Such must be certainly true of emergent species such as *Ricinodendron heudelotii, Chlorophora excelsa, Nauclea diderrichii, Ceiba pentandra*, and the forest species of *Terminalia* which cannot survive in shade, and which must grow rapidly in the gaps where they germinated if they are to escape shading.

Height growth of trees is not often recorded, except in some seedling studies, and in plantations, but the pattern of growth may be inferred from comparisons of girth and height of established trees. Figure 4.22 shows that height growth may be reduced while girth continues to increase. It would appear that, after the maximum height is achieved, the crown of the tree expands horizontally, rather than vertically, thus increasing the leaf area of the tree and allowing further girth increment. Except for young trees, then, height is a poor guide to tree age. What determines the maximum height that a tree may achieve is not known with certainty, but available evidence suggests that it is determined partly genetically (most microphanerophyte species are incapable of reaching the upper canopy even under apparently ideal conditions) and partly environmentally (*Ceiba pentandra* trees may reach 60 m in height under moist conditions, but in dry areas rarely exceed 25–30 m (Hall and Swaine, 1976)).

Some trees, notably those which colonize gaps and open spaces in the forest, show spectacular height growth. The short-lived pioneer tree *Musanga*

Table 4.5. Diameter increments and mortality of saplings in the understorey of forest in Nini-Suhien National Park, Ghana. All trees between 1.3 m tall and 32 mm dbh (10 cm gbh) were recorded on a 450 m^2 transect in December 1978 and in January 1980

	1978	1980
Number of trees measured	406	395
Number dying	11	
Annual mortality (%)	1.35	
Mean annual diameter increment (mm ± s.e)	0.21 ± 0.05	
Maximum	4.20	
Minimum	−1.55	

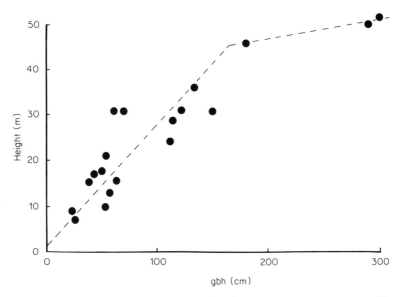

Fig. 4.22. Relation between tree height and girth in *Pterygota macrocarpa* (Sterculiaceae), Bia National Park, Ghana.

cecropioides, for example, has been seen to increase in height by as much as 5 m in one year. Similar performance is seen in many of the large pioneer species such as *Terminalia ivorensis* (Swaine and Hall, 1983). Such rapid height growth has led people to assume that these species have an intrinsically high photosynthetic capacity, but work by Coombe (1960), Coombe and Hadfield (1962), and Okali (1972) has cast doubt on this assertion. Small pioneers, such as *Musanga cecropioides* and *Trema orientalis*, and large pioneers including *Ceiba pentandra*, *Terminalia ivorensis*, and *Chlorophora excelsa*, have lower carbon fixation rates than many herbaceous species. Direct comparisons of fixation rates between pioneers and shade-tolerant species have only recently been made, but for Central American species (Oberbauer and Strain, 1984). This work shows for the first time, that pioneers do have higher rates than shade tolerant species.

4. Longevity and mortality

The age achieved by large forest trees is still a matter of conjecture. Estimates based on the summation of passage times through girth classes are unreliable as individual trees vary so greatly. Using the average tree in each size class for such calculations suggests ages of 200–300 years for trees with girth exceeding 3 m; but if the fastest-growing individuals in each class were used for the same calculations, ages an order of magnitude less result. In plantations, fast-growing pioneers such as *Terminalia ivorensis* have reached 36.5 m in height and 2.4 m in girth after only 22 years (Lamb and Ntima, 1970). Carbon dating of very large trees in tropical forest, on the other hand, seems to confirm that ages of hundreds of years are possible (Redhead and Taylor, 1970; Whitmore, 1984, pp. 110–1), but whether such isolated records represent the general rule is yet to be determined (Ashton, 1981).

The age and growth rate of forest trees together determine the size-class structure of the forest described earlier (see Fig. 4.5). The number of large canopy trees in an area will be relatively small, limited by available space, but the number of younger trees potentially capable of growing to occupy a place in the canopy will be very much greater. It follows that many of these smaller trees must die before they reach full size. Mortality rates among forest trees have been measured by monitoring the fate of individual trees over a number of years. All the evidence presently available suggests that, seedlings apart, the chance of death is independent of tree size (and thus of age); in any size class, the proportion of the trees which die in a given period is the same (Fig. 4.23). Constant mortality rates mean, of course, that the survivorship curve for forest trees will be a negative exponential, just as is seen in the distribution of size classes for natural forest.

Although the majority of forest tree species show size distributions of this form, it has often been observed that certain species do not conform. Shade-intolerant species which need a gap for germination and growth, such as *Ceiba pentandra*, often display unusual distributions, with most of their

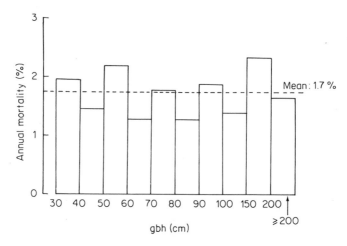

Fig. 4.23. Mortality of trees in different girth classes at Kade, Ghana, calculated from observations of 233 deaths over 12 years in 2 ha of forest. There is no significant difference in the mortality rates in different size classes.

members concentrated in the larger size classes, for example *Campnosperma brevipetiolatum* in Whitmore (1974), and the 'heliotrope' class of Rollet, (1974). These species, as has already been said, must grow fast to achieve a place in the upper canopy. It follows therefore, that they will pass relatively quickly through the smaller size classes, and we can expect to see relatively few of the smaller trees at any one time.

5.   Regeneration and recruitment

In a stable forest system, tree mortality is balanced by recruitment into the smaller size classes (Table 4.6) which ultimately depends on seed dispersal, germination, and establishment.

After dispersal as seeds, most plants seem to suffer a period of intense mortality. The evidence for this in tropical trees is mostly indirect: many trees produce vast quantities of seed from time to time, but the number of seedlings found under the forest canopy is usually very small in comparison. In the life-cycle of a plant from seed to mature, reproducing tree, this first stage is the most uncertain and the probability of a seed surviving to become an established seedling is very small (see p. 74 and Sarukhan, 1980). The causes of this massive loss of seed are not known in detail for more than a few species, but likely reasons include seed predation by insects and small mammals, the lack of suitable stimuli or conditions for germination, the competition for light, nutrients, and root space among germinating seedlings, and predation by herbivores of the young seedlings.

The fate of young seedlings of forest species has been monitored in several studies. T. Synnott (personal communication) found for *Entandrophragma utile* in Uganda that only 1.3 per cent of an original cohort of 2800 planted seeds

Table 4.6   Mortality and recruitment of trees > 30 cm gbh in eight 0.25 ha plots of forest at Kade, Ghana, between 1970 and 1982

	Plot								Total (2 ha)
	A	B	C	D	E	F	G	H	
Number of deaths (d)	20	29	21	29	50	31	30	23	233
Number of recruits (b)	21	27	18	32	29	36	29	19	211
Mean tree density (n)	139	142	130	157	142	132	138	144	1124
Turnover rate (yr)[a]	81	61	80	62	43	47	56	82	61
Mortality rate (% yr^{-1})	1.2	1.7	1.3	1.5	2.9	2.0	1.8	1.3	1.72
Recruitment rate (% yr^{-1})	1.3	1.6	1.2	1.7	1.7	2.3	1.8	1.1	1.56

[a] Calculated as $12n/\frac{1}{2}(d+b)$.

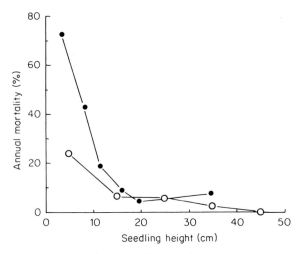

Fig. 4.24. Mortality of tree seedlings in different height classes: (●) Pinkwae sacred grove, Ghana (from Lieberman 1980); (○) Shai Hills Game Production Reserve, Ghana. Mortality is very high among the smaller seedlings, but slight among established seedlings.

survived beyond two and a half years. Large numbers were lost as seeds to rodents, and the remainder to herbivores. In dry forest in Ghana two studies (Fig. 4.24) have shown that such mortality is most intense in the smallest seedlings, with a marked increase in survivorship with increasing seedling height (Lieberman, 1980). In these forests at least, it would appear that once the seedlings have achieved about 20–30 cm in height, their mortality is scarcely greater than that of large trees.

6.   Forest succession

So far in this chapter events in the forest which occur on a relatively short time-scale have been considered. Germination, growth, predation, flowering, dispersal, and death are all events which may be observed even over fairly short periods of a few months to a year. With succession the concern is with long-term changes following the creation of a substantial opening in the forest and the initiation of a 'secondary' forest. Our use of the terms 'secondary' and 'primary' in relation to forests and forest species needs some explanation. By definition, secondary forest is what grows up when primary forest is destroyed. The new forest which eventually replaces this regrowth through succession is expected to be much like the original primary forest, but could be called secondary because of its past history. Since the past history of any area of forest is rarely known reliably, it is wise to use the term 'secondary' with some caution. In this chapter, we mean by forests called 'primary', or 'closed-canopy', areas dominated by tall shade-tolerant species (Fig. 4.25). By

Fig. 4.25. Interior of primary forest in Nini-Suhien National Park, Ghana. The buttressed tree is *Cynometra ananta* (Caesalpiniaceae). All size classes of tree are well represented.

secondary forests we mean areas recently regrown from cleared land which contain no large shade-tolerant trees but an abundance of pioneer species (Fig. 4.26). The distinction between the two kinds only becomes difficult in the later stages of succession, which are not well known if only because they are difficult to distinguish from fully developed, stable primary forest. Primary forest is sometimes called 'virgin' forest, meaning forest which has never been disturbed. Since there is a continual turnover of trees even in primary forest, the term 'virgin forest' seems to have little meaning and is impossible to apply in practice.

Mention has already been made of pioneers or shade-intolerant species which are found as young plants only in gaps. These, like the primary, shade-tolerant species, can be divided into life-form classes based on size. One can recognize large pioneers which can grow large enough to form part of the upper canopy, and small pioneer trees, such as *Musanga cecropioides*, *Rauvolfia vomitoria*, *Harungana madagascariensis*, and *Macaranga* spp., which rarely exceed 20 m in height. Small pioneers are usually regarded as short-lived (c. 20 years for *Musanga* – Ross, 1954), often dying even before they are overshadowed by taller species.

Pioneer species share a number of characteristics which set them apart from primary forest species. Their dependence on high light intensity for growth has already been noted, but the higher light intensity (or changes in its spectral composition) in gaps is also necessary to induce germination of the seeds of some pioneers. Other conditions which change on the formation of a gap, and which may be involved in breaking dormancy in pioneer seeds, include higher maximum and lower minimum temperatures (in both air and soil), and increases in the diurnal range of atmospheric humidity and soil moisture.

Pioneer trees typically grow and mature very rapidly, setting seed when quite young (e.g. six months in *Trema orientalis* (Swaine and Hall, 1983)). Their seeds are often very small, produced more or less continuously and in large numbers. They are generally either wind-dispersed or dispersed by birds and bats. Their seeds become very widely dispersed and can be found in large numbers (Table 4.7) in the soils under closed-canopy forest (Keay, 1960; Hall and Swaine, 1980), where they remain dormant until a gap is created and suitable conditions arrive for germination. Whitmore (1983) has reviewed this topic.

Little is known of the rate at which such seeds are dispersed into forest, but the most likely sources of the large numbers of seeds found in forest soils are either a slow accumulation from the seed 'rain' over many years, or a relict population of seeds deposited by a previous stand of mature pioneers at the same spot.

Pioneer species form part of the forest only when there is a gap in the canopy. If it is only a small opening, especially if it does not greatly expose the soil, the gap may be closed by the accelerated growth of existing trees and

Fig. 4.26. Interior of secondary forest near Kibi, Ghana. The trees with stilt roots are *Musanga cecropioides* (Moraceae), probably about eight years old. Note the absence of large trees and the dense ground flora.

Table 4.7. The number of seeds germinating from two $\frac{1}{2}$ m² soil samples collected to a depth of 4 cm from primary forest in different parts of Ghana. (Adapted from Hall and Swaine, 1980)

	Sample				Mean
	1	2	3	4	
Secondary forest species (number seedlings m^{-2})	373	694	632	163	466
Primary forest species (number seedlings m^{-2})	11	2	1	0	4

climbers. Larger openings of 100 m² or more usually permit the establishment of a few pioneer trees. The largest gaps, created, for example, by the simultaneous fall of several large trees during a storm, or more commonly nowadays by forest clearance by man, allow large stands of pioneers to establish. In many parts of West Africa, *Musanga cecropioides* may dominate the regrowth within a year, with varying proportions of other pioneer species (Swaine and Hall, 1983; Ross, 1954). Once the cover of vegetation has been re-established, no further recruitment of pioneers is possible unless further disturbance occurs, but the shade-tolerant primary species can continue to invade. Small plants of primary species are evident in secondary forest in considerable numbers, even in the first year, either as survivors of the original forest, or as seedlings established subsequently. Their growth is slow compared with the pioneers, but they are more persistent, surviving beyond the death of the small pioneers to form the beginning of new primary forest along with the larger pioneers.

Forest areas destroyed by windthrow or felling are thus expected to regrow into something very like the original forest, though the process of recovery may be slow, and will depend on the severity of the original disturbance. Frequent and long-continued farming, for example, may so deplete soils of tree seed, that forest cannot re-establish, and a weedy grassland or 'derived savanna' may take over. Isolated forest patches which are cleared may be so distant from a source of seed to replenish the primary species complement that they remain floristically impoverished for many years. The later stages of forest regrowth following clearance are, however, rather poorly known because the time-span involved exceeds the life of most research programmes.

References

Adam, J. G. (1970). Etat actuel de la végétation des Monts Nimba au Liberia et en Guinée. *Adansonia*, n.s., **10**, 193–211.

Ahn, P. M. (1970). *West African Soils*. Oxford University Press, London.

Alexandre, D. Y. (1977). Régéneration naturelle d'un arbre caracteristique de la forêt équatoriale de Côte d'Ivoire: *Turreanthus africanus* Pellegr. *Oecologia Plantarum*, **12**, 241–262.

Alexandre, D. Y. (1978). La rôle disseminateur des éléphants en forêt de Taï, Côte d'Ivoire. *La Terre et la Vie*, **32**, 47–62.

Alexandre, D. Y. (1980). Caractère saissonair de la fructification dans une forêt hygrophile de Côte d'Ivoire. *Rev. Ecol.* (Terre Vie), **34**, 335–350.

Ashton, P. S. (1981). The need for information regarding tree age and growth in tropical forest. *Yale Univ. School of For. Environ. Studies Bull.*, **94**, 3–6.

Aubréville, A. (1959). *La Flore Forestière de la Côte d'Ivoire*, 3 vols. Centre Technique Forestier Tropicale, Nogent-sur-Marne.

Baidoe, J. F. (1970). The selection system as practised in Ghana. *Comm. For. Rev.* **49**, 159–165.

Chachu, E. O. (1976). An investigation of toposequences and associated forest cover in Ijaiye Forest Reserve. Ph.D. Thesis, University of Ibadan, Ibadan, Nigeria.

Coombe, D. E. (1960). An analysis of the growth of *Trema guineensis*. *J. Ecol.*, **48**, 219–231.

Coombe, D. E., and Hadfield, W. (1962). An analysis of the growth of *Musanga cecropioides*. *J. Ecol.*, **50**, 221–234.

Faegri, K., and van der Pijl, L. (1971). *Principles of Pollination Ecology*. Pergamon Press, London.

Gartlan, J. S., McKey, D. B., Waterman, P. G., Mbi, C. N., and Struhsaker, T. T. (1980). A comparative study of the phytochemistry of two African rain forests. *Biochem. System. Ecol.*, **8**, 401–422.

Guillaumet, J–L., and Adjanohoun, E. (1971). La végétation de la Côte d'Ivoire. *Mem. Off. Rech. Scient. tech. Outre-mer*, **50**, 157–263.

Hall, John B. (1973). Vegetational zones on the southern slopes of Mount Cameroon. *Vegetation*, **27**, 49–69.

Hall, John B. (1977). Forest types in Nigeria: an analysis of pre-exploitation forest enumeration data. *J. Ecol.*, **65**, 187–199.

Hall, John B., and Medler, J. A. (1975). Highland vegetation in south-eastern Nigeria and its affinities. *Vegetatio*, **29**, 191–198.

Hall, J. B., and Swaine, M. D. (1976). Classification and ecology of closed-canopy forest in Ghana. *J. Ecol.*, **64**, 913–951.

Hall, J. B., and Swaine, M. D. (1980). Seed stocks in Ghanaian forest soils. *Biotropica*, **12**, 256–263.

Hall, J. B., and Swaine, M. D. (1981). *Distribution and Ecology of Vascular Plants in a Tropical Rain Forest: Forest Vegetation in Ghana*, W. Junk, The Hague, 383 pp.

Janzen, D. H. (1970). Herbivores and the number of tree species in tropical forest. *Amer. Nat.*, **104**, 501–528.

Janzen, D. H. (1972). Protection of *Barteria* (Passifloraceae) by *Pachysima* ants (Pseudomyrmecinae) in a Nigerian rain forest. *Ecology*, **53**, 885–892.

Jeník, J. (1969). The life-form of *Scaphopetalum amoenum* A. Chev. *Preslia*, **41**, 109–112.

Jeník J., and Hall, J. B. (1966). The ecological effects of the harmattan wind in Djebobo massif, Togo. *J. Ecol.*, **54**, 767–779.

Johansson, D. (1974). Ecology of vascular epiphytes in West African rain forest. *Acta phytogeogr. suec.*, **59**, 1–129.

John, D. M. (1973). Accumulation and decay of litter and net production of forest in tropical West Africa. *Oikos*, **24**, 430–435.

Keay, R. W. J. (1957). Wind-dispersed species in a Nigerian forest. *J. Ecol.*, **45**, 471–478.

Keay, R. W. J. (1959). *An Outline of Nigerian Vegetation*, 3rd edn. Federal Ministry of Information, Lagos.

Keay, R. W. J. (1960). Seeds in forest soils. *Niger. For. Inf. Bull.*, n.s., **4**, 1–12.

Lamb, A. F. A., and Ntima, O. O. (1970). *Terminalia ivorensis*. Fast-growing timber trees of the lowland tropics. *Commonw. For. Inst. Paper*, No. 5, Oxford.

Lawson, G. W., Armstrong-Mensah, K. O., and Hall, J. B. (1970). A catena in tropical moist semi-deciduous forest near Kade, Ghana. *J. Ecol.*, **58**, 317–398.

Longman, K. A., and Jeník, J. (1974). *Tropical Forest and its Environment*. Longman, London, 196 pp.

Lieberman, D. D. (1980). Dynamics of forest and thicket vegetation on the Accra Plains, Ghana. Ph.D. Thesis, University of Ghana, 222 pp.

Medway, Lord (1972). Phenology of a tropical rain forest in Malaya. *Biol. J. Linn. Soc.*, **4**, 117–146.

Monk, C. D. (1966). An ecological significance of evergreenness. *Ecology*, **47**, 504–505.

Moore, H. E. (1973). Palms in the tropical forest ecosystems of Africa and South America. In B. J. Meggars, E. S. Ayensu, and W. D. Duckworth (Eds), *Tropical Forest Ecosystems of Africa and South America: a Comparative Review*, pp. 63–88. Smithsonian Institution Press, Washington.

Nicholson, D. I. (1965). A study of virgin forest near Sandakan, North Borneo. *Symposium on Ecological Research in Humid Tropics Vegetation, Kuching 1963*, pp. 67–87. Unesco, Paris.

Obaton, M. (1960). Les lianes ligneuses à structure anomale des forêts denses d'Afrique occidentale. *Annls. Sci. nat.*, ser. 12, **1**, 1–219.

Oberbauer, S. T., and Strain, B. S. (1984). Photosynthesis and successional status of Costa Rican rain forest trees. *Photosynthesis Research*, **5**, 227–237.

Ogden, J. (1981). Dendrochronological studies and the determination of tree ages in the Australian tropics. *J. Biogeogr.*, **8**, 405–420.

Okali, D. U. U. (1972). Growth rates of some West African forest-tree seedlings in shade. *Ann. Bot.*, **36**, 953–959.

Papadakis, J. (1966). *Crop Ecologic Survey in West Africa (Liberia, Ivory Coast, Ghana, Togo, Dahomey, Nigeria)*. FAO, Rome, 103 pp. + atlas.

Purseglove, J. W. (1972). *Tropical Crops. Monocotyledons*. Longman, London, 607 pp.

Raunkiaer, C. (1934). *The Life-forms of Plants and Statistical Plant Geography*. Oxford University Press, London.

Redhead, J. F., and Taylor, D. A. H. (1970). The age of *Entandrophragma cylindricum* (Sprague). *J. W. Afr. Sc. Ass.*, **15**, 19.

Richards, P. W. (1952). *The Tropical Rain Forest*. Cambridge University Press, Cambridge.

Richards, P. W. (1963). Ecological notes on West African vegetation III. The upland forests of Cameroons mountain. *J. Ecol.*, **51**, 529–554.

Rollet, B. (1974). *L'architecture des Forêts Denses Humides Sempervirentes de Plaine*. Centre Technique Forestier Tropicale, Nogent-sur-Marne.

Ross, R. (1954). Ecological studies on the rain forest of Southern Nigeria III. Secondary succession in Shasha Forest Reserve. *J. Ecol.*, **42**, 259–282.

Sandford, W. W. (1968). Distribution of epiphytic orchids in semi-deciduous tropical forest in southern Nigeria. *J. Ecol.*, **56**, 697–705.

Sandford, W. W. (1969). The distribution of epiphytic orchids in Nigeria in relation to each other and to geographic location and climate, type of vegetation and tree species. *Biol. J. Linn. Soc.*, **1**, 247–285.

Sarukhan, J. (1980). Demographic problems in tropical systems. In O. T. Solbrig (Ed.), *Demography and Evolution in Plant Populations*, pp. 161–188. Blackwell Scientific Publications, Oxford.

Schnell, R. (1952). Végétation et flore de la region montagneuse du Nimba. *Mém. Inst. fr. Afr. noire*, **22**, Dakar.

Spichiger, R., and Pamard, C. (1973). Recherches sur le contact forêt-savane en Côte d'Ivoire; étude du recrû forestier sur les parcelles cultivées en lisière d'un îlot forestier dans le sud du pays Baoulé. *Candollea*, **28**, 21–37.

Swaine, M. D., and Beer, T. (1977). Explosive seed dispersal in *Hura crepitans* L. (Euphorbiaceae). *New Phytol.*, **78**, 695–708.

Swaine, M. D., and Hall, J. B. (1976). An application of ordination to the identification of forest types. *Vegetation*, **32**, 83–86.

Swaine, M. D., and Hall, J. B. (1977). The ecology and conservation of upland forests in Ghana. *Proceedings of Ghana SCOPE's Conference on Environment and Development in West Africa*, September 1974, pp. 151–158. CSIR, Accra.

Swaine, M. D., and Hall, J. B. (1983). Early succession on cleared forest land in Ghana. *J. Ecol.* **71**, 601–627.

Swaine, M. D., Hall, J. B., and Lock, J. M. (1976). The forest–savanna boundary in west-central Ghana. *Ghana J. Sci.* **16**, 35–52.

Waterman, P. G., Mbi, C. N., McKey, D. B., and Gartlan, J. S. (1980). African rain forest vegetation and rumen microbes: phenolic compounds and nutrients as correlates of digestibility. *Oecologia*, **47**, 22–33.

Waterman, P. G., Meshal, I. A., Hall, J. B., and Swaine, M. D. (1978). Biochemical systematics and ecology of the Toddalioideae in the central part of the West African forest zone. *Biochem. System. Ecol.*, **6**, 239–245.

White, F. (1979). The Guinea–Congolian Region and its relationships to other phytochoria. *Bull, Jard. Bot. Nat. Belg.*, **49**, 11–55.

White, F. (1983). *The Vegetation of Africa*, Unesco, Paris, 356 pp.

Whitmore, T. C. (1974). Change with time and the role of cyclones in tropical rain forests on Kolombangara, Solomon Islands. *Commonw. For. Inst. Paper*, No. 46.

Whitmore, T. C. (1983). Secondary succession from seed in tropical rain forests. *Forestry Abstracts*, **44**, 767–769.

Whitmore, T. C. (1984). *Tropical Rain Forests of the Far East*. (2nd edn), Oxford University Press, Oxford, 352 pp.

Zeven, A. C. (1967). *Agricultural Research Report*, No. 689. Centre for Agricultural Publications and Documents, Wageningen.

Plant Ecology in West Africa
Edited by G. W. Lawson
© 1986 John Wiley & Sons Ltd

CHAPTER 5

Savanna

William W. Sanford and Augustine O. Isichei
Department of Botany, University of Ife, Nigeria

1. Definition of savanna

1. Vegetation

Fosberg (1961), in an attempt to set up a standard, international set of ecological terms, called savanna any 'closed grass or other predominantly herbaceous vegetation with scattered or widely spaced woody plants' and subdivided this into tall savanna, in which the herbaceous cover is at least 1 m high, and low savanna in which it is less. Savanna is separated from 'steppe' on the basis of the herbaceous cover being closed in savanna but not closed in steppe, the latter thus including 'desert scrub'. This is essentially a broad growth-form definition and is unsatisfactory on two counts; savanna as now generally conceived of has a cover predominantly of grass, not of any other herbaceous material; savanna may or may not have woody plants, and if they are present they may be either scattered or fairly dense.

Other attempts to define savanna by its vegetation have been made. Beard (1967) reviews the terms agreed upon at Yangambi by the Scientific Council for Africa South of the Sahara (CSA, 1956). These are more elaborate and exact than the simple definition stated above but also broadly depend on growth form, together with structural features. Savanna is described as a 'formation of grasses at least 80 cm high, forming a continuous layer dominating a lower stratum. Usually burnt annually. Leaves of grasses flat, basal, and cauline. Woody plants usually present.' This is divided into savanna woodland — 'trees and shrubs forming a canopy which is generally light'; tree and shrub savanna — 'savanna as defined, with trees and shrubs scattered'; and grass savanna — 'trees and shrubs generally absent'. Steppe is defined: 'Open herbaceous vegetation, sometimes also with woody plants. Usually not burnt. Perennial grasses usually less than 80 cm high, widely spaced. Leaves of grasses narrow, rolled, or folded, mainly basal. Annual plants very often are abundant between perennials.' Formations associated with savanna are also defined. Woodland:

'Open forest; tree stratum deciduous of small or medium-sized trees with crowns more or less touching, the canopy light; grass stratum sometimes sparse, or mixed with other herbaceous and suffrutescent vegetation. Sometimes evergreen or partly evergreen.' Thicket: 'Shrubby vegetation, evergreen or deciduous, usually more or less impenetrable, often in clumps, with grass stratum absent or discontinuous.' Scrub: 'Open shrub-land, as opposed to closed thicket.' A major difficulty here is the lack of a clear functional separation between 'savanna woodland' and 'woodland'.

2. Climate

Not all definitions of savanna, however, have been based on vegetation. Jaeger (1945) followed by Troll (1952) attempted to define savanna on climatic grounds with little reference to vegetation. 'Humid savannah' encompasses the climatic zone of tropical summer rain with a dry period of from 2.5 to 5 months; 'dry savannah', the regions of summer rain with a dry period of from 5 to 7.5 months; and 'thorn-scrub savannah', the regions with a dry period of from 7.5 to 10 months. Troll has pointed out that savanna always reflects a combination of edaphic, biotic, and anthropogenic factors, and therefore includes all forms, from treeless grasslands to forest. This definition of savanna rather mixes causative factors and appearance.

Trewartha (1954) in his modification of Koppen's world map, also views savanna fundamentally from the aspect of climate; he classes as tropical savanna areas with a distinct alternation of wet and dry seasons with rain in the summer generally exceeding 600 mm and sometimes up to 2000 mm. Walter (1973), however, in a world-wide review of savanna, places the approximate boundary of tree savanna at 400 mm, stating that at below this level only grass or grass and small woody plants occur.

His views (Walter, 1971, pp. 238–239), are similar to those of the early German plant geographers Schimper and Drude, as he looks upon savanna as 'the natural homogeneous zonal vegetation of the tropical summer rain zone showing a closed grass cover and scattered, individual woody plants, either shrubs or trees. This climatically conditioned savannah probably occurs in Africa only in areas with rainfall below 600 mm and corresponds to the "thorn-scrub" of Jaeger.' Where anthropogenic factors are important he uses the term 'secondary savannah'. He excludes 'woodland' and considers 'parkland savannah' (grassland with widely spaced trees, often selected for their economic importance) to be a mosaic of different communities. His climatic savanna is, then, comparable to the Sahel in West Africa which, however, does not always have a continuous cover of grass. Furthermore the grasses are predominantly annual and often patchy in distribution.

3. Soil

Still other workers have attempted to demarcate savanna according to edaphic considerations. Thus one of the conclusions of the 1964 IGU–Unesco symposium in Venezuela as reported by Adejuwon (1970) was that there are 'natural savannas which can be considered as edaphic climaxes.

It is certainly obvious in any West African region of forest margin or savanna that patches of open vegetation and forest or woodland occur in relation to soil character. Chachu (pers. comm.) and Milligan (1979) have pointed out how frequently the vegetation of the Kainji Lake National Park region in northwestern Nigeria seems to be determined by soil depth. Associated with shallow soil are decreased moisture availability, increased leaching of nutrients, and increased soil temperature, all of which affect vegetation.

Ramsey and de Leeuw (1965a, b) have studied vegetation and soil parent material in northern Nigeria and have concluded that 'only in extreme cases does the parent material of the soil influence the composition of the arboreal vegetation and then mainly through the soil–water relationship'. Ahn (1970), however, reported that the geological boundary between Pre-Cambrian Birrimian formations and sandstone appears to determine the position of the forest–savanna boundary in Ghana. Hall (1977) has clearly demonstrated a demarcation of forest type and species composition in relation to broad soil type (tropical ferruginous–ferralitic).

However, such soil–vegetation relationships, as well as relationships between climate and soil and vegetation, obviously confuse cause and result: a definition of savanna should describe the physical thing that we are going to call 'savanna' rather than list the factors which we believe have brought it about.

4. A synthetic definition

From published descriptions of savanna and direct observation of savanna in West Africa, a definition may be arrived at: seasonal tropical vegetation in which there is a closed or nearly closed cover of grasses at least 80 cm high with flat, usually cauline, leaves; usually burnt annually; trees and shrubs at various densities most often present.

The fundamental quality which provides whatever physiognomic, structural, and physiological unity the concept of savanna may have is the presence of a nearly continuous ground cover of grass. Recently Menaut (1983a) has suggested that the most useful characterization of savanna is the ratio between above- and below-ground woody and grass biomass. Such a characterization would not only be useful on a local basis but also on a world-wide basis.

Seasonality, while having climatic implication of a wet and dry period, also implies physical change of the vegetation: the grasses mature, dry, and brown;

many of the trees and shrubs are deciduous. Annual burning indicates that we would not expect to find any considerable amount of dead vegetation accumulating on the ground. The use of the terms 'tree' and 'shrub' indicates that woody plants of at least two height classes are normally present. Trees are understood to be woody plants which at maturity are at least 3 m high, while shrubs are less than 3 m high. (It is more satisfactory to use height as a criterion rather than such often-times stated, but in practice ambiguous, criteria as 'having only one or several stems emerging from the ground'.)

For a discussion of the abiotic environment associated with such vegetation, see section III, below.

II. Fundamental processes of savanna systems

1.   Stability as a functional aim

The definition of savanna adopted above is rightly based on gross observable features of the vegetation and necessarily ignores the functional dynamics of savanna as an ecosystem type. Before any discussion of the causes of savanna and before detailed description of West African savanna types and before consideration of their study and management, it may be helpful to summarize briefly the main processes of savanna as a system.

How much of a *system*, in the strict sense, savanna is may be debated, but if any assemblage of plants remains for an appreciable period of time, there must be a satisfactory relationship between the plants of the assemblage and between the plants and the abiotic environment and to this extent a *system* may be said to exist. This is not really saying anything except that if a group of plants survives, its survival must be possible. It is possible, however, to go further and agree that such assemblages are not static in species abundance and composition but change over time, the rate of change being inversely proportional to the adaptive fitness of the individual plants to each other and to their environment of animals, soil, and climate. In other words, environmental pressure and variation in organisms, brought about by immigration and by phenoplastic and genotypic variability, act together in such a way that the most successful forms (i.e. the forms most productive of propagules able to establish themselves and in turn reproduce) of the most successful species replace less successful forms or/and less successful species.

If, barring change in environment, plants are continually being replaced by better adapted ones, the rate of change will decline until it is so slow that the system is more or less stable (using 'stable' to mean 'constant' or 'unchanging'). But a complicating factor spoils this simple picture: each plant species, indeed each individual plant, changes the environment to varying degrees and in varying ways; thus the only condition under which a stable system can develop is that of the development of an assemblage of perfectly adapted plants which do not further change the environment.

The dynamics of the savanna ecosystem thus depend upon the stability of the abiotic environment and the level of adaptive fitness of the plants in that environment.

The climate of West Africa in recent geological times has been relatively unstable or 'capricious' as Livingstone has termed it in a recent review (1975). No part of the African continent has escaped serious climatic change during the past 20 000 years. During the Holocene, 10 000 – 7000 BP, there was great forest expansion, although the continent was never covered with moist forest, and large areas of grassland were present in East Africa very long ago. Dry conditions existed before the period of moist forest expansion (about 13 000 BP) and a gradual drying again occurred from 5000 to 3000 years BP. Livingstone considers the present biota to represent a lag concentrate of species that were able to persist through drier and cooler conditions than those of today. Thus, in the savanna, dry-adapted plants have been gradually infiltrated by some plants adapted to somewhat moister conditions. During the last 1000 years or less, much of the savanna has been subject to annual dry-season burning and in very recent times (perhaps only during the last 300–400 years) to extensive clearing and farming. Annual burning has constituted a stabilizing pressure by selecting species of a certain physiological/morphological fire-resistant type, and by tending to limit vegetation in number of individuals and to control the size-class distribution of individuals. On the other hand, cultivation, and to a lesser extent livestock grazing and browsing, not being regular in occurrence, have been destabilizing in effect. The repeated disturbance of natural vegetation at irregular intervals has destroyed much of the vegetation and has opened the area for immigration of colonizers or weeds, temporary inhabitants which are supplanted by other species when the disturbance is discontinued. It is thus obvious that any piece of savanna will vary in composition and structure according to its history of land use. Such local variations may be ignored for the time being and an attempt made to generalize the tendencies of change leading towards stabilization, and to the major processes of a more or less stable savanna vegetation.

2. Mineral cycling

Crucial for an assemblage of plants which long survives is efficient mineral cycling. If essential minerals in available form are present in finite supply and are not made available for reuse, the community will decline in productivity and eventually starve to death. The tropical climate of high temperature which increases the rate of mineralization of organic matter in the soil and of heavy seasonal rain which may bring about erosion, runoff, and leaching conspires to make tropical soils especially vulnerable to deterioration. In West African savanna regions, cultivated crops decline drastically in yield after only three or four years, because there is almost no nutrient cycling in a maize or millet or guinea-corn field. Surviving plant assemblages must avoid this. The main factors in nutrient conservation and cycling which allow a community of plants to persist are a multilayered canopy cover to lessen the impact of rain, a close network of roots to absorb nutrients as soon as they go into solution to prevent their loss by leaching or runoff, and adequate litter fall and decomposition to return nutrients to the soil for reuse. The ideal savanna would, then, have at least two strata of woody plants and a cover of grass and forbs.

Typical Guinea savanna may roughly be described as having an A stratum of large trees at least 7 m in height and a B stratum of woody plants less than 7 m high. Tree canopy cover in northern Guinea savanna in Nigeria ranges, on an average, from about 25 to 35 per cent. (Such cover means a density of trees at least 30 cm in girth at breast height of from 0.010 to 0.020 m $^{-2}$.) Woodland in this region ranges from about 50 to 80 per cent in canopy cover (with a tree density of from 0.040 to 0.20 m $^{-2}$). Southern Guinea may have slightly more canopy cover but is often essentially the same in cover as northern Guinea. In the derived or transition zone, tree cover may be higher in woodland which approaches forest in structure. The herbaceous cover during the rainy season in all regions is well over 200 per cent, often over 400 per cent, being composed of several layers.

The extent of the root network can be approximated by measurement of below-ground biomass. The most precise figures available are those of Cesar and Menaut (1974) for derived savanna in Ivory Coast for eight categories of vegetation, ranging from grass savanna with less than 1 per cent cover by woody plants to woodland with over 50 per cent woody cover. The seasonal mean below-ground biomass was reported to vary from 10.1 t ha^{-1} (1010 g m^{-2}) to 19.0 t ha^{-1} (1900 g m^{-2}), with the ratios of above-ground/below-ground biomass ranging from 0.22 to 0.51. Variation over the year was considerable, with maximum below-ground biomass as high as 28.8 t ha^{-1}. Isichei (1979) has estimated below-ground biomass in derived and Guinea savanna in Nigeria. At the peak period of herbaceous standing crop (October–November) total below-ground biomass ranged from 1322 to 4334 g m^{-2} in derived savanna and from 402 to 1972 g m^{-2} in Guinea Savanna.

Somewhat more information is available for dead material above ground and litter fall. Total standing dead matter above ground in the Ivory Coast study ranged from 0.28 to 0.67 t ha^{-1} (28 to 67 g m^{-2}). Leaf litter fall has been reported by Isichei to range from 63 to 317 g m^{-2} yr^{-1}, and wood litter fall from 223 to 387 g m^{-2} yr^{-1} in derived savanna; 118 to 241 g m^{-2} yr^{-1} leaf fall and from 70 to 672 g m^{-2} yr^{-1} wood fall in Guinea savanna. Hopkins (1966) has reported 90 g m^{-2} yr^{-1} leaf fall in derived savanna (Olokemeji Forest

Reserve, Nigeria), and Collins (1977) has reported 239 g m^{-2} yr^{-1} leaf fall and 139 g m^{-2} yr^{-1} wood fall in Guinea savanna.

As important as the amount of litter fall is the rate of disappearance of the litter. If litter is collected in plastic bags, termites and the larger soil fauna are excluded and disappearance is brought about mainly by bacteria and fungi. This, however, gives an unrealistic estimate of disappearance. Collins (1977) has estimated that the termite, *Macrotermes bellicosus*, alone removes 60.1 per cent of the annual wood fall and 2.9 per cent of the leaf fall in his study area in Nigerian Guinea savanna (an area with a very high density of this termite). The fungus-growing Macrotermitinae were reported to remove 35.5 per cent of the annual leaf litter fall in this region. Isichei (1979), working in Guinea savanna further north, as well as derived and southern Guinea savanna, estimated litter disappearance under completely natural conditions with the litter unenclosed. His reported disappearance rates may be somewhat too low because he was not able to collect data at frequent enough intervals, but they give an extremely good idea of the rapidity of litter disappearance in savanna. Using the exponential growth equation, as suggested by Olson (1963):

$$N_t = N_0 e^{-rt}$$

where N is the litter remaining after time t and N_0 the litter at the beginning of the time interval; rates, r, estimated on a yearly basis were found to range from 1.68 to 4.62 (mean 2.72) for leaf disappearance, with the highest rate of disappearance in northern Guinea where termites were abundant. This means that on the average, 50 per cent of the annual leaf fall disappears within a little over three months. The values of r for wood disappearance were much more variable, probably being more dependent on termite density, and ranged from 0.32 to 1.16 (mean 0.59). Thus, 50 per cent of the annual wood fall disappears in about one and a half years.

Some idea of the amount of nutrients returned with litter is obtained by examination of Isichei's figures for nitrogen content of litter: leaf litter contained from 0.81 to 1.22 per cent nitrogen (mean 0.98 per cent) while wood litter ranged from 0.50 to 0.61 per cent nitrogen (mean 0.57 per cent) by weight of oven-dried material.

3. A nitrogen model

As stated at the beginning of this section, if an assemblage of plants is to survive, nutrients must be conserved and cycled so that a nearly constant level of availability is maintained. Figure 5.1 illustrates nitrogen stocks and flows in savanna woodland (as a composite of the early burnt plot in derived savanna at Olokemeji Forest Reserve and a 1 ha research plot in northern Guinea

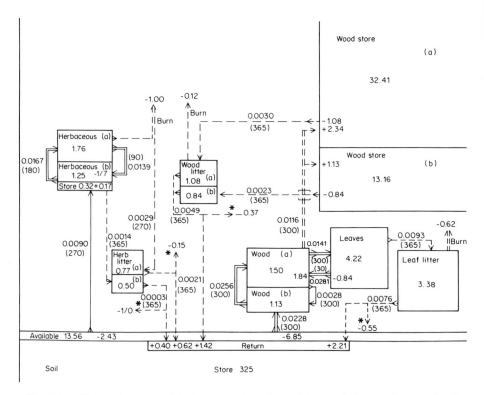

Fig. 5.1. Compartment model of estimated stocks and flows of nitrogen in woodland savanna (composite of the early plot in derived savanna of Olokemeji Forest Reserve, western Nigeria, and *Isoberlinia–Afzelia* woodland in Kainji Lake National Park, north-western Nigeria). *Notes*: (1) Values in brackets indicate number of days in the year a flow is operative; (2) arrows indicate direction of movement of nitrogen; (3) compartments are approximately proportional in size and values are in g N m^{-2} yr^{-1}; (4) decimal values beside flow lines indicate nitrogen flow rates in g m^{-2} day^{-1}; (5) a = above ground; b = below ground; (6) the soil returns shown are estimates of nitrogen reaching the soil after loss of nitrogen by nitrification and ammonification during litter decomposition.

savanna, Kainji Lake National Park, both Nigeria). From soil analyses, Isichei estimated that the soil sink to a depth of 45 cm holds about 325 g N m^{-2}. From tissue and soil analyses and from controlled nutrient feeding studies (Opakunle, 1978) he has estimated that 2.5 per cent of the soil nitrogen becomes available each year for plant uptake. All of this is taken up by the vegetation in this model. Woody vegetation stores 45.6 g N m^{-2} and adds 3.5 g N m^{-2} in new wood each year, while returning 3.6 g N m^{-2} in litter and dead below-ground material. The herbaceous biomass takes up 2.4 g N m^{-2} annually, loses 1 g m^{-2} by volatilization with burning, and returns 1.3 g m^{-2}

in litter and dead below-ground material. This means that a deficit of about 1 g N m^{-2} yr^{-1} is incurred. Such a situation represents as nearly a balanced system as one is apt to find — there is always some 'leakage' of nutrients in any cycle. Open savanna shows a greater deficit. Isichei's composite model indicates that only 61 per cent of the available soil nitrogen is taken up by the vegetation; 39 per cent is subject to leaching, runoff, and use by micro-organisms. Woody vegetation utilizes 2.2 g N m^{-2} and returns 1.8 g m^{-2}, while herbaceous vegetation takes up 2.3 g m^{-2}, loses 1.4 g m^{-2} from burning, and returns 1.1 g. The debit balance is 1.6 g N m^{-2}, yr^{-1} plus up to 5.3 g N m^{-2} not taken up from the soil by higher plants and possibly lost.

These model estimates, while crude, illustrate that the most efficient mineral uptake and cycling occurs in woodland, and that considerable annual loss ('leakage') may occur in more open savanna. Such loss may, however, be made up by inputs. Both woodland and open savanna will receive nutrients in rain. West African savanna, for example, probably receives between 0.39 and 0.75 g N m^{-2} yr^{-1} (Adeniyi, 1982), with the most likely mean value about 0.45 g m^{-2} (Jones and Bromfield, 1970). Further addition comes from N_2-fixation by free-living soil micro-organisms (*Azotobacter* and *Clostridium*), from *Rhizobium*–Leguminosae associations and from the loose association between grass roots and the bacteria *Azospirillum lipeferum* and *A. brasilense*. An additional input for open savanna is N_2-fixation by soil crusts of blue-green algae (largely *Scytonema*) which may fix from 0.33 to 0.92 g N m^{-2} yr^{-1} (Isichei, 1980).

Of nutrients other than nitrogen, only sulphur is subject to loss through volatilization by burning; the others are returned to the soil in ash as well as by litter (ash is, however, most often moved by wind and early rains away from its site of origin). All nutrients are present in considerable amounts in rain-water. Most often limiting in West African savanna is phosphorus; most easily leached in cultivated land is potassium.

4. Energy utilization

Just as it may be argued that a community of plants tends to develop such that mineral cycling improves in efficiency, it may be argued that a community develops towards increased efficiency of energy utilization. Ecological efficiency, E, may be defined:

$$\frac{Y}{I} = E$$

where Y is yield of dry matter and I energy input — light in the case of plants, food in the case of animals (Conrad, 1977).

The logic of this argument is simple: if there is any unused light — i.e. any light not reflected back into space or absorbed — such light represents a free resource for primary producers, part of an unfilled niche. Because of plant competition and reproductive expansion, any unfilled niche tends, in time, to become filled. Thus, as shading occurs from the growth of large plants, smaller plants with low light compensation points enter the system.

Working in a region of northern Guinea savanna in Nigeria, where the mean daily global radiation during the growing season (May to November, about 200 days) is approximately $16\,242$ kJ m^{-2} day^{-1}, Isichei found the mean dry-matter production in open grassland savanna to be 725 g m^{-2} yr^{-1} and in a nearby woodland, 1295 g m^{-2} yr^{-1}, representing ecological efficiencies of 0.4 and 0.70 per cent respectively.

As has been seen, savanna woodland is more efficient in mineral cycling; it can now be seen that it is more efficient in energy utilization. This strongly suggests that savanna, if undisturbed except by annual burning, will tend to develop into woodland savanna (see Figs. 5.2 and 5.3).

III. Causes of savanna

It has long been argued that savanna is *caused* by something, the 'something' sometimes held to be climate (low moisture availability for at least part of the

Fig. 5.2   Woodland savanna: *Isoberlinia–Afzelia*, Kainji Lake National Park, north-western Nigeria. (Photograph W. W. Sanford.)

Fig. 5.3. Juvenile bushland (*Burkea–Detarium–Terminalia avicennioides* Savanna) Kainji Lake National Park, north-western Nigeria. (Photograph by W. W. Sanford.)

year), sometimes soil (shallow or/and infertile soil) or, a more recent favourite, anthropic disturbance, especially burning. We will examine these hypothetical causes, although stating at the outset that we do not believe savanna has a *cause* but rather than this broad vegetation system develops in response to a cluster of interacting factors.

1. Climate and soil

Causation by climate or/and soil is usually linked with the old climax or poly-climax theory. According to the classical view of Clements (1916), still favoured by some geographers, vegetations develop directionally through a number of successional phases until a stable and ecologically balanced system, the climax, is achieved. Successions are held to be convergent, all ending up in a state determined by climate according to the climax theory, or by climate, soil, and possibly other factors such as anthropic disturbance according to the poly-climax theory.

As the introductory remarks to section II, above, indicate, we do not subscribe to any simple view of succession and climax. In the first place, vegetation varies over time species by species. Such an individualistic understanding of community change, sometimes considered modern and controversial, was, as Horn (1975) points out, already accepted as common knowledge by Thoreau in 1860. Secondly, there is good evidence (e.g. Matthews, 1979) that successions do not converge, that they strongly tend to diverge.

It cannot be argued, however, that climate and soil do not have great importance in determining the composition and structure of vegetation. As already indicated, the moisture regime is fundamental. It hardly needs expatiation that some plants are adapted to low moisture or/and extended dry periods and some are not. The question is, rather, can the definite structure which we have described as savanna be definitely associated with the moisture regime?

Taking Keay's (1959a) savanna zonation for Nigeria as an example, definite ranges of the ratio of annual precipitation/potential evapotranspiration can be seen in Table 5.1.

Table 5.1.

Vegetation zone	Precipitation/evapotranspiration ratio
Derived savanna	0.75–1.0
Southern Guinea savanna	0.65–0.88
Northern Guinea savanna	0.40–0.66
Sudan savanna	0.21–0.40
Sahel (thorn scrub)	< 0.21

But this hardly answers the question, as Keay may have as likely defined the zones by climate as the vegetation of the zones may have been determined by climate.

From the definition of savanna (Section 1.4, above), it is clear that moisture availability must be seasonal, that a dry season must be long enough and severe enough for most herbaceous vegetation to dry thoroughly to ground level. The major problem is whether or not a grass cover would develop in response to such a climatic regime rather than a dry forest — without appreciable grass. This question at once brings soil and topography into the picture. Riverine or gallery or riparian forests (or woodlands) are frequent in savanna regions. Such bands of vegetation follow the course of seasonal or perennial streams and often approximate to rain forest in structure and physiognomy. Grass is often absent. Such vegetation is, however, largely restricted to the more humid Guinea savanna, although trees and shrubs are taller and denser along watercourses even in semi-arid regions. The species composition of such woodlands, while containing species found in rain forest, is distinctive, not mirroring rain forest but rather containing a few widespread forest species (e.g. *Anthocleista vogelii, Antiaris africana, Ceiba pentandra,*

Cola gigantea), together with larger savanna trees (e.g. *Vitex doniana, Khaya senegalensis*) and trees found only along streams in either forest or savanna (such as *Berlinia grandiflora, Irvingia smithii, Pterocarpus santalinoides*). It is likely that the water-table, soil fertility, and atmospheric humidity in such relatively narrow bands along streams, often in steep valleys, approximate the abiotic environment of seasonal forest regions, and that such vegetation would not survive at a distance from the stream whether or not there was anthropic disturbance.

More intriguing are patches of dry forest, usually dominated by *Diospyros mespiliformis*, found in the midst of Guinea savanna on more or less level ground away from streams. (Fig 5.4). Our recent examination of such dry forests in northern Guinea savanna in Nigeria shows that *Diospyros* saplings of all size classes develop under *Diospyros* canopy, while other species are rare; thus the next generation is very apt to have the same composition and structure as the present one. Such one-species-dominated forests appear to conform to classical views of climax (especially as defined for tropical rain forest by Connell, 1978). While moisture availability is probably a contributing factor to the development of such vegetation patches, soil depth (associated with moisture availability and nutrient store) and character are perhaps more important. Even in the absence of disturbance, the requirements of *Diospyros* when occurring in drier regions (as opposed to rain-forest areas) seem to be specialized enough to preclude the spread of such dry forests over any extensive area.

Superficially similar, except for the presence of a grass stratum, are fairly dense woodlands dominated by *Isoberlinia* spp. or by *Isoberlinia* and *Afzelia africana*. Our observations so far, however, lead us to conclude that such woodlands may not be replaced by the same species composition in the next generation.

Less obvious than the variation in savanna brought about by such woodland patches is the gradual variation in structure and species composition with toposequence: the vegetation at the top of a hill is different from that on the lower slopes or at the base. Such variation is often associated not only with drainage and water-table but also with soil structure, almost impenetrable ironstone often lying sufficiently near the surface to restrict root development.

It is apparent, then, that soil characters — depth, structure and texture, fertility — strongly influence vegetation on a local scale. This still does not answer the question of whether or not savanna would develop in response to soil and climate without anthropic interference. The conclusion of the 1964 IGU–Unesco symposium in Venezuela, referred to above (section I.3) was that: (1) savanna is not a climatic climax; (2) a majority of the units of this vegetation can be considered as anthropic — existing because of the activities of man; (3) there are also natural savannas that can be considered as edaphic climaxes. Acceptance of this without considerable reservation seems rash. It is

Fig. 5.4.   *Diospyros mespiliformis* with a dense undergrowth of saplings growing on an old termite mound in *Isoberlinia-Afzelia* woodland, Kainji Lake National Park, north-western Nigeria. (Photograph by W. W. Sanford.)

known that extensive grasslands, probably fitting our definition of savanna, existed in East Africa long before there was either burning to any extent or cultivation at all (Livingstone, 1975). Furthermore, such grasslands were too extensive to be edaphically developed. A much more reasonable approach is that of Sarmiento and Monasterio (1975) who put forth a holocentric interpretation of savanna as a broad ecosystem type, the multiple factors leading to its development varying in different climatic–vegetation zones. They point out that in regions of rain forest in tropical rainy climates (such as tropical America, where they worked), the type of forest (species composition, structure, physiognomy) clearly varies with the rainfall regime; and where the rainfall is variable from year to year or is slightly low or where the dry season is prolonged, savanna vegetation may develop and supplant forest in local areas of sandy soil, senile soils of latocolic evolution, or on very shallow soils. In drier tropical forests, savannas develop first in areas of poorer and/or shallower soils, but with extreme drought stress they may develop even over fertile soils. Thus savannas are seen to develop in response to a dynamic interplay of soil and climate — and species available for establishment. Clearly, burning may augment any of the conditions tending to bring about savanna development and may lead to its extension.

Menaut (1983a) has recently reviewed probable causes of savanna and concluded that the superior adaptation of grasses to extreme seasonality and low or/and variable moisture supply largely accounts for grass formations replacing or occurring with woody formations. He points out that at the regenerating stage woody plants and grasses are in competition and grasses have the advantage in infertile soils and regions of drought stress. Medina (1983) has also emphasized the importance of adaptation to low soil nutrient status of many grasses.

That all West African savanna represents a replacement of forest is questionable. As Livingstone (1975) has pointed out, Africa was never covered by forest, although there was great forest expansion 10 000–7000 years BP so that much of West Africa may have been forested and this may have been replaced by savanna during subsequent drier periods. For about the last 3000 years, however, moister conditions have prevailed and savanna regions have become hospitable to a somewhat wider range of plants.

2. Anthropic disturbance

The question of whether or not savanna as defined would be present except in patches, where poorer soil and climate worked together for its development, without anthropic disturbance has still not been definitely answered. If savanna is undisturbed it moves towards woodland, would such woodland have a grass cover and fit our definition of *savanna* woodland or would grass disappear and the vegetation be dry forest?

Menaut (1977), working in Ivory Coast, reports that woodland savanna cannot normally develop unless the area has had some fire protection, but once this has been afforded 'transformation of an open shrub savanna into a dense savanna woodland seems irreversible whatever might be the future of the plots without any major climatic change'. Extensive savanna woodlands in West Africa which are annually burnt and yet still remain woodland bear out Menaut's latter contention. The question remains, however, whether grass would disappear from such woodland if burning ceased.

The Olokemeji Fire Plots in derived savanna of south-western Nigeria were established in 1929–30. The fire-protected plot is now without grass and is a regrowth forest. This does not provide us with information as to what would happen in Guinea or Sudan savanna. The Red Volta Fire Plots in north-eastern Ghana are on the border of northern Guinea and Sudan savanna with a mean annual rainfall of c. 1100 mm. They were established in 1947 and the most recent report (Brookman-Amissah *et al.*, 1980) shows that in 1977 the fire-protected plots had a density of trees at least 30 cm in girth of 0.02 m^{-2} which is comparable to the density of many areas of annually burnt savanna in Nigerian northern Guinea savanna. These plots had a grass biomass (182 g m^{-2} dry matter) greater than that of the late burnt plots and greater than that of the annually burnt plots at the border of northern Guinea and Sudan savanna in north-western Nigeria as monitored by the Nigerian Man and Biosphere team in 1979 (Sanford *et al.*, 1982b). Whether or not grass will continue in this plot cannot be known. It will depend on which is able to withstand full canopy shading better: grass species or woody plant saplings. Results of our current research indicate that several species of annual grasses (e.g. *Pennisetum pedicellatum, P. subangustum, Brachiaria* spp.) and a few perennials (.e.g. *Andropogon tectorum, A. gayanus, Beckeropsis uniseta*) tolerate or prefer some shade and will continue growth under full canopy if the canopy is high. Small woody plants, however, and grasses are serious competitors. Thus, whether or not grass will continue in the fire-protected plot will probably depend on the size-class distribution of woody plants. At present, 84 per cent of the woody plants are less than 30 cm in girth in the Ghana plots; a stable situation for Guinea savanna is 82 per cent of plants less than 40 cm in girth (Sanford *et al.*, 1982a). The Ghana plot does not appear to be far from a stable distribution, and canopy shading by the larger trees may limit expansion of the percentage of saplings. Ultimately, whether or not grass continues, and whether or not the density of small woody plants will be limited and more or less stable, will depend upon an interplay of soil, climate, and species composition of the area.

If land is cleared and cultivated, then abandoned to nature, the process of vegetational change, termed 'secondary succession', begins. In humid regions of rain-forest potential, such an open area is first dominated by annual forbs (e.g. *Ageratum conyzoides, Synedrella nodiflora, Spigelia anthelmia*), annual sedges such as *Mariscus alternifolius*, and annual grasses such as *Brachiaria deflexa*. Perennial forbs then begin to increase in number. In one of our tests in south-western Nigeria the mean number of species in five 2500 m^2 plots was 19.6 the first year and 21.8 the second year with the mean rooted shoot density per square metre increasing from 123 to 273 and the percentage of broad-leaved plants increasing from 47.0 to 86.8. With further time, annual grasses such as *Pennisetum subangustum* and perennials such as *Panicum maximum* and *Rottboellia exaltata* dominated some plots while in others the broad-leaved *Melanthera scandens* and *Ipomoea involucrata* remained most abundant. In all cases the broad-leaved composite *Eupatorim odoratum* became dominant and nearly the sole species four or five years after the succession began, to be gradually replaced by woody plants (first *Trema orientalis* and *Musanga cecropioides*). Only if the cleared area is annually burned will grasses continue to dominate after the third or fourth year. In such cases, *Andropogon tectorum* often becomes dominant in deep fertile soils and *Monocymbium ceresiiforme* and *Loudetia* spp. in shallow soil.

In drier, more northern regions, the pattern of succession is quite different. Annual grasses dominate the first year with only a scattering of forbs (such as *Blumea aurita, Borreria* spp., *Combretum sericeum*). The only detailed record available for an appreciable period of time is for the north-eastern Ghana fire plots, referred to above, where grass continued to be dominant in the fire-protected area. The number of species of forbs, after 30 years of protection, was 20 compared to 16 in the early burnt and 15 in the late burnt area. These results suggest that climate without burning may bring about predominantly grass as the herbaceous vegetation in drier regions, but that in moister regions burning is necessary. In all cases either anthropic disturbance or cultivation or intense burning or locally very poor or shallow soil prevent savanna from moving towards woodland savanna as opposed to open or bush savanna. (Fig. 5.5).

IV. Effects of fire

1. *Time of burning and general effects*

Although burning has already been mentioned as a frequent contributory cause in the development and extension of savanna, it is necessary to discuss its effects further as it is almost invariably a management practice throughout West Africa. How long burning has been regularly practised is unknown. Man probably evolved in East African grasslands very long ago, but his transition from hunting and gathering to farming began only about 10 000 years ago (Washburn, 1978) and burning before and considerably after that must have been very rare. It is of course true that burning may result from natural causes, but we have never accepted as in any way likely Rose Innes's (1972) too often quoted suggestion that fires may start from sparks caused by boulders rolling

Fig. 5.5 A region of northern Guinea savanna where cutting and burning over the years have led to a degraded Shrub Grassland. (Photograph by W. W. Sanford.)

down slopes. Lightning does, on occasion, cause fires, but as Walter (1971, p. 239) remarks, fires caused by lightning are much less frequent than those caused by man and usually do not spread very far in natural woodland.

Savanna is most often burnt, as Menaut (1977) remarks, in the middle of the dry season — that is, usually during January, although some burning begins in mid or late December and extends through February. In areas of cattle migration, burning occurs shortly after the arrival of herds from more northerly ranges. As vegetation in the north dries and streams and water-holes evaporate, livestock move into more humid areas, first making use of crop residues and areas of still green grass along stream beds. Then, as drying continues, the savanna is burnt to clear away dead standing material and to stimulate new growth (flushing) of the grass. Tentative work by Opakunle, unpublished, indicates that either burning or clipping at ground level and removal of grass will bring about flushing but burning is more effective. (See Fig. 5.6).

Such savanna burning besides removing, on the average, about 82 per cent of the herbaceous above-ground biomass (Isichei and Sanford, 1980) causes leaf shedding of most trees. This is followed by leaf flush, normally occurring in the late mid dry season (February in the Nigerian Guinea zone). Brookman-Amissah et al. (1980) report, for the north-eastern Ghana experiment, that by March, 57 per cent of the trees in the fire-protected plot were in leaf, 76 per cent in the early burnt (burnt four months earlier), and 43 per cent in the late burnt area, as yet unburnt. These data suggest that fire has some stimulating effect on tree flush, perhaps comparable to its effect on grass. That the effect is not brought about alone by difference in species composition is shown by the same difference in flushing time occurring for *Entada africana* and *Combretum glutinosum* which occur in all plots.

2. Soil

Soil organic matter is of great importance in improving soil texture and its water-holding capacity, in increasing cation exchange capacity, and in providing nitrogen. Because of high soil temperature, organic material is very rapidly mineralized in the tropics so its concentration in the soil is generally low. In some situations the carbon content of soil at a depth of 30–45 cm is greater than at 0–15 cm. This condition, especially frequent in savanna woodland, may be brought about both by more rapid mineralization of humus near the soil surface, where temperatures are higher, and by the presence at lower levels of increased underground biomass of woody vegetation. Because of mineralization near the surface, litter fall is of surprisingly little importance in adding humus to the soil. Even though only about 18.4 per cent of the year's

Fig. 5.6   Burning in early January in northern Guinea bushland, northern Nigeria. (Photograph by P. J. Newbould.)

leaf litter fall and 11.7 per cent of the wood fall is burnt by annual fires (Isichei and Sanford, 1980) there is not a significant positive relationship between litter fall and soil organic matter content (although there is between tree density and soil organic content at the 30–45 cm depth). It is very unlikely that organic matter in the upper level of soil is oxidized by the annual fires, as soil temperature at a depth of 2 cm is raised by 14 °C at most and often by as little as 3–4 °C, and then only for a few minutes during burning (Ramsay and Rose Innes, 1963). However, removal of shading material by burning will lead to higher soil temperatures during the remainder of the dry season and this will lead to more rapid mineralization. Brookman-Amissah et al. (1980) found a significant decrease of organic matter with burning at the 0–5 cm depth but not at lower depths, although a statistically insignificant trend towards more organic matter beneath fire-protected vegetation can be seen from their data. Moore (1960), working at the Olokemeji Fire Plots 30 years after the commencement of controlled burning, found the humus content of the upper 20 cm of soil higher under early burnt vegetation than under the fire-protected plot, but lower under the late burnt vegetation than under that which was fire protected. His data were not, however, subject to statistical analysis so it is impossible to know whether or not the differences reported are significant. Isichei (1979) has reported soil carbon concentrations of the same plots 50 years after controlled burning began and found the same relationships as Moore at the 0–15 cm depth.

Moore (1960) and Isichei (1979) reported the same trend for soil total Kjeldahl nitrogen at Olokemeji as for carbon except that Isichei found little variation from plot to plot at the 30–45 cm depth. Brookman-Amissah et al. (1980) found significant difference only at the 0–5 cm level.

The two elements, carbon and nitrogen, are most likely to be affected by burning as they are volatilized in material burnt and lost from the system. Other nutrients, excepting sulphur, are returned to the soil in the ash, although they may be displaced by wind and water. Considerable attention has been given to nitrogen loss through burning. Values as high as 28.02 kg ha^{-1} or 0.28 g m^{-2} (25 lb per acre) for tall grass in the derived zone of Ghana have been reported (Nye and Greenland, 1960, p.54), but Isichei and Sanford (1980) estimate that in natural (unimproved) savanna only about 13 kg ha^{-1} yr^{-1} (range of from 11 to 16 kg ha^{-1} or from 0.11 to 0.16 g m^{-2}) are lost by burning. Clearly, an equivalent amount must be added to the system if deterioration is not to occur. As we have shown above, considerable input of nitrogen comes from rain and from biological fixation and this explains why there is little if any difference in nitrogen content of the soil with annual burning.

Sanford (1982a) in a review of burning has stated: 'on the whole, climate appears more important than burning in limiting the concentration of organic matter in the soil through its effect on the rate of mineralization. Most soil carbon appears to be derived from underground biomass.' Nitrogen more or less follows carbon.

Burning temporarily raises soil pH slightly. In some soils, especially in the derived savanna, texture may be improved by burning (Fagbenro, 1982) but in others, especially in northern Guinea regions, surface compaction may result, as indicated by the increase in bulk density reported by Brookman-Amissah *et al.* (1980). Considerable redistribution of non-volatile minerals occurs on a micro-scale with wind movement of ash and also some, via rain, on a larger scale. All in all, however, annual burning has little effect on the soil and whatever effects are observed are more dependent on vegetational change brought about by burning than by burning itself.

3. Vegetation

(a) Physiognomy and structure

If one looks at the Olokemeji Fire Plots today, 51 years after the controlled burning experiment started, one sees, regrowth rain forest in the fire-protected plot, while the early burnt plot superficially has the structure and physiognomy of southern Guinea savanna and the late burnt plot looks like northern Guinea savanna. Using only physiognomic characters Isichei carried out principal components analysis of the three Olokemeji plots together with 14 sites including all types of savanna. The results are shown in Figure 5.7. Clear clumping of the Olokemeji plots at the extreme right is seen on the ordination of the first and second axes. These results indicate that while annual burning affects structure and physiognomy of vegetation, 50 years is not enough to turn derived savanna into Guinea savanna.

The characters most affected by burning are shown in Table 5.2 (taken from Sanford, 1980). These characters are useful in classifying vegetation, as will be discussed below. As can be seen from Table 5.3, the percentage of woody legumes increases with burning in derived savanna. This was also true in the north-eastern Ghana experiements. Here the highest percentage of legumes was found in the early burnt plots in 1960, but by 1976 the percentage of woody legume individuals was 73 per cent in the late burnt plots, 66 per cent in the early burnt plots, and 52 per cent in the fire-protected plots.

Of particular interest is the effect of burning on succession and stability in savanna. The demographic changes in woody plant populations at Olokemeji are shown in Table 5.4 (also from Sanford, 1982a). From these results it can be seen that burning, by removal of small saplings, results in a larger mean girth and hence a distinctive size-class distribution. A stable girth size distribution is reached in less time in burnt than in unburnt vegetation, and probably in less time in late burnt than in early burnt vegetation (Sanford, 1982a). One reason for this is the restricted number of species which can survive the more intense

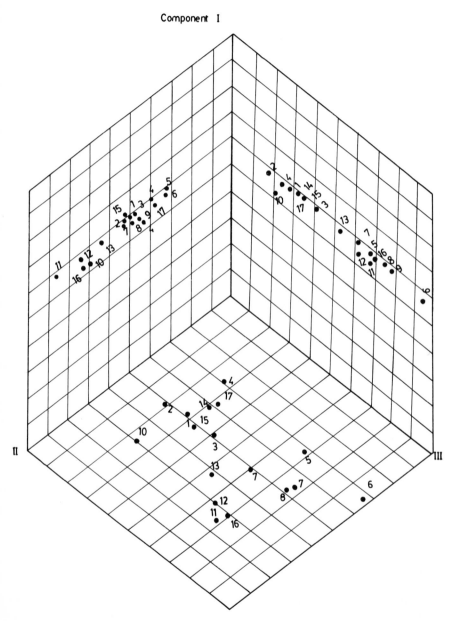

Fig. 5.7. Principal components ordination of 17 savanna sites in Nigeria using characters given in Table 5.2, and listed by site name and location and vegetation type (Keay classification).

(1)	Igbeti: 08°46'N, 04°09'E;	southern Guinea.
(2)	Borgu, Oli River: 10°06'N, 3°38'E;	northern Guinea.
(3)	Borgu, Tungan Giwa: 10°06'N, 3°38'E;	northern Guinea.
(4)	Olokemeji Plot A: 7°26'N, 3°32'E;	derived savanna.
(5)	Olokemeji Plot B: 7°26'N, 3°32'E;	derived savanna.
(6)	Olokemeji Plot C: 7°26'N, 3°32'E;	derived savanna.
(7)	Near Gidan Waya: 09°37'N, 08°24'E;	southern Guinea.
(8)	Near Tungan Giwa: 10°06'N, 3°38'E	northern Guinea.
(9)	Oli River bank: 10°6'N, 3°38'E;	northern Guinea.
(10)	Vom Hills, Plateau: 09°41'N, 08°44'E;	plateau savanna.
(11)	Near Gusau: 13°24'N, 05°10'E;	Sudan savanna.
(12)	Km 313 Sokoto–Zaria Road: 11°07'N, 07°44'E;	Sudan savanna.
(13)	Near Zaria: 11°07'N, 07°44'E;	Sudan savanna.
(14)	Km 132 Kaduna–Jos: 09°55'N, 08°53'E;	Sudan savanna.
(15)	Near Keffi: 10°17'N, 08°40'E;	southern Guinea savanna.
(16)	25 miles N of Sokoto: 13°04'N, 05°16'E;	Sudan savanna.
(17)	Gurara falls near Suleja: 08°12'N, 06°41'E;	southern Guinea.

late burning together with discouragement of the immigration of new species (Fig. 5.8).

Some idea of the relative stability of vegetation may be gained from species diversity estimates, although this relationship is still controversial. It has usually been claimed that diversity increases with successional age and that maximum diversity is associated with climax. North American studies of old fields (e.g. Bazzaz, 1975) have generally shown that there is an initial rapid increase in diversity followed by an abrupt drop (as in the example given earlier, where species diversity increased until an abrupt drop was brought about by dominance of *Eupatorium odoratum*), then a slow rise (in our example as woody plants began to supplant *Eupatorium*). Odum (1969), however, has suggested that theoretically species diversity should decline somewhat at climax, and Connell (1978), at least for tropical forest, has presented evidence that at climax the vegetation is dominated by one or a few most successful species so diversity is low. This is in logical agreement with May's (1973) contention that increased diversity confers increased fragility upon a system — at least as mathematically modelled.

Species diversity, evenness, and richness are shown for the Olokemeji and Red Volta (Ghana) Fire Plots in Table 5.5 (Sanford 1982a from data of Isichei, 1979, and Brookman-Amissah *et al.*, 1980). Increased evenness is probably positively related to increased stability, and a tendency for evenness to increase with burning is clear in both data sets. Species diversity, on the other hand, is consistently highest in fire-protected plots as is species richness.

(b) Life-form

Considerable controversy has been engendered concerning the effect of late and early burning on the proportion of perennial to annual grasses. Examina-

Table 5.2.   MAB-3: Field data sheet. Physiognomic/Structural Savanna Vegetation
Survey
Site: ....   Locality: ....   Date: ....   Enumerator: ....

Attribute	Rank
	1 2 3 4

A. Tree height and canopy
  1.   Canopy: no closure; patchy cover; *c.* 50% closure;
      almost complete closure
  2.   Trees over 10 m: none, very few; 10–25%; 25–50%;
      over 50%
  3.   Trees/shrubs 3–6 m: (rank as above)
  4.   Trees/shrubs under 3 m: (rank as above)
  5.   Trees/shrubs under 3 m: mostly scattered; some in
      clumps or thickets; moderately clumped; extensive
      thickets

B. Leaf/leaflet characters of woody elements
  6.   Compound: none, very few; 10–25%; 25–50%; >50%
  7.   Less than 5 cm long (rank as above)
  8.   Over 10 cm long (rank as above)
  9.   Pubescent, hairy or glaucous (rank as above)
 10.   Tip acuminate (rank as above)

C. Tree/shrub habit and appearance
 11.   High-branched, first branch over 6 m from ground (rank
      as above)
 12.   Low-branched, first branch less than 2 m from ground
      (rank as above)
 13.   Branches of trees over 5 m tall at an angle of *c.* 45° or
      less with bole (rank as above)
 14.   Branches crooked or twisted (rank as above)
 15.   Boles slanting, crooked or twisted (rank as above)
 16.   Bark dark and prominently fissured or flaky (rank as
      above)

D. Taxonomic features of woody elements
 17.   *Elaeis guineenis*: none; few; moderate; many
 18.   Leguminosae: none or very few; 10–25%; 25–50%;
      over 50% of woody plants present

E. Herbaceous vegetation: life-form and taxonomic features
 19.   Grass over 2 m high (rank as above, on cover basis)
 20.   Grass under 1 m high (rank as above, on cover basis)
 21.   Broad-leaved herbs (rank as above, on cover basis)
 22.   Leguminosae: none or very few; 10–25% of broad-
      leaved herbs present; 25–50%; over 50% broad-leaved
      herbs present
 23.   Cyperaceae: none or very few; some in small patches;
      moderate amount, spread; many spread or in large
      patches

Table 5.2. (*continued*)

Site: .... Locality: .... Date: .... Enumerator: ....

Attribute	Rank			
	1	2	3	4

24. Aloe (rank as above, No. 23)
25. Succulent Euphorbiaceae (rank as above, No. 23)

F. Surface soil and topography
26. Slope: none; gentle slope; moderate; steep
27. Bare ground: not visible; in few small patches; scattered
      patches; large patches present
28. Surface soil: red-brown; light brown; grey; dark grey-
      black
29. Surface soil: gravelly; sandy; some clay; much clay
30. Termite mounds: none; few; moderate; many (all types
      of above-ground mounds)

G. The eight (8) most abundant woody plants, in order of abundance
  1.
  2.
  3.
  4.
  5.
  6.
  7.
  8.

H. The four (4) most abundant grasses, in order of abundance
  1.
  2.
  3.
  4.

I. Comment

tion of available quantitative data has led us to conclude that in the relatively moist derived savanna areas, burning time and perennial : annual grass ratios are not significantly related. Isichei has shown that in the late burnt Olokemeji plot about 92 per cent of the yield was of perennial grasses and in the early burnt plot about 91 per cent (both on the basis of dry weight). In drier, more marginal regions there does not seem to be any relationship of the ratio with burning time, but more annuals are found in fire-protected than in burnt areas in the Ghana experiments. Afolayan (1979), working in northern Guinea savanna in Nigeria, has reported a tendency for an increase of perennials with late burning, but statistical significance is lacking.

Table 5.3. A comparison of physiognomic/structural attribute scores of the early and late-burnt fire plots at Olokemeji Forest Reserve with those of the fire-protected plot

Attribute	Attribute scores expressed as percentage of the fire-protected plot score		
	Plot B (early burnt)	Plot A (late burnt)	Difference between A and B
Group I			
Canopy closed	86	53	33
Leaves/leaflets glabrous, glossy	95	75	20
Trees high-branched; above 6 m	85	46	39
Boles/branches ± straight	76	41	35
Bark smooth, light coloured	76	47	29
Leaves/leaflets over 12 cm long	76	76	0
Group Ib			
Branches acutely angled with the bole	100	63	37
Trees over 9 m tall	106	69	37
Leaves compound	150	75	75
Group II			
Bark fissured, dark coloured	133	167	34
Grass over 1.5 m high	271	285	15
Leaves/leaflets pubescent, hairy	100	120	20
Trees branched at above 1.5 m	138	175	37
Branches, boles crooked or twisted	162	188	26
Group IIb			
Woody Leguminosae	220	180	40

Table 5.4. Demography of the three Olokemeji Fire Plots, south-western Nigeria

	Condition in 1976			Population change between 1969 and 1976%			
	Number of individuals	Species	Mean girth	Dead or missing	New individuals	New species	Species extinctions
Plot[a] A	133	14	34.6	12	15	0	14
B	236	24	34.2	11	25	4	8
C	456	33	27.4	15	26	5	18

[a] A—late burnt; B—early burnt; C—fire protected. The percentage of species with dead individuals and no new individuals is 29% in A, 25% in B, and 37% in C.

Fig 5.8.   Fire scar on the bole of *Crossopteryx febrifuga* in northern Guinea bushland. Both this species and *Afzelia africana* almost invariably bear fire scars and probably a high percentage of the young saplings are killed by fire. (Photograph by W. W. Sanford.)

Table 5.5. Species diversity, evenness, and richness of the fire plots of Olokemeji, Nigeria, and Red Volta, Ghana

	Diversity		Evenness[a]		Richness
	H'	$e^{H'}$	$(l/C \cdot e^{H'})$	$H'/\log_e N$	Spp. number/ $\log_e$ area
Olokemeji 1976					
Late burnt	2.32	10.2	0.82	0.88	1.88
Early burnt	2.36	10.6	0.59	0.74	3.22
Not burnt	2.82	16.8	0.72	0.78	5.09
Ghana					
Late burnt					
1950	1.70	5.47	0.76	0.68	1.30
1977	1.88	6.55	0.70	0.73	1.41
Early burnt					
1950	1.91	6.75	0.85	0.80	1.19
1977	1.58	4.85	0.62	0.60	1.52
Not burnt					
1950	1.82	6.17	0.79	0.76	1.19
1977	2.58	13.19	0.74	0.74	3.47

[a] C = the Simpson coefficient: $1/\Sigma p_i^2$, where p_i is the percentage of individuals of the *i*th species; N = total number of species.

Species distribution of grasses appears to be largely controlled by soil fertility and depth, tree canopy cover (i.e. shading), and mechanical disturbance. Some species such as *Loudetia* spp., *Monocymbium ceresiiforme*, and *Schizachyrium exile* can manage in poor, shallow soils, while others such as *Andropogon tectorum* require deep, fertile soils. Species such as *Hyparrhenia* spp. and *Schizachyrium* spp. do best in full sun, while other such as *Andropogon* spp. and *Beckeropsis uniseta* do best in light shade. In general, frequent mechanical disturbance leads to an abundance of annuals and such weedy perennials as *Rottboellia exaltata* and *Sporobolus pyramidalis*.

In summary, annual burning leads to a relatively stable vegetation from the standpoints of structure and size-class distribution of trees and general physiognomy. Species diversity decreases with burning as does species richness, but evenness increases. Various physiognomic characters are infiuenced by burning, especially canopy closure which is decreased, due to the prevalence of high branching; mean tree height is decreased, mean girth is increased. With burning, species are selected which have smaller leaves; pubescent or glaucous leaves are more common as are compound leaves. Taxonomically, the percentage of woody legumes increases with burning. Time of burning does not appear to influence the proportion of annual to perennial grasses. Grass species distribution is influenced rather by soil, mechanical disturbance, and shading; the latter is, however, strongly affected by the burning regime.

4. Primary production

(a) Woody plants
Most data relating burning to primary production consist of stem counts or basal area estimates of woody plants. A few examples make the general trend clear. After 47 years of controlled burning at Olokemeji, the late burnt plot contained 16 per cent of the total number of stems of the three fire plots combined, the early burnt 29 per cent, and the fire protected 55 per cent. An estimation of wood volume (Isichei, 1979), however, gives a different picture: the late burnt plot contained 15 per cent of the total volume of the three plots, the early burnt 48 per cent, and the fire protected 37 per cent. The high volume of the early burnt plot is accounted for by the larger mean girth of the trees in this plot. Other data from derived savanna are those from the fire experiments at Kokondekro in Ivory Coast (reported by Ramsay and Rose Innes, 1963). After 25 years of controlled burning, the late burnt plot contained 17 per cent of the total stems, the early burnt 36 per cent, and the fire protected 47 per cent. These figures are not so different from those from Olokemeji, even though Kokondekro was not initially clear-felled as was Olokemeji. The only detailed data from 'true' savanna are those from the Red Volta (Ghana) Plots. Here, after 27 years, the late burnt plot contained only 7.5 per cent of the stems over 30 cm in girth, the early burnt 15.6 per cent, and the fire protected 77 per cent, indicating a considerably greater effect of burning in the drier region than in the more humid areas.

(b) Herbaceous plants
Other than firewood, the major products of the savanna are livestock fodder, thatching, and mat-fencing materials — all herbaceous. Isichei (1979) has compared herbaceous maximum standing crop in the three fire plots at Olokemeji over a three-year period. The late burnt plot averaged 496 ± 68 g m^{-2} dry weight while the early burnt averaged 400 ± 43 g m^{-2} — not a comfortably large difference, while the fire-protected plot produced virtually no herbaceous crop. The late burnt yearly mean was greater than that of the early burnt two years out of three.

Afolayan (1977, 1978) has compared maximum herbaceous standing crop in savanna of three vegetation types at Kainji Lake National Park, north-western Nigeria, ranging from very open savanna to woodland. The areas had previously been burnt annually, probably in January, for many years. Without clear felling, Afolayan subjected the plots to controlled burning for two years. Grass dry weight ranged from 88 to 264 g m^{-2} in the fire-protected plots, from

110 to 393 g m^{-2} in the early burnt, and from 244 to 622 g m^{-2} in the late burnt plots, with late burnt plots in comparable vegetation always having the greatest yield. Isichei (1979) reported two-year mean values of 324 g m^{-2} for open savanna and 259 g m^{-2} for woodland, both burnt in January. Very interestingly, Afolayan's data allow estimation of grass growth rates, using the exponential equation

$$r = \frac{\log_e N_t - \log_e N_0}{t}$$

and analysis of variance shows significant difference between rates with burning treatment at $P<0.01$. Growth rates were highest in unburnt and early burnt plots (0.49, 0.47) and lowest in late burnt (0.42) where the greatest yield was achieved. Examination of regression equations (time : yield) show that the highest Y-intercept occurs with the late burnt plots, indicating that the most grass is already present at the beginning of the growing season. This initial grass may represent the flush stimulated by late burning, whereas the flush stimulated by early burning may have been unable to withstand the long period before the rains and so dried away. That this might be so is suggested by there being little difference between late and early burning yields in derived savanna (Olokemeji) where, because of moister conditions, early flush may survive until the rains.

On the other hand, Opakunle's current work (unpublished) shows that the growth rate itself is greater at least in the first month in unburnt grass.

The Red Volta (Ghana) data differ sharply from both the derived (Olokemeji) and the northern Guinea (Kainji Lake) data in that the grass biomass (dry weight) is significantly the greatest in the early burnt plots (260 g m^{-2}) and greater in the fire protected (182 g m^{-2}) that in the late burnt plots (144 g m^{-2}). This suggests that in the drier, northern savanna shading may be conducive to grass growth, possibly through amelioration of temperature with consequent reduction of transpiration and respiration (Sanford et al., 1982b).

(c) Summary
Very obviously, fire decreases the density of woody stems and late (intense) burning has a greater effect than early burning. In very humid savanna (as in derived savanna regions) the wood volume may sometimes be greater in early burnt plots than in fire protected ones in early successional stages, as smaller mean girth in the fire-protected area may not be compensated for by the increase in stem density.

The effect of fire on herbaceous production varies greatly with climate. In drier, more marginal regions early burning or even fire protection may result in higher grass yields than late burning. In the southern derived savanna, time of burning makes little difference in yield, but somewhat greater yield is achieved with late burning. In intermediate (Guinea savanna) regions, late burning usually appears to increase grass above-ground biomass, although no results of long-term experiments are available.

V. The Savanna Zones of West Africa

So far, an attempt has been made to define savanna and to review its probable causes. The zonation scheme of Keay (1959) and Aubréville *et al.* (1958) has been consistently used. The most difficult task remains: the attempt to express what is actually meant by these zones.

Papadakis (1965) has published a useful synthesis of the vegetation maps of West African countries. He uses the terms coastal savanna, transition forest (between rain forest and Guinea savanna), Guinea savanna, Sudan savanna, thorny tree or shrub (Sahel) savanna and montane savanna. This differs from Keay, but is in agreement with Chevalier (1900) in not separating Guinea savanna into southern and northern Guinea. Keay argues for this separation in his *Outline of Nigerian Vegetation*, but we have found, at least from the standpoints of management and agricultural potential, that the separation is not particularly useful: in practice it is certainly often difficult to make the distinction except perhaps on purely climatic grounds. (It may be argued that Clayton's (1957) separation of Sudan into sub-Sudan and Sudan savannas is more meaningful, but this has not been found useful in practice either.) Papadakis's use of 'montane' is somewhat looser than we would accept. We find Boughey's (1955) classification for tropical Africa more realistic in which 'montane' refers to vegetation found above 3000 m. According to this usage, montane savanna would be found in West Africa only on Mount Cameroun and on the island of Bioko. Hall and Medler (1975a, b) use the term 'highland' for vegetation above 1220 m (4000 ft). Most West African high areas could thus well be termed either highland or sub-montane savanna and woodland. We find that vegetation changes appreciably even when the elevation reaches only 760 m; this corresponds with the demarcation between Boughey's lowland and foothill zones. The factor largely responsible for change, as one of us has discussed previously (Sanford, 1974) is moisture availability. (It should be realized, however, that the classification of vegetation according to elevation alone is fraught with many of the same difficulties as classifying it according to climate or soil alone.)

1. Conventional zonation

Most recently, White (1983) has completed the *UNESCO/AETFAT/UNSO Vegetation Map of Africa* (Fig. 1.1). 'Savanna' is not used at all because the author not only finds its definition lacking in precision but also considers it a foreign word to Africa. (It comes via the Spanish from the language of the

Caribbean Indians and was first used in 1535 by Oviédo to designate clearings
in the forest. It is difficult to see that this term is any more foreign to Africa than
the English and French terms so widely used.) The approximate equivalents
from this most recent work will be added at appropriate points in this chapter.

The conventional zonation most current in West Africa will very briefly be
reviewed and then a summary of our suggested, more detailed classification of
savanna more or less as we presented it at the Nigerian Man and Biosphere
Workshop in New Bussa (Sanford, 1982b) will be proposed.

(a) Guinea Savanna
Southern Guinea and transition or derived savanna are classed by White (1983)
as mosaic lowland rain forest and secondary grassland with most of the region
being Guineo-Congolian in vegetation with the more northerly region a mosaic
of lowland rain forest, *Isoberlinia* woodland and secondary grassland. This
implies that the entire region is 'derived' — a position with which we disagree.

The northern Guinea zone is termed Sudanian woodland with abundant
Isoberlinia.

Guinea savanna (the zone of gallery forest of Rousseau (1932), 'forêt
clairierée tropophile', Jacques-Felix, 1950, 1951) is a region of high grass at
least 1.2 m tall; characterized by such trees as *Lophira lanceolata, Daniellia
oliveri, Parkia clappertoniana, Isoberlinia doka* and *I. tomentosa*, various
Combretum spp., and *Terminalia* spp. These may be close enough together to
form a closed or nearly closed — although light and irregular — canopy
(woodland), or scattered (bushland or open savanna), or species of economic
use, particularly *P. clappertoniana, Butyrospermum paradoxum* and *Adansonia digitata*, may be scattered at some distance apart in open grassland almost
free of shrubs or smaller trees (parkland). Keay (1959a) describes southern
Guinea as open woodland with high grasses from 1.5 to 3 m tall, trees up to
12–15 m (rarely up to 30 m) high with rather short boles and broad leaves. The
trees *Lophira lanceolata, Daniellia oliveri, Afzelia africana, Hymenocardia
acida, Piliostigma thonningii*, and *Vitex doniana* are especially common. The
grasses are mainly *Andropogon* spp., *Hyparrhenia* spp., *Schizachyrium* spp.,
and *Pennisetum* spp. Transition or derived savanna (Fig. 5.9) is intermediate
between lowland rain forest and southern Guinea savanna and is considered
anthropogenic both in origin and maintenance — in other words, it is largely
defined by its presumed cause. This becomes a sticky problem in logic when
we consider that southern Guinea savanna is also often thought to be caused
and/or maintained by man. One can try to fall back on age and say that derived
savanna is more recently anthropogenically formed. In practice, however, the
difference is broadly physiognomic and structural, with derived savanna being
largely in mosaic patches or in very irregular bands containing a number of
forest trees and many oil-palms (*Elaeis guineensis*); the grass, *Andropogon*

Fig. 5.9. Transition or derived savanna, with a dense stand of *Andropogon tectorum* in the foreground and a mixture of small savanna and forest trees in the background. (Photograph by W. W. Sanford.)

tectorum, is especially abundant, and this gradually almost drops out as one proceeds northward into Guinea savanna.

Coastal savanna is in our opinion a form of Guinea or humid savanna closely allied to transition (derived) savanna. Keay (1959b), in his explanatory notes for the vegetation map of Africa south of the Tropic of Cancer, terms this formation 'coastal forest-savanna mosaic' or, in the Ghana region, 'coastal scrub and grassland zone', noting that because of the high atmospheric moisture derived from the nearby sea, the vegetation is somewhat different from inland regions of the same annual rainfall. The species found tend to be a mixture of inland savanna species and forest species. White (1983) terms this 'West African coastal mosaic' occupying an anomalous dry area of the extensive zone of transition between Guineo-Congolian and Sudanian regions. Such vegetation occupies a strip about 25 km wide along the coast from about 1° W to 2° 40′ E, although any exact border is impossible to set as it is continuous with what can be termed typical Guinea savanna (*sensu* Keay, 1959a). (It should be noted that this latter type of vegetation dips southward from western Nigeria through Ghana.) Coastal savanna has been largely determined by the moisture regime in the first place, and secondarily by very long-term grazing, land clearing, etc. The more or less open grasslands extending all the way to the sea in this area long ago very probably contributed to the early rise and power of the kingdom of Dahomey and later to foreign colonization and the slave trade. Savanna is much more hospitable to the deeds of men, whether they be peaceful or violent, than is forest. (And of course man arose from hominoid ancestral animals in the arid and semi-arid savannas of East Africa around Lake Turkana rather than in forested Africa or moist, fertile Mesopotamia.)

The rainfall/evapotranspiration ratio begins to decrease as we move from central Nigeria westward at about Ibadan (ratio of 1.00). South-west of this city, at Pobé on the Nigeria–Republic of Benin border, the ratio is also 1.00 even though this site is much nearer the sea and rainfall generally increases seawardly. The ratio at Ondo, Nigeria, at the same latitude but eastward is 1.49. Lagos, Nigeria, on the sea, has a ratio of 2.10, while Cotonou, Benin, also on the sea, has a ratio of 1.38; further west at Ouidah, Benin, the ratio has dropped to 0.99 and at Lomé, Togo, to 0.89. At Accra, Ghana, the ratio is only 0.70. These ratios correspond to those found in typical derived and southern Guinea savanna zones (see Section III. 1, above). Westward from here, the ratio again rises so that at Takoradi, Ghana, it is 1.36 and at Axim, 2.60.

This strange climatic dip has lead to the vegetational anomaly long termed the 'Dahomey Gap' which is ancient enough to have affected the West African distribution of birds (Moreau, 1966, 1969) and permitted some speciation in plants as well. In general, the coastal vegetation of Ivory Coast is similar to that of Cameroun and the extreme south-east of Nigeria, but a number of differences occur. The Accra plains have a remarkable concentration of endemic and disjunct species (Jeník and Hall, 1976).

East of Accra, there are extensive regions of impacted soils with such poor drainage that low-lying pockets are seasonally waterlogged, while higher ground is occupied by steppe-like vegetation — low grasses, often with rolled or cylindrical leaves, and sparsely scattered small shrubs. Some areas are almost completely bare. Slightly further inland, hills are covered with woody vegetation typical of transition savanna woodland of Nigeria and are surrounded by open grass plains, maintained in such an open condition by semi-nomadic grazing. This is in contrast with the more humid plaints west of Accra where farming is common. Thus the Ghana coastal plains east of Accra are very different from the western portion (often differentiated as the Winneba Plains). The coastal savanna of Benin and Togo varies from littoral sand communities of coconut palms and brackish marshes to scrub vegetation resembling highly disturbed transition savanna.

While the savanna of the coastal regions of Benin, Togo, and central and eastern Ghana have been brought about primarily by the climate and secondarily by man who, acting as positive feedback, has followed this more open vegetation to the sea and made it more open still by his activities, soil has had considerable effect as well. The soils of the Dahomey Gap are in general poor. Both ferrisols and ferrallitic soils occur (Ahn, 1970). The ferrisols tend to be richer and better structured than most tropical soils of comparable depth, while the ferrallitic soils have a low cation exchange capacity and are characterized by advanced leaching and weathering. In the area of lowest rainfall/evapotranspiration ratios, the better ferrisols may possibly somewhat compensate for the unfavorable moisture regime so that small patches of dense vegetation are not uncommon and many forest species manage to survive. Usually, however, forest remnants are associated with topography, occupying valleys which not only accumulate richer, deeper soil but also are sheltered from drying winds. On the other hand, richer soils have lead to more intensive agriculture. That woody vegetation is often almost entirely missing is the result of man's continual interference. The most detailed studies of the vegetation of the Accra plains are those of Jeník and Hall (1976) and Liebermann (1982).

Northern Guinea savanna is more similar to the East African 'miombo' and is a generally open woodland or 'bushland' with grasses somewhat shorter than in southern Guinea savanna. Trees with compound leaves become relatively more abundant. Particularly frequent species are *Isoberlinia doka* and *I. tomentosa*, *Terminalia avicennioides*, *Detarium microcarpum*, *Piliostigma thonningii*, *Parinari polyandra* and *P. curatellifolia* (=*Maranthes* spp.), *Burkea africana*, many *Combretum* spp. Grasses are mainly *Andropogon*, *Hyparrhenia*, and *Schizachyrium* spp. (Fig. 5.10).

(b) Sudan Savanna

This zone is not differentiated from northern Guinea by White (1983) but is said to be composed of Sudanian woodland with abundant *Isoberlinia*, Sudanian undifferentiated woodland with islands of *Isoberlinia* and, in the northernmost areas, undifferentiated woodland. Keay's Sahel savanna becomes Sahel *Acacia* wooded grassland and deciduous bushland.

The Sudan zone is termed 'savane typique' and said to be characterized by *Butyrospermum paradoxum* by Rousseau (1932), and 'forêt clairierée tropophile' characterized by *Butyrospermum* and *I. doka* by Jacques-Felix (1950, 1951). It is often open grassland or parkland with scattered *Butyrospermum* or *Adansonia* trees or low woodland or bushland. Grasses are lower, usually under 1.2 m tall. Cultivation is very extensive, particularly of millet and guinea-corn, and this is probably responsible for the greater prevalence of bushland and parkland as opposed to woodland. The lower grasses may also derive from cultivation, often being colonizing weedy species, with annuals abundant. Woody plants are *Acacia* spp., many Combretaceae (especially noticeable is *Guiera senegalensis*), *Piliostigma thonningii* and now *P. reticulatum*, *Balanites aegyptiaca*, *Adansonia digitata*, *Capparis* spp., *Monotes kerstingii*, and *Anogeissus leiocarpus*, which is distributed from the fringing rain forest to Sahel scrub but is often particularly noticeable here as a larger tree, many times in clumps of several. It is interesting to note that *Anogeissus*,

Fig. 5.10. Mature bushland in northern Guinea savanna after January burning. (Photograph by Sheila Jeyifo).

Balanites, Monotes, and occasional *Adansonia* extend into the Sahel area to a limited degree. (It should be emphasized, in case anyone thinks taxonomic descriptions of vegetation zones too easy, that most of the woody species mentioned extend from the derived savanna zone throughout the Guinea savanna and into the Sudan zone but vary in abundance.) (Fig. 5.11).

(c) Upland Savanna

Highland or submontane grasslands (montane elements of White, 1983) are well developed in the Moka area of the island of Bioko, in the western Cameroun highlands and in the Obudu and Mambila plateaus of Nigeria: all of these highlands are related in geological formation, having been formed by volcanic action. In all of the areas, there is sufficient moisture for forest growth and often fertile soil as well, although because of very heavy rainfall leaching is extensive wherever vegetation cover has been removed. Open grassland has apparently been maintained largely by animal overgrazing and cutting and burning by man, but in some local areas, leaching and erosion have been extensive enough in the past to create edaphic grasslands. Species lists of a typical West African highland (the Obudu Plateau, Nigeria) are given by Hall and Medler (1975a, b; see also Hall, 1971). These highland areas are usually open grassland in which the grasses are less than 1 m tall; most common species include *Eragrostis* spp., *Hyparrhenia* spp., *Sporobolus* spp. — espe-

Fig. 5.11. On the border between Sudan savanna and Sahelian scrub, near Maiduguri, north-eastern Nigeria, in early July. (Photograph by W. W. Sanford.)

cially *S. africanus*, with a very few woody plants except in the ravines where forests or woodlands commonly develop. Woodland in such local areas is probably accounted for both by deeper and more fertile soil and by less disturbance from man because of its inconvenience for farming, etc. Tree species found are those of montane and submontane forests. The common temperate zone fern, *Osmunda regalis*, is often found along watercourses where there is no tree shading, and *Pteridium aquilinum*, the temperate zone 'bracken', may occur in extensive, almost closed stands (Fig. 5.12).

(d) A proposed savanna classification
One of us (Sanford, 1982b) has worked out a scheme for savanna classification and mapping which is currently in the experimental stage. The aim is to provide a feasible system from the standpoint of labour input which at the same time provides practical information on a scale of detail suitable for management planning. The conventional zonation terms are retained as the framework, but they are defined climatologically. The second hierarchical step concerns soil and topography; the third, physiognomic and structural features of the vegetation; the fourth, taxonomic features. This scheme is outlined below.

1. Climatic-ecological zone	Length of rainy season (days)	Precipitation/ evapotranspiration
(a) Transition savanna	> 200	0.66–1.0
(b) Guinea savanna	150–200	0.40–0.75
(c) Sudan savanna	110–150	0.21–0.40
(d) Sahel short grassland	< 110	< 0.21

2. Topography and soil
Topography: Broken (i.e. ironstone ridges, inselbergs, small hills and valleys, mountains) with local slopes: (1) steep, (2) moderate, (3) gradual; undulating; plain.
Soil: Ferrallitic (Food and Agriculture Organization ferrasols) (1) yellow-brown; (2) red, ferruginous (FAO ferric luvisols) (1) on sandy parent material, (2) on crystalline acid rocks, (3) non-differentiated. Ferrisols (weakly developed soils and rocky areas) (1) young soils, (2) lithosols, (3) soils on iron-pan crusts. Brown and red soils of arid and semi-arid regions. Volcanic soils or soils with volcanic elements: sand/clay ratio; percentage organic matter (or C).

Fig. 5.12. Montane savanna in the Cameroun Highlands. (Photograph by W. W. Sanford.)

3. *Physiognomic and Structural Features of the Vegetation*

Overall physiognomy based on woody plant density:

(a) Woodland: Woody plant density — one tree per less than 40 m^2, of which at least 35 per cent are over 30 cm in girth.

(b) Tree savanna: Woody plant density — one tree per from 40 to 100 m^2 of which over 35 per cent are over 30 cm in girth.

(c) Bushland: Woody plant density — one tree or shrub per less than 40 cm^2 of which less than 35 per cent are over 30 cm in girth.

(d) Wooded savanna: Woody plant density — one tree per over 100 m^2 of which over 25 per cent are over 30 cm in girth.

(e) Shrub savanna: Woody plant density of one per from 40 to 100 m^2 of which less than 35 per cent are over 30 cm in girth.

(f) Open or shrub grassland: Woody plant density of one per over 100 m^2 of which less than 25 per cent are over 30 cm in girth.

(g) Permanent or seasonal savanna swamp (subdivided as (a–f) above; see Fig. 5.13).

(These formations are prefixed by 'highland' when found above 1220 m.)

Size-class distribution of woody plants

Each of the seven formations is classed according to the distribution of girth sizes of all woody plants over 1 m high and at least 1 cm in girth (at the

Fig. 5.13. Surface iron-pan with *Loudetia* cf. *arundinacea* growing in crevices, and abnormally luxuriant vegetation growing in the drainage area of the pan (Jos Plateau, Nigeria.)

mid-point if under 3 m in height, at breast height otherwise). (Goodness of fit may be tested by chi-square or G tests).

Mature — 59 per cent of the woody plants from 1 to 20 cm in girth; 21 per cent from 21 to 40 cm; 10 per cent from 41 to 60 cm; 5 per cent from 61 to 80 cm; 3 per cent from 81 to 100 cm; 3 per cent over 100 cm in girth.

Juvenile — if there is significant deviation from the above distribution as increase in number of stems in size classes below 61 cm.

Old — if there is significant deviation from the above distribution as increase in number of stems above 60 cm.

4.   Taxonomic composition of the vegetation
Each formation is characterized by the two to four most abundant woody plants and by the two to four most abundant grasses.

VI.   Management problems

1.   Conservation and improvement of soil fertility

(a) Clay in the soil
While savanna regions may suffer drought for part of the year, for another part rainfall is so intense that leaching and erosion may occur. Leaching is only

prevented by ion-adsorbing properties of the soil and by plant uptake of soluble nutrients and their storage in organic molecules. Soil properties preventing leaching are the presence of clay and organic matter. Clay particles, being negatively charged, hold cations. The most abundant clay in tropical soils is kaolinite, which is, unfortunately, relatively inefficient as a cation adsorber because of its large particle size (up to 1 μm in diameter). Much more efficient are the rarer montmorillonite clays. Jones and Wild (1975) report data from various sources concerning the mean percentage and range of clay in various tropical soils. The two most common soils in West Africa are ferruginous tropical and ferrallitic soils, making up approximately 58 and 10 per cent respectively of West African savanna soil. Both contain only about 9 per cent clay, with a range from 0.0 to 34 per cent. Vertisols are richest in clay (mean 52 per cent, range 22 – 79 per cent, but they make up only about 1.9 per cent of West African savanna soils. The clay of these soils is about 63 per cent montmorillonite, whereas ferruginous and ferallitic soils contain mainly kaolinite.

While clays are helpful in holding ions, they are very fine particled and tend to lead to soil compaction with the result that rain, particularly the first heavy rains of the season, do not penetrate but run off, carrying soil particles and dissolved nutrients. Soils tending to become compacted must be kept well covered by vegetation and/or litter. As mentioned earlier (Section IV.1) burning may increase soil compaction as well as remove plant cover and litter and should, therefore, be carried out cautiously over heavy soils.

Not much can be done to change the clay content of soil. While addition of organic matter is effective in lightening and improving the texture of soils in temperate zones, rapid mineralization makes this relatively ineffective in the tropics. The best means of conserving and utilizing heavy clay soils appears to be to keep them covered with grass under a light, high tree canopy. Leaf fall from the trees will protect the soil after grass burning and underground biomass will add some humus to the soil.

'Fadama', seasonally inundated land along streams and in floodplains, contains on an average 24 per cent clay (range 1.5–71.5 per cent) and the clay is often montmorillonite (Jones and Wild, 1975). Primary production is potentially very high in such restricted regions (e.g. Afolayan (1977) reported the maximum standing crop from fadama along the Oli River in Kainji Lake National Park, northern Nigeria, to be 172 per cent of the maximum achieved in nearby woodland savanna and 323 per cent of that achieved in almost adjacent *Burkea–Terminalia* open savanna). Because of this production potential, such restricted regions are increasingly being utilized for agriculture. This has disturbed nomadic and transhumance movements of livestock, as formerly such valley bottoms were grassland refuges for the dry season. Clearly, dry season substitutes for *fadama* grazing must be found (Fig. 5.14).

Fig. 5.14. Savanna marsh along a river. The very tall grass is *Pennisetum purpureum*, 'elephant grass', (Photograph by W. W. Sanford.)

(b) Organic matter in the soil

The importance of organic matter in soil has been discussed in sections II.2 and IV.2, but a few additional comments are needed. The main organic component of soil is humus, a collective term for an amorphous group of colloids containing about 60 per cent carbon and 5 per cent nitrogen. According to Ahn, (1970, p.123), West African savanna soils contain from 1 to 3 per cent organic matter. Micro-variation from metre to metre is often greater than mean variation from hectare to hectare as the presence of unmineralized organic matter so much depends on vegetation, especially trees.

While it has been shown above (section IV.2) that the organic matter content of soil is more dependent on below-ground biomass than on surface litter, litter is extremely important in protecting soil from runoff and erosion and, by shading, lowering soil temperature and so slowing the rate of mineralization. Mulching with crop residue has proved very effective in accomplishing these aims as well as in controlling weeds, but crop residues are becoming increasingly important as livestock fodder, and this trend will continue as pressures against nomadism increase (see section VIII.3, below).

(c) Vegetation cover

(i) *In relation to water* In section II.2 we showed how important a multi-layered above-ground cover of vegetation and a below-ground root network are in maintaining an ecosystem. It is the disturbance of such a vegetation system which most frequently leads to deterioration of the soil — often irreversibly. For example, in Ivory Coast savanna, bare soil with a slope of 7 per cent lost from 24 to 32 per cent of the rain as runoff and from 10 477 to 11 770 t $km^{-2} yr^{-1}$ (10.4–11.8 g m^{-2}) of soil; an experiment in Senegal showed 40 per cent loss of rain as runoff from bare soil and 17 per cent loss from rather sparse grassland, with soil loss of 2313 t $km^{-2} yr^{-1}$ (2.3 g m^{-2}) from bare ground and only 488 t $km^{-2} yr^{-1}$ (0.49 g m^{-2}) from the grassland (CTFT, 1969). Relative erosion has been estimated by Roose (1975); if erosion from bare ground is rated 1, erosion from forest is 0.001; from savanna in good condition, 0.01; from palms, coffee, or cocoa with cover plants growing beneath, 0.1; from yams, cassava, 0.2–0.8; from ground-nut, 0.4–0.8; from maize, sorghum, millet, from 0.3 to 0.9 (Fig. 5.15).

Leaching from well-developed savanna is very little, but may be considerable from farmland. For example, in northern Nigeria, with a mean annual rainfall of 1114 mm, leaching losses from grain fields range from 1.6 kg ha^{-1} yr^{-1} of sodium to 2.1 and 4.9 kg of magnesium and calcium respectively, and from 6.3 to 7.4 kg of potassium and nitrogen respectively (Jones and Wild, 1975, p. 141).

Fig. 5.15. Tree savanna in a *fadama* along the Oli River in Kainji Lake National Park, north-western Nigeria. The large trees are *Terminalia macroptera*. (Photograph by W. W. Sanford.)

The factors which conspire to deplete soil resources are heavy rains (both amount per unit time and drop size are important parameters), land slope and fine soil particle size. Because of the first two factors, highlands may be especially subject to leaching, runoff, and erosion. For example, the Obudu Highlands of eastern Nigeria receive about 4300 mm of rain during 275 days. The soil is leached and acidic. The area is so erosion prone that where grass has been trampled by human footpaths or cattle paths, erosion channels soon start. We have seen some that are yet only the width of the path (about 50 cm) but already 1.5 m deep.

Very often sheet erosion may remove even more soil than gully erosion, but it is not as readily discerned. This is true of lowlands as well as highlands. For example, it has been stated (Brammer, 1962) that erosion was not a serious problem in Ghana, but in 1972, Adu (1972) described erosion in the Navrongo–Bawku area of northern Ghana and estimated that 40 per cent of the land was affected by sheet erosion; as much as 8 per cent of the land in this area had lost both A and B profiles.

(ii) In relation to wind Water is not the only important factor in causing erosion: wind is also a menace. Each year, throughout West Africa, a haze of dust, largely fine diatomaceous earth, is brought down from the Sahara. European explorers and settlers of the eighteenth century referred to this, the harmattan, as 'the smokes', a visually apt term. A small amount of nutrient material is deposited with the dust, but this is of little importance.

Of considerable importance is the wind removal of loose soil from Sahelian and Sudan zones within West Africa. We can find no reliable figures of soil removal, but some idea of the global extent of such removal is given by the study of Prospero and Nees (1976). They found that most of the solid aerosol of the North Atlantic trade wind belt is mineral dust originating from the arid and semi-arid regions of West Africa. Mean concentration in the atmosphere above Barbados, West Indies, during June, July, and August ranged from 6.2×10^{-6} g m^{-3} in 1965 to 26.5×10^{-6} g m^{-3} in 1973. The authors relate this great increase to the Sahelian drought.

2.   Conservation of vegetation

(a) Gene conservation
Obvious necessities for conservation of vegetation are the prevention of erosion, runoff, and leaching, the maintenance of watersheds and implementation of efficient mineral cycling (see also section VII.4, below). Other than these, there are the less obvious needs of gene conservation for future use in plant breeding and the conservation of attractive scenery associated with quality of life and recreation.

Plants of savanna areas are often subject to marginal environments as regards moisture regime, temperature range, soil nutrient status, and herbivore and disease predation and thus may possibly provide valuable genetic material for improving resistance and habitat amplitude of economic plants. This is perhaps most obviously true of the Gramineae, most of which have C_4 metabolism and are thus adapted to high temperatures and light intensity. (The first product of CO_2 fixation is a 4-carbon molecule instead of the usual 3-carbon phosphoglycerate; plants having this metabolic pathway do not have appreciable, light-induced respiration and so do not waste carbon substrate under conditions of high light intensity; at the same time the photosynthetic mechanism is less sensitive to heat and so continues to function efficiently at much higher temperatures than is the case with C_3 plants.)

Besides such high-temperature tolerance, savanna plants often show remarkable tolerance of drought conditions and, in some cases, of high soil/water salinity. For example, some hydromorphic soils in northern Nigeria have pH values as high as 9.6 in the middle horizon where there are high concentrations of calcium and sodium salts (Jones and Wild, 1975, p. 41, citing Pullan, 1962). Plants growing in such an environment must be unusually tolerant of high salt concentrations. Areas around Lake Chad are also halomorphic, and with long irrigation many savanna soils may accumulate salts to a toxic level.

An interesting example of insect resistance is found with the mature seeds of tropical American *Mucuna* (Papilionaceae). (Some tropical American *Mucuna* spp. are grown as pasture plants cover crops in West Africa: also there are six species of this genus found in West Africa, one of which, *M. pruriens*, is also found in America.) Several species of this genus have sufficiently high concentrations of L-dopa (3,4-dihydroxy-phenylalanine) in the mature seeds to prevent insect attack (Rehr *et al.*, 1973).

(b) Scenic amenity

Obviously, no one enjoys viewing bare earth for long, but some types of vegetation are more aesthetically pleasing and/or more suited to recreational activities than others. In general, forest and woodland are considered more pleasing than bushland, scrub, or open grassland. As discussed earlier, woodland is the most nearly stable (perhaps 'climax') savanna structure and so is desirable from more utilitarian considerations as well. While it is very resistant to natural environmental fluctuations and even to annual burning, it is, of course, quickly destroyed — and for many, many years — by cutting. Increased firewood demands throughout West Africa are making serious inroads on natural woodland. Even more disastrous is large-scale 'Green Revolution' cultivation. Clearly, it is necessary to determine which areas are suitable for cultivation without endangering future soil fertility through erosion and leaching and which areas should be left as woodland. Unfortunately, such decisions are often made according to criteria far from ecological considerations.

3.   Nomadism, transhumance, and settled agriculture

Throughout the savanna regions of West Africa, conflict between traditional patterns of nomadism and transhumance livestock management and settled agriculture is increasing in range and intensity (Figs. 5.16 and 5.17). Man-made lakes such as the Volta in Ghana and Kainji Lake in Nigeria not only flood grazing land but bring about land-use transformation through irrigation over large surrounding areas. Growing populations and increasing demands for a higher standard of living have forced all countries to implement various agricultural expansion schemes and put more land under monoculture and intensive agriculture. Milligan and Sule (1982), of the International Livestock Centre for Africa, have recently discussed this problem. Possible solutions are: (1) settlement of the traditional herdsmen with managed pastures or ranches; (2) restricted movement with supplementary livestock feeding from such agricultural by-products as cottonseed cake, or from hay or insilage prepared from natural savanna grasses; or (3) planned transhumance with natural savanna management. Quite clearly, no one alternative provides a satisfactory solution and all must be judiciously used. From economic and logistic considerations, the third choice will probably be the most extensively applied

Fig. 5.16. Gully erosion beginning where the grass cover has been removed. (Photograph by A. O. Isichei.)

Fig. 5.17. Dry season shelters of Fulani herdsmen in southern Guinea bushland. Modified transhumance is practised with these herdsmen taking their livestock further north soon after the rainy season beings. (Photograph by W. W. Sanford.)

for a long time. Supplementary feeding is, in the case of products such as seed cake, very expensive, and in the case of hay and insilage requiring of labour at times and locations where it is not available. In all cases the logistics of such feeding is a serious constraint. It is very likely, however, that soon the tall grasses of the humid savanna must be utilized more efficiently than at present. Cutting during the mid-growth period (about three to three and a half months after the start of the rainy season, usually late June to mid-July) would provide a large bulk of material with fair protein content. (Protein concentration declines almost exponentially from the period of flushing to the time of maturity and maximum standing crop.) Cutting would induce regrowth and at least two, in some areas three, crops could be harvested for hay or insilage per year. This raises an ecological question requiring urgent investigation: What will be the effect on the soil of such intensive harvesting and removal of nutrients from the system?

4. Desertification

A practical definition of desert is any temperate or tropical region where almost no primary production occurs. This situation can be brought about by either drought or mineral depletion of the soil or toxification of the soil as by salt or heavy metal accumulation. The savanna is particularly subject to all three conditions and so must be considered an ecosystem of considerable fragility.

(a) Climatic fluctuation and change
The recent seven-year Sahelian drought in Africa turned the attention of many climatologists to the possibility of climate change, but no clear-cut evidence has yet emerged. Changes several thousand years ago are known (and briefly discussed in section III.1), but more recent changes are problematical. The possibility which has received the most attention is the increase of carbon dioxide in the atmosphere. Gates (pers. comm., 1977) has reported that atmospheric CO_2 is increasing at about 1.0 ppm yr^{-1}. This would result in a doubling of the pre-industrial CO_2 concentration by about 2025–2050 and would lead to an average global temperature increase of about 2.0 C. Such an increase in marginal areas, such as the drier savannas, would result in drought and probable desertification.

Of less controversial importance are the year-to-year climatic fluctuations and possible short-term (most often from 10 to 60 years) cyclic tendencies of rainfall. In areas of marginal water availability, such as the entire Sahelian area and at least part of the Sudan zone, such variation means that in some years crops will fail and available fodder will not be enough to support the livestock population. In effect, such areas are utilized beyond their carrying capacity for some years and 'crash' in other years. The recent Sahelian drought was such a crash. An idea of yearly variation is given by the mean precipitation and standard deviation for 10-day periods at Maiduguri in north-eastern Nigeria on the border of Sudan savanna and Sahelian short grassland (11° 51′ N, 13° 05′ E): the standard deviation averages 114 per cent of the mean rainfall.

(b) Anthropogenic effects
Even more important than climatic fluctuations are the disruptive activities of man. These include burning of vegetation, heavy grazing by domestic livestock, wood-cutting, and cultivation. All of these have been mentioned before so their probable impact on savanna desertification will now be only summarized. Grove (1973) has stated that 'burning of the vegetation is possibly not a very important agency in the process of desertification. . . . An exception to this general rule might be woodland alongside watercourses.' We agree with this assessment, but would like to amplify slightly the exception. The land along watercourses is often steeply sloping so that if vegetation is removed, erosion may be drastic. We have seen very recent evidence of this in the Nigerian Obudu Highlands where recent fires have destroyed woodlands along steep ravines; the topsoil has already nearly disappeared.

Heavy grazing, while extremely serious in many regions of East Africa, is not extensive in West Africa. One reason is the seasonal livestock migration enforced by custom, tsetse-fly, and the unpalatable character of tall grasses as the season progresses (Fig. 5.18). Exceptions do occur in some Sudan regions, in some highland areas (the Jos Plateau in Nigeria, for example), in the Sahelian region in general, and around boreholes in particular. In the latter case, water, brought by modern technology to alleviate chronic water shortage, has brought about intensive settlement near water resulting in overgrazing to such an extent that near-desert surrounds many boreholes. We have seen this from the ground in driving from Maiduguri, Nigeria, to Lake Chad, and a remote-sensing specialist told us how puzzled he was at first by the blank white areas surrounding boreholes. Besides this extreme case of livestock concentration, the changes in land use mentioned in section VI.3 have brought about more intensive grazing of the ever-shrinking free-range areas. In the near future this may lead to desertification in many areas now well covered with vegetation.

The recent increase of firewood cutting has already been mentioned (Section VI.2b). Mortimore and Wilson (1965) have estimated that nearly three-quarters of Kano (Nigeria) city's firewood consumption of some 75,000 t yr^{-1} is brought in by donkey from within a radius of c. 20 km. As Kano and similar cities are

Fig. 5.18. Cattle moving southward into Guinea savanna in January (mid-dry season). The vegetation is mature bushland, near New Bussa, north-western Nigeria. (Photograph by Sheila Jeyifo.)

Fig. 5.19. Firewood collected near Jebba in the southern Guinea zone of Nigeria ready to be transported out. Donkeys are being replaced by lorries and the radius of collection around the large towns grows ever wider. (Photograph by Sheila Jeyifo.)

growing at a rate of from 5 to 10 per cent annually, the firewood problem will become great unless other fuels can be substituted (Fig. 5.19). Once woodland has been cut, its regeneration is difficult. As discussed earlier (Section IV.2), it is probably almost always necessary to provide a considerable number of years of fire protection before woodland can be established. In areas of heavy grazing by either domestic livestock or wild herbivores, an only slightly less long period of fencing is necessary.

The relationship between desertification and cultivation is too obvious to require amplification, although it should be pointed out that the importation of foreign, large-scale agricultural technology is increasing faster than the know-ledge of how to manage it. Such management practices as no or minimum tillage, use of legume cover crops, crop rotation, and judicious use of fertilizers will certainly be necessary if disaster is to be avoided.

VII. Coda

Ultimately the fate of the people of West Africa rests upon the fertility of savanna soil. Savanna, covering up to nearly 80 per cent of the land area of many West African countries, must supply most of the food. A few years ago it might optimistically have been believed that no matter what was done to this land, European and American agricultural technology, improved breeds of crops and livestock, massive doses of fertilizer, and frequent spraying with pesticides and herbicides would solve all problems and not only maintain but increase food production. Such blind optimism is still harboured in the minds of a number of politicians, but citizens of developing countries have recently seen the possibility for massive physical loss of soil by water and wind, together with depletion of soluble minerals by leaching; they have seen that new breeds of plants and animals may not be suited to local conditions and that uniformity of breeds anywhere may lead to destruction by pests and disease; they have seen that pesticides and herbicides may have disastrous side-effects besides costing more money than is often available; they have seen that world-wide inflation may make the purchase of sufficient quantities of fertilizers impossible except by the richest countries (which perhaps need them the least); and that the existence of some crucial nutrients is finite. This knowledge is increasingly leading men everywhere to realize that the land and its ecosystems must be handled with the utmost care and that such care can only be learned by the local study of local conditions by men and women who have a tangible stake in the form of children and grandchildren in the future of the land.

References

Adejuwon, J. O. (1970). The ecological status of coastal savannas in Nigeria. *J. Trop, Geogr.*, **30**, 1–10.

Adeniyi, F. (1982). Mineral input in rain free-fall. In W. W. Sanford, H. M. Yesufu and J. S. O. Ayeni (Eds.), *Nigerian Savanna: Proc. Man and Biosphere State-of-Knowledge Workshop: Kainji Lake Research Inst.*, New Bussa, 20–24 April 1980, pp. 352–376. Unesco/Kainji Lake Research Inst., New Bussa.

Adu, S. V. (1972). Eroded savanna soils of the Navrongo–Bawku area, northern Ghana: *Ghana J. Agric. Sci.*, **5**, 3–12.

Afolayan, T. A. (1977). Savanna structure and productivity in relation to burning and grazing regimes in Kainji Lake National Park. Ph.D. thesis, University of Ibadan.

Afolayan, T. A. (1978). Grass biomass production in a Northern Guinea Savanna ecosystem. *Oecol. Plant.*, **13**, 375–386.

Afolayan, T. A. (1979). Change in percentage ground cover of perennial grasses under different burning regimes. *Vegatatio*, **39**, 35–41.

Ahn, P. M. (1970). *West African Soils*. Oxford University Press, London.

Aubréville, A., Duvigneaud, P., Hoyle, A. C., Keay, R. W. J., Mendonça, F. A., and Pichi-Sermolli, R. E. G. (1958). *Vegetation Map of Africa South of the Tropic of Cancer* (Map sheet 1 : 10 000 000). Clarendon Press, Oxford.

Bazzaz, F. A. C. (1975). Plant species diversity in old-field successional ecosystems in southern Illinois. *Ecology*, **56**, 485–488.

Brammer, H. (1962). Soils. In J. B. Wells (Ed.), *Agriculture and Land Use in Ghana*, pp. 88–126, Oxford University Press, London.

Beard, J. S. (1967). Some vegetation types of tropical Australia in relation to those of Africa and America. *J. Ecol.*, **55**, 271–290.

Boughey, A. S. (1955). The nomenclature of the vegetation zones on the mountains of tropical Africa. *Webbia*, **11**, 413–423.

Brookman-Amissah, J., Hall, J. B., Swaine, M. D., and Attakorah, J. Y. (1980). A reassessment of a fire protection experiment in north-eastern Ghana savanna. *J. Appl. Ecol.*, **17**, 85–100.

Cesar, J., and Menaut, J. C. (1974). Analyse d'un ecosysteme tropicale humide: la savane de Lamto (Côte d'Ivoire). II. Peuplement vegetal. *Bull. de liaison des Chercheurs de Lamto*, Numero special, 1974, Fasc. II.

Chachu, R. E. O. (1982). A tentative hypothesis of plant succession in Kainji Lake National Park. In W. W. Sanford, H. M. Yesufu, and J. S. O. Ayeni (Eds.), *Nigerian Savanna: Procs. Man and Biosphere State-of-Knowledge, New Bussa, 20–24 April 1980*, pp. 406–417. Unesco/Kainji Lake Research Inst., New Bussa.

Chevalier, A. (1900). Les zones et les provinces botanique de l'A.O.F. *c.r. hebd. séanc. Acad. Sci.*, Paris, **130**, 1205–1208.

Child, G. S. (1974). *An Ecological Survey of the Borgu Game Reserve*. FAO, Rome.

Clayton, W. D. (1957). A preliminary survey of soil and vegetation in Northern Nigeria. *Regional Res. Station, Samaru, Zaria. Northern Nigeria: Departmental Report.*

Clayton, W. D. (1961). Derived savanna in Kabba province, Nigeria. *J. Ecol.*, **49**, 595–604.

Clements, F. E. (1916). *Plant Succession. An Analysis of the Development of Vegetation*, No. 242. Carnegie Inst. Washington, DC.

Collins, N. M. (1977). Vegetation and litter production in Southern Guinea Savanna, Nigeria. *Oecologia (Berl.)*, **28**, 163–175.

Connell, J. H. (1978). Diversity in tropical rain forests and coral reefs. *Science*, **199**, 1302–1310.

Conrad, M. (1977). The thermodynamic meaning of ecological efficiency. *Amer. Natur.*, **11**, 99–106.

Centre Technique Forestier Tropical (CTFT). (1969). *Conservation des sols au sud du Sahara*, Nogent-sur-Marne, 211 pp. (Techniques rurales en Afrique, No. 12).

CSA (Scientific Council for Africa) (1956). *CSA Specialist Meeting on Phytogeography*, Yangambi, No. 53, p. 35. CCTA, London.

Fagbenro, J. A. (1982). Effects of fire on soil with special reference to Tropical Africa. In W. W. Sanford, H. M. Yesufu, and J. S. O. Ayeni (Eds.), *Nigerian Savanna: Proc. Man and Biosphere State-of-Knowledge Workshop: Kainji Lake Research Inst., New Bussa, 20–24 April 1980*, pp. 128–143, Unesco/Kainji Lake Research Inst., New Bussa.

Fosberg, F. R. (1961). A Classification of vegetation for general purposes. *Trop. Ecol.*, **2**, 1–28.

Geerling, C. (1973). Vegetation map of Borgu Game Reserve. Unpublished; Dept. Forest Resources Management, Univ. of Ibadan, Nigeria.

Grove, A. T. (1973). Desertification in the African environment. In *Drought in Africa*, pp. 33–45. Report of the 1973 Symposium, Centre for African Students, School of Oriental and African Studies, University of London.

Hall, John, B. (1971). Environment and vegetation on Nigeria's highlands. *Vegatatio*, **3**, 339–359.

Hall, John, B. (1975). Fire as an experimental variable at Auno. *Ife Univ. Herbarium Bull.*, No. 10.

Hall, John, B. (1977). Forest-types in Nigeria: An analysis of pre-exploitation forest enumeration data. *J. Ecol.*, **65**, 187–199.

Hall, John, B., and Medler, J. A. (1975a). Nigeria and its affinities. *Vegatatio*, **29**, 191–198.

Hall, John, B. and Medler, J. A. (1975b). The flora of the Obudu Plateau and associated highlands: an annotated species list. *Ife Univ. Herbarium Bull*, No. 9.

Hill, M. O. (1973). Reciprocal averaging: an eigenvector method of ordination. *J. Ecol.*, **61**, 237–249.

Hopkins, B. (1966). Vegetation of the Olokemeji Forest Reserve, Nigeria IV. The litter and soil with special reference to their seasonal changes. *J. Ecol.*, **54**, 687–703.

Horn, H. S. (1975). Succession. In R. M. May (Ed.), *Theoretical Ecology*, Ch. 10. Blackwell, Oxford.

Howell, J. M. (1968). The Borgu Game Reserve of Northern Nigeria. Part 1, *Nig. Field*, **38**, 39–116; Part 2, 147–165.

Isichei, A. O. (1980). Nitrogen fixation by blue-green algal soil crusts in Nigerian Savanna. In T. Rosswall (Ed.), *Nitrogen Cycling in W. African Ecosystems, Proc. SCOPE/UNEP Workshop, IITA, Ibadan, Dec. 1978*, pp. 191–198. SCOPE/UNEP, Uppsala.

Isichei, A. O. (1979). Elucidation of stocks and flows of nitrogen in some Nigerian savanna ecosystems. Ph.D. thesis, University of Ife, Nigeria.

Isichei, A. O. and Sanford, W. W. (1980). Nitrogen loss by burning from Nigerian grassland ecosystems. in T. Rosswall (Ed.), *Nitrogen Cycling in W. African ecosystems, Proc. SCOPE/UNEP, Workshop, IITA, Ibadan, Nigeria. Dec. 1978*, pp. 325–332. SCOPE/UNEP, Uppsala.

Jacques-Felix, H. (1950). Geographie des denudations et degradations du so au Cameroun. *Minist. France outre-mer, Direction Agric. Elev. For., Bull. Sci.*, No. 3, 1–84. Paris.

Jacques-Felix, H. (1951). Regions naturelles et paypagos vegetaux, *Encycl. Afr. Jr. Cameroun-Togo*, 15–29. Paris.

Jaeger, F. (1945). Zur Gliederung und Benennung des tropischen Grasslandgürtels. *Verh. Naturf. Ges. Basel*, 56, 509–520.

Jeník, J., and Hall, J. B. (1976). Plant communities of the Accra Plains, Ghana. *Folios Geobot, Phytotax. Praha*, **11**, 163–212.

Jones, M. J. and Bromfield, A. R. (1970). Nitrogen in rainfall at Samaru, Nigeria. *Nature (London)*, **227**, 86.

Keay, R. W. J. (1959a). *An Outline of Nigerian Vegetation*, 3rd edn. Government Printer, Lagos.

Keay, R. W. J. (1959b). *Vegetation Map of Africa, Explanatory Notes*. Oxford University Press, London.

Liebermann, D. (1982). Seasonality and phenology in a dry tropical forest in Ghana. *J. Ecol.*, **70**, 791–806.

Livingstone, D. A. (1975). Late Quaternary climatic change in Africa. *Ann. Rev. Ecol. Systematics*, **6**, 249–280.

Matthews, J. A. (1979). A study of the variability of some successional and climax plant assemblage-type using multiple discriminant analysis. *J. Ecol.*, **67**, 255–272.

Medina, E. (1983). Organic matter production of the grass cover in S. American savanna, with emphasis on nutritional aspects. In *Report of Symp. on Savanna and Woodland Ecosystems in Trop. America and Africa: A Comparison, Univ. of Brasilia, Brazil, 2–7 Oct. 1983*, pp. 19–20. Internat. Biosciences Network, Paris.

May, R. M. (1973). *Stability and Complexity in Model Ecosystems*. Princeton University Press, New York.

Menaut, J. C. (1977). Evolution of plots protected from fire since 13 years in a Guinea Savanna of Ivory Coast. *Proceedings 4th Intern. Symp. Tropical Ecol.*, Panama.

Menaut, J. C. (1983a). A tentative synthetic approach to the functioning of savanna ecosystems. In *Report of Symp. on Savanna and Woodland Ecosystems in Trop.*

America and Africa: A Comparison, Univ. of Brasilia, Brazil, 2–7 Oct. 1983, pp. 19–20. Internat. Biosciences Network, Paris.

Menaut, J. C. (1983b). The vegetation of African savannas. In F. Bouliere, (Ed.), *Tropical Savannas.* Elsevier, Amsterdam, Oxford, New York.

Milligan, Kevin (1979). An ecological basis for the management of Kainji Lake National Park. Ph.D. thesis, University of Ibadan.

Milligan, K., and Sule, B. (1982). Natural forage resources and their dietary value. In W. W. Sanford, H. M. Yesufu, and J. S. O. Ayeni (Eds.), *Nigerian Savanna: Proc. Man and Biosphere State-of-Knowledge, Workshop, Kainji Lake Res. Inst. New Bussa, 20–24 April 1980,* pp. 190–207. Unesco/Kainji Lake Res. Inst., New Bussa.

Moore, A. W. (1960). The influence of annual burning on a soil in the derived savanna zone of Nigeria. *7th Internat. Congr. Soil Trans.,* **4**, 257–264.

Moreau, R. E. (1966). *The Bird Fauna of Africa and its Islands.* Academic Press, New York and London.

Moreau, R. E. (1969). Climatic changes and the distribution of forest vertebrates in West Africa. *J. Zool. (London),* **158**, 39–61.

Mortimore, M. J., and Wilson, J. (1965). *Land and People in the Kano Close-settled Zone.* Occasional Papers 1, Ahmadu Bello Univ., Dept. of Geography, Zaria, Nigeria.

Nye, P. H., and Greenland, D. J. (1960). *The Soil Under Shifting Cultivation.* Tech. Commun. No. 51, Commonwealth Bureau of Soils, Bucks, England.

Odum, E. (1969). The strategy of ecosystem development. *Science,* **164**, 262–270.

Olson, J. S. (1963). Energy storage and the balance of producers and decomposers in ecological systems. *Ecology,* **44**, 323–331.

Opakunle, J. S. (1978). Nitrogen partitioning in relation to season and nitrogen content of the growth medium in two Nigerian occurring grasses — *Andropogon gayanus* Kunth. and *Schizachyrium sanguineum* (Retz.) Alston. M.Sc. dissertation. Univ. of Ife.

Papadakis, J. (1965). *Crop Ecologic Survey in West Africa.* FAO, Rome.

Prospero, J. M., and Nees, R. T. (1976). Dust concentration in the atmosphere of the equatorial North Atlantic: possible relationship to the Sahelian drought. *Science,* **196**, 1196–1198.

Pullan, R. A. (1962). A report on the reconnaissance soil survey of the Nguru–Hadejia–Gumel area with special reference to the establishment of an experimental farm. *Samaru Soil Survey Bull.,* 18. Inst. Agric. Res., Samaru, Nigeria.

Ramsay, D. McC., and de Leeuw, P. N. (1965a). An analysis of Nigerian Savanna III. The vegetation of the middle Gongola region by soil parent materials. *J. Ecol.,* **53**, 643–660.

Ramsay, D. McC., and de Leeuw, P. N. (1965b). IV. Ordination of vegetation developed on different parent materials. *J. Ecol.,* **53**, 661–677.

Ramsey, J. M. and Rose Innes, R. (1963). Some quantitative observations on the effects of fire on the Guinea Savanna vegetation of Northern Ghana over a period of eleven years. *African Soils,* **8**(1), 41–85.

Rehr, S. S., Janzen, P. H., and Feeny, P. P. (1973). L. Dopa in legume seeds: A chemical barrier to insect attack. *Science,* **181**, 81–82.

Roose, E. (1975). Paper presented at the workshop on *Soil Conservation and Management in the Humid Tropics,* Ibadan, Nigeria, 30 June – 4 July, 1975, as reported in *Tropical Ecosystems,* p. 264, Unesco/UNEP/FAO, Paris.

Rose Innes, R. (1972). Fire in West African vegetation. *Proc. Tall Timbers Fire Ecology Conf.,* **11**, 147–173.

Rousseau, J. A. (1932). Mission d'études forestières dans la region du Nord-Cameroun. *Bull, Agence. Gener. Colonies,* Extrait No. 285, 1766–1823.

Sanford, W. W. (1974). The use of epiphytic orchids to characterize vegetation in Nigeria. *Bot. J. Linn, Soc.*, **68**, 291–301.

Sanford, W. W. (1982a). Savanna burning: A review. In W. W. Sanford, H. M. Yesufu, and J. S. O. Ayeni (Eds.), *Nigerian Savanna: Procs. Man and Biosphere State-of-Knowledge Workshop, Kainji Lake Res. Inst., New Bussa, 20–24 April 1980*, pp. 160–188, Unesco/Kainji Lake Res. Inst., New Bussa.

Sanford, W. W. (1982b). Savanna: an overview. In W. W. Sanford, H. M. Yesufu, and J. S. O. Ayeni (Eds.), *Nigerian Savanna: Procs. Man and Biosphere State-of-Knowledge Workshop, Kainji Lake Res. Inst., New Bussa, 20–24 April 1980*, pp. 3–23, Unesco/Kainji Lake Res. Inst., New Bussa.

Sanford, W. W., Obot, E. A., and Wari, W. (1982a). Savanna vegetational succession: a means of assessment by the distribution of girth size of woody elements. In W. W. Sanford, H. M. Yesufu, and J. S. O. Ayeni (Eds.), *Nigerian Savanna: Procs. Man and Biosphere State-of-Knowledge Workshop: Kainji Lake Res. Inst., New Bussa, 20–24 April 1980*, pp. 418–432. Unesco/Kainji Lake Res. Inst., New Bussa.

Sanford, W. W., Usman, S., Obot, E. A., Isichei, A. O., and Wari, W. (1982b). The relationship of woody plants to herbaceous production in Nigerian savanna. *J. Trop. Agric. (Trinidad)*, **59**, 315–318.

Sarmiento, G., and Monasterio, M. (1975). A critical consideration of the environmental conditions associated with the occurrence of savanna ecosystems in Tropical America. In F. B. Golley and E. Medina, *Tropical Ecological Systems*, Ch. 6, pp. 223–250. Springer-Verlag, Berlin, Heidelberg, New York.

Thoreau, Hod. (1860). The succession of forest trees. In *Excursions* (1863). Houghton Mifflin, Boston.

Trewartha, G. T. (1954). *An Introduction to Climate*. McGraw-Hill, New York.

Troll, C. (1952). Das Pflanzenkleid der Tropen in seiner abhangigkeit von Klima, Boden und Mensch. *Dtsch Geogr.*, **28**, 35–66.

Walter, H. (1971). *Ecology of Tropical and Subtropical Vegetation*. Oliver and Boyd, Edinburgh.

Walter, H. (1973). *Vegetation of the Earth in Relation to Climate and the Ecophysiological Conditions* (trans. from 2nd edn. by J. Wiesar). English Universities Press, London.

White, F. (1983). *Unesco/AETFAT/UNSO Vegetation Map of Africa*. Unesco, Paris, 356 pp. and map.

Washburn, S. L. (1978). The evolution of man. *Scient. Amer.*, **239** (3), 146–155.

Plant Ecology in West Africa
Edited by G. W. Lawson
© 1986 John Wiley & Sons Ltd

CHAPTER 6

Desert and Sahel

Hubert Gillet*
Laboratoire d'Ethnobotanique, 57 Rue Cuvier, Paris,
France

I. The desert zone

Between the Mediterranean zone to the north and the Sahelian zone to the south lies the desert zone occupied by the immense Sahara Desert. It is the realm of sand (erg), pebbles (reg), rocky plateaus (hamadas), and mountains (Tassili des 'Ajjers, Hoggar, Tibesti): the domain of extreme aridity. Vegetation is absent, except on favoured sites (wadi beds, high ground) simply because of the lack of rain (Fig.6.1). The absence or rarity of precipitation is what causes this dryness. When it does rain heavily, although that would be not more than once in 10 years, a crop of fine grass grows up (*Eragrostis, Aristida, Tribulus*, etc.) and covers the gentle hollows with a green carpet. It is the period of the *acheb*. The Sahara turns green for a period of a few weeks while the ground remains moist. Then the remorseless sun parches the little plants which would have taken the opportunity to scatter their tiny seeds on the surface of the ground. Vegetative activity is not completely absent, but it is certainly not in evidence every year.

Despite the uniform nature of this drought which stamps its powerful imprint over the desert, there is, if one considers the rhythm of vegetation activity, need to distinguish between the northern and southern desert zones. Thus, the Sahara might be divided into two by a transitional zone which runs through it very close to the middle, equidistant from the Mediterranean and from the Sahelian zone.

---

* Professor Gillet's chapter was written in French and I am much indebted to Dr Roger Gravill, History Department, University of Lagos who produced a first draft in English, also to Mr Cline-Cole, Department of Geography, Bayero University who read through a later version and made many useful suggestions, though I myself take responsibility for the final English interpretation of Professor Gillet's contribution. G.W.L.

Fig. 6.1. Koboué Gorge (Ennedi, Chad). Note wadi with scanty vegetation at bottom of gorge.

1. The northern desert zone

Here, when the rains are of Mediterranean origin, they are associated with powerful depressions which come from the north and penetrate more or less as far as the northern half of the Sahara. Though they may occur at any time, they nearly always come during the cool season between October and the end of April. The plants follow a Mediterranean type of rhythm and it is not surprising that under such conditions the commonest Northern Saharan species should be of Mediterranean origin (*Cupressus dupreziana* of the Tassili des 'Ajjers for example).

In this zone life concentrates where water exists permanently, that is, in the cases where the water-table lies near the surface. In such places the date-palm (*Phoenix dactylifera*) is king. There is truth in the popular saying, 'it has its feet in water and its head in the sun'. It creates a cool and protective shade beneath which can be established two levels of agriculture, firstly of the most varied vegetables, such as beans, aubergines, tomatoes, peppers, assorted greens, radishes, and secondly of fruit-trees such as various citrus fruits, pomegranate, almond, and fig trees.

The palm-tree is a large consumer of water and it is said that in full production it can take up $3 m^3$ of water a week, but it is very tolerant of salt being able to withstand concentrations of up to 6 g l^{-1}. The species is dioecious and to avoid the growth of useless male stems — one male branch to 50 females is enough to ensure pollination — the growers, under certain fixed conditions, prune the rejected shoots or *djebars* from the bases of the female trunks.

But in modern times life in the palm-groves tends to be in jeopardy. Untreated palm-trees or those weakened by the lack of water fall victims to a fungus disease, the *bayoud* which is difficult to stem, and the inhabitants are now seeking employment in public works or oil exploration where they find more remunerative occupations than the hard life of a cultivator, who has to draw well-water every day.

2.   The tropical desert zone

This zone is distinguished from the northern desert zone by the mode of origin of the rains. These, when they come, are carried by the tropical monsoon, the air masses being of equatorial origin, in other words by the extreme northward extension of the tropical front. This difference is fundamental since the rains arrive under the high summer conditions from the end of July to the end of August, i.e. during the hot season. Warmth and humidity thus unite to produce a vigorous and explosive take-off of vegetation.

The rains are fierce and in a short while, from 10 to 30 mm of water can inundate the soil. A few days (generally three or four) afterwards the ground cracks open and the first plantlets appear. The plants which emerge under these conditions are of tropical origin. They are found mostly to the south where they are incorporated in the main body of Sahelian plants.

Some plants represent a genuinely South Saharan endemic element: they are scarce and in danger of disappearing, resisting only with difficulty the prevailing extreme dryness. Certain of them find refuge in the South Saharan Mountains where, due to runoff, a few millimetres of rainfall may suffice to irrigate a few privileged positions. We have had the opportunity to study these at Termit, a small mountain about 700 m high located some 220 km south-east of Aïr. Here an annual grass *Coelachyrum brevifolium* has established itself on the banks of the wadi and a perennial stoloniferous grass, *Eleusine compressa*, creeps and attaches firmly in the substratum. The characteristic plants of this zone are of tropical origin. They belong to the genera whose species constitute the basic Sahelian flora. In general they are non-demanding, widely distributed species which cover a large part of the tropical zone.

Thus one may regard the Sahara as being crossed through its middle by an ill-defined imaginary line separating a northern type desert zone to the north from a tropical desert zone to the south. These two zones, although they both belong to the same desert scenario, are rendered fundamentally distinct by the rhythm of vegetational activity. The first belong to the *Holarctis* empire and the second to the *Paleotropis*. These two zones come into contact and even overlap in the areas of high relief (Hoggar, Aïr, and Tibesti).

II. The Sahelian zone — general scheme

As soon as the rains attain 100 mm and become relatively regular in their onset there is enough moisture to sustain an ill-nourished herbaceous cover of steppe-like aspect but of permanent occurrence. This mostly open herbaceous layer is a source of nutrition for wild and domesticated herbivores and constitutes a natural pasture. Further to the south, rainfall totals of 500–550 mm permit large-scale rain-fed cereal and food-crop cultivation. Here permanent farming populations install themselves in settled villages. Thus, betwen the desert zone to the north, where the constant dryness does not allow the maintenance of permanent vegetation, and the Sudanese zone to the south which is the domain of agriculture, is situated the Sahelian zone, a region of extensive animal husbandry. It is thus bounded by two well-defined biogeographical frontiers.

The general physiognomy of the Sahelian zone becomes progressively modified along a north–south gradient. Little by little, the vegetation comes to occupy more of the terrain, total plant biomass increases, and eventually a point is reached close to the 300–350 mm isohyet when trees move away from watercourses and spread out over the entire countryside. A treeless sahel gives place to one with trees. Thus, one can distinguish a northern Sahel type comprising a single herbaceous stratum and a southern Sahel type comprising two strata — a herbaceous layer and a tree layer. Naturally, the dividing line between these two is not abrupt. The trees first leave the wadi banks to group themselves in clusters in depressions which become increasingly dense. Individual trees then appear on the fringes of these clusters. Gradually the vegetation distribution changes from concentration in clusters to one of complete dispersal. At the southern limit of the Southern Sahel the number of trees per hectare increases noticeably.

This division into two Sahels is generally recognized but some authors, notably Boudet (1972), divided the Sahel into three zones:

(a) The northern or desert Sahel with an annual rainfall of from 50 to 200 mm.

(b) The true Sahel which receives between 200 and 400 mm of rainfall per year.

(c) The southern or Sudanese Sahel with rainfall of between 400 and 550 mm per annum, where the more regular rains permit regular cultivation of millet.

III. The northern Sahel zone

The 100 mm rainfall isohyet used in delimiting the northern boundary of the Sahelian zone forms an important biological frontier because it fixes the northern limit of widespread tropical species such as 'cram-cram' or *Cenchrus biflorus*, easily recognizable by its spikelets surrounded by a husk of prickly bracts, the *Aristida* annuals common to the Sahel (*Aristida mutabilis*, *Aristida adscensionis* etc.), and *Schoenefeldia gracilis*, in other words the annual grasses which make up the Sahelian summer 'prairies'. This 100 mm of rainfall represents the northern limit of summer pastures.

1. *Structure of the vegetation*

It is convenient to distinguish two herbaceous strata.

(a) The perennial species of steppe-like distribution present in isolated tufts scattered several metres apart. Two main species share the sandy stretches, *Panicum turgidum* and *Cyperus conglomeratus*.

Panicum turgidum, a large species, prefers light sands. It is characterized by dormant buds positioned at the nodes and protected by envelopes of foliage and it is capable of producing vegetative growth following the lightest of showers when the buds and the branches turn green and become assimilators.

Cyperus conglomeratus is a species in low tufts, whose leafy branches arise from a hypogeous caespitose basal part. The plant is astonishingly drought-resistant and provided that the rhizomes stay buried in sand, it can survive in a dormant state for years. It may disappear during a series of dry years, but it conserves the capacity to rise and prosper when conditions become favourable.

These two species are much appreciated by animals. The *Panicum* is sought by gazelles and even more by the oryx and the addax, which consume great quantities during autumn and winter. These antelopes regale themselves on the shoots which emerge from the sand at the base of the tuft.

(b) The stratum of little annual plants which occur between the tufts.

These plants effect their vegetative cycle rapidly while the soil retains surface humidity. They do not display marked drought-resistant characters for they develop precisely during the time when the air is humid and the climate is not desert-like. Even so, certain species do show features of adaptation to dryness such as microphylls and the lower surface of the leaves covered with divergent white hairs (*Indigofera sessiliflora* and *Indigofera argentea*).

Others fold the upper surface of their leaves against one another, leaving only their pubescent undersides exposed to the air (*Tribulus ochroleurus*). In yet other cases the two halves of the leaves are folded with the midrib serving as a hinge (*Monsonia senegalensis*).

The density of this herbaceous covering varies with the amount of rain as does the size of individual plants. Certain species grow while conditions are favourable, and the size attained by the stalks is proportionate to the amount of rain received. Such plants as *Sesamum alatum* and *Ceratothera sesamoides* are veritable biological rain-gauges, as are also to some extent the less-branched praucicaulous Gramineae (*Cenchrus prieurii*).

Other plants reduce to the minimum their vegetative parts and the individual plant then becomes simply a link between the seeds of one generation and those of the next. In such cases the stalks, branches, and peduncles are very fine, like capillaries, the flowers very small, and the seeds numerous and minute. The most representative example is certainly *Mollugo cerviana*, which can perform its vegetative cycle in less than three weeks. If it rains only a little, one can see puny *Mollugo* plants a few centimetres high producing a limited number of flowers and seeds; if, on the other hand it is a rainy year, the same *Mollugo* can produce branches in all directions in a luxuriant growth with a cover of hundreds of flowers. These so-called 'short-cycle' plants make the optimum use of climatic variations which are never the same from one year to the next.

Certain plants with large leaves and tender stems are evident only in the years when the rain is above average. They are totally absent in years of poor rainfall. The most striking case is that of the wild colocynth, *Colocynthis vulgaris*, which needs plenty of water to develop its fruits. These are 10–12 cm in diameter and consist of 90 per cent water. When the dry season comes, the colocynth drops its globular fruits on the ground where they gradually dry out. In certain favourable low-lying zones, they are so thick on the ground that it is difficult to walk without treading on some of them.

2.   Dominant species of the herbaceous layer

After the rains the desolate dunes are covered in a few days with a vegetation composed essentially of short grasses with slender leaves. The exact composition of this herbaceous cover varies according to the soil's moisture-retention capacity, but the short *Aristida* annuals, which are the ones demanding the least water, predominate.

On the soils where vegetation has to grow between the pebbles, *Aristida funiculata* and *Aristida adscensionis* constitute the dominant species. In the driest places, the plants even extend their branches as far as the humid soil in order to absorb the maximum humidity (*Boerhaavia coccinea*, *Tribulus*).

In the sandy places where livestock roam the flora is composed of species which are favoured by organic matter such as *Cenchrus biflorus*, *Brachiaria deflexa*, *Gynandropsis gynandra*, *Ipomoea dissecta*, etc.

In places where the water lingers on for some time after the rains, and where the soil is enriched with fine clay particles, annual plants establish a cover over the soil of more than 90 per cent and may number up to 1100 individuals (*Echinochloa colona*), or even 1500 to the square metre (*Panicum laetum*); sometimes the long leaves of certain leguminous plants are superimposed in several levels (*Cassia tora*).

IV. The southern Sahel zone

As one passes into the southern Sahel the tree becomes a predominant element of the countryside. Its presence alone is a determining factor in the composition and dynamics of the vegetation. Little by little as one gradually proceeds towards the south of the southern Sahel zone, the trees become more numerous. As their foliage becomes more luxuriant the lower vegetation also diversifies: leguminous plants become more abundant and replace the grasses at the end of their period of growth. The active period of vegetation growth is extended; while it was only of a few weeks' duration in the northern Sahel zone, it runs to three months in the extreme south. The lower vegetation becomes organized into several strata.

1.   The structure of the vegetation

(a) The tree layer
The density of this layer depends on the amount of rainfall: thus, in the north of the zone there are only a few trees per hectare, while there are up to 50 ha^{-1} in the south. It follows that the tree biomass increases regularly along a north to south gradient.

The first trees to appear over the whole terrain, are the very same ones that were confined to watercourses in the northern Sahel zone: *Maerua crassifolia* and, especially *Acacia tortilis* subsp., *raddiana. Acacia tortilis* subsp. *raddiana* is the typical tree, and by its spreading posture and entangled branches covered with abundant foliage which give each other mutual protection, it forms an effective screen against the rays of the sun. In the shade of the tree, evaporation is greatly diminished and herbaceous vegetation benefits from more moderate conditions.

Thus, one sees in the shelter of the trees, a true herbaceous flora, which appears early, is very luxuriant, and survives a long time after the end of the rains before finally succumbing to the dry conditions. What is involved is a flora composed of species which find the shade beneficial and are termed sciaphiles. Among them we can also number some North Sudanese species, which are here, however, in a position of resistance rather than hospitality. To put it another way, the trees shade allows a 'biological counter-attack' harbouring those plants, which may be regarded both as an advance guard and also a rearguard. The large-leaved grasses such as *Urochloa lata*, the composites such as *Sclerocarpus africanus* and *Blainvillea gayana*, etc. may be cited as examples. The quick-growing tropical creepers find favourable conditions there and hitching themselves to the lower boughs, rise to invade the tree, entangle with its branches, and thus increase the denseness of the foliage. Certain of these are annual such as *Momordica balsamina* or *Coccinia grandis*, while others are perennial such as *Leptadenia heterophylla* and *Leptadenia hastata*.

In actual fact, the tree creates within its sphere an environment where the conditions of life are easier and the extremes of temperature are moderated. The soil benefits from a regular supply of litter and thanks to the cover of such dense annual vegetation it is efficiently protected from scorching due to the sun. The lower temperatures slow down the mineralization of organic matter while at the same time enhancing the mobilization of nitrogen. It is easy to see, therefore, why the herbaceous vegetation cover is more luxuriant. Again, physiologists working at the station of Fété Olé (northern Senegal) and Niono (near Bamako in Mali) have demonstrated that photosynthetic efficiency is four times greater in the shade than in the sun and this is just as true under 350 mm of rainfall as under 600 mm. It has thus been established that in the radiation-rich countries of the Sahel, the sun's intensity checks vegetative growth. Plants have to mobilize a part of their energy to struggle against thermal excess.

The role of the tree may also be measured by other activities. The roots which can penetrate down more than 2 m, get at trace elements, which are rare or non-existent at the surface of the soil (copper, cobalt, manganese, molybdenum, magnesium, boron, etc.) and recycle them in the general system, through the agency of the leaves, which fall to the ground.

Finally, it can be said that each tree represents an oasis of greenness, where the last plants find refuge before being eliminated by the advance of the desert, or where newcomers can find a footing. The tree is the last bastion of greenness and its disappearance signifies the irremediable stranglehold of the desert.

(b) The annual herbaceous layer
This benefits from the more prolonged humidity than prevails at similar strata in the non-Sahel regions. Thus it is organized in three levels as given below.

(i) Prostrate plants. These spread themselves on the soil surface by the elongation of their branches. Some of these plants proliferating in this way may provide a good cover to the soil surface, especially if the rains have been well distributed. Included here are several species of Convolvulaceae, such as *Ipomoea aitonii*, which is able to grow upwards and ramify in the trees, *Merremia pinnata, Jacquemontia capitata,* etc. and Commelinaceae such as *Commelina forskalaei*, not to mention certain euphorbias which remain prostrate on the ground such as *Euphorbia scordifolia*, also *Zornia glochiata* and *Alysicarpus ovalifolius*, which are much liked by animals.

(ii) Erect plants of medium size. One finds here a group of plants which are much better developed than in the North Sahel zone. The graminaceous ones such as *Aristida mutabilis, Schoenefeldia gracilis,* and *Digitaria gayana* throw out suckers from the start of their growth and thus multiply their leafy stems.

Also included in this group are such diverse plants as *Phyllanthus*, *Cassia mimosoides*, *Polygala erioptera*, *Monechma ciliatum*, etc.

(iii) Taller erect plants.   These become noticeable when the rains have been particularly heavy. Under such conditions certain grasses proliferate to the extent of exceeding 1 m in height. *Cenchrus biflorus* and *Cenchrus prieurii* attain such a height and become so extensive, that when in fruit they make walking difficult for the traveller. Other specifically South Sahel annual plants of which *Aristida stipoides*, *Tephrosia linearis*, and *Hibiscus asper* constitute examples, can reach heights of 1.50 m.

Under certain conditions, at the end of the rainy season, while the grasses are fading out one after another, annual leguminous plants push through in such a profuse manner that they are bunched together in compact, monospecific tufts completely covering the soil over considerable stretches. Thus, may be seen pure stands of *Crotalaria podocarpa* or of *Indigofera astragalina* or again of *Indigofera secundiflora*. Certain leguminous plants may remain quite green even into the dry season, for example *Tephrosia obcordata*.

V.   Xeromorphic characteristics of woody vegetation

The Sahel's ligneous vegetation has to face a formidable ordeal, that of surviving the prolonged dry season of nine or even ten months, without succumbing. During this long period, or at least during the hottest and driest season, which runs from March to June the sensitive organs are exposed to a burning sun. Such plants must survive by reducing water loss through evaporation to the minimum, come what may. The tree must protect itself at all costs. Several procedures are open to it.

(a) The deciduous type

Like trees in temperate countries which lose their leaves and enter a period of rest during the unfavourable season, i.e. during the cold season, certain Sahel trees shed their leaves at the end of the rains and remain dormant during the dry season (but compare *Acacia albida*, see Fig.6.2). They are active only in the rainy season. Only a few Sahel trees follow this procedure and their assimilation period is short. An example is *Commiphora africana*, typical of the northern part of the South Sahel whose tripartite leaves turn yellow in the autumn before falling. It is true that to continue in leaf for the longest possible time, *Commiphora* comes into foliage at the first signs of approach of the rainy season, often at the first moist winds from the south-west. It is evident that with such tender leaves, *Commiphora* is ill-adapted to the Sahel's aridity and it would appear that this tree is very gradually adapting itself to the extreme climate of the Sahel. *Commiphora quadricincata* of the North Sahel follows the

Fig. 6.2. Sahel savanna in the rainy season with a tree of *Acacia albida* in the foreground. Note that though *Acacia albida* remains in leaf during the dry season, thereby representing a valuable source of dry-season fodder, it is shown here in a leafless condition in the rainy season.

same rhythm but its leaves are somewhat thicker. Others that may be cited are *Combretum aculeatum, Grewia,* etc.

(b) Microphylls and assimilating branches
Here the leaves look rather like accessory organs. They are green, small, and deciduous, appearing for only a few weeks while humid conditions prevail. In the end it is the green branches which carry out most of the photosynthesis. Such branches are thin, puny, and greatly subdivided. The reduction of leaf surface is compensated for by a concurrent proliferation of branchlets. The most typical example here is *Capparis decidua* (also known as *Capparis aphylla*). The very common *Leptadenia pyrotechnica* with rush-like stems is another example.

The leaves of the *Acacia* are equally microphyllous (*Acacia tortilis* subsp. *raddiana, Acacia seyal, Acacia senegal,* etc.), but here the branches are covered with a thick bark.

(c) Coriaceous leaves
One efficient way of reducing evaporation is by the thickness and impermeability of the leaf surface. Such leaves become thick, leathery, and

persistent. They are tender only in their infancy and then gradually they harden. Many Sahel shrubs and trees belong to this category: *Boscia senegalensis* (Fig.6.3), *Maerua crassifolia*, *Balanites aegyptiaca*, *Piliostigma reticulatum*, *Salvadora persica*.

(d) Hairy or glandular leaves

In this case the epidermis of the leaf is covered by a dense padding of frequently stellate hairs as in the cases of Malvaceae or Sterculiaceae (*Waltheria indica*) or more often forming a tomentum (*Cadabre farinosa*).

These frequently glandular hairs make the leaves sticky to the touch (*Cadabre glandulosa*).

(e) Spiny branches

The young branch gradually becomes lignified and assumes a rigid consistency without getting any thicker. This occurs after the fall of the leaves in *Maerua crassifolia*, but this hardening of the branches is often a reaction to overgrazing. It is a defence mechanism (*Grewia tenax*, *Combretum aculeatum*, etc.).

(f) Thorns

The thorns are often modified stipules, which have lost their foliage-like character to assume a hard consistency. A thorn is an organ from which practically no evaporation occurs.

Fig. 6.3.   An old tree of *Boscia senegalensis*.

The straight and sharp stipular thorns of certain acacias are noteworthy. Because of their ivory white colour they reflect the light: *Acacia tortilis* subsp. *raddiana* (Fig. 6.4), *Acacia seyal, Acacia nilotica*. The thorns of *Balanites aegyptiaca* (Fig.6.5) are impressive (up to 10 cm long). Occasionally, thorns are curved into a semicircle *Acacia mellifera, Acacia senegal*).

(g) Associated characteristics
Certain trees particularly well adapted to drought combine several xeromorphic features simultaneously for example, succulent leaves, thorns, and assimilating branches in *Balanites aegyptiaca*; leathery, glutinous leaves in *Combretum glutinosum*; microphylls and thorny branches in *Maerua crassifolia*, etc.

1. Remarks

In the extreme south of the southern Sahel, where the rainy season lasts for nearly three months, the xeromorphic characteristics become less marked both for herbaceous and woody plants. For example, the leaves of some grasses increase in size and surround the base of the plant (*Urochloa lata*); *Brachiaria* with flexible leaves gradually replaces *Aristida* which has leaves bent into a rigid V; the herbaceous carpet is richer and the leaves become superimposed in several storeys.

Fig. 6.4.   *Acacia tortilis* subsp. *raddiana* dislodged by a sudden flood.

Fig. 6.5. *Balanites aegyptiaca*, Wadi Rimé Ranch, Chad.

The trees described as platyphyllous, with long but leathery leaves, make their appearance (*Combretum, Terminalia*). Sometimes the leaves are not only flexible but compound and deciduous (*Stereospermum kunthianum*).

VI. Way of life: nomadism and transhumance

The Sahel zone, which extends across Africa south of the Sahara, from the Atlantic Ocean to the Red Sea and covers more than 4 million km^2 is thinly populated: the number of inhabitants is estimated at 25 millions with a density of six persons per square kilometre. If one omits townspeople the density falls to two or three inhabitants to the square kilometre. Yet it must not be thought that the Sahel is a human desert. You can stop anywhere you like in this immense no man's land, and believe yourself alone for a radius of several kilometres, but after a few moments to your great astonishment there appears a walker or a camel rider from somewhere who has made a detour to come and greet you!

Sahel life is conditioned to the highest degree by climatic conditions and more particularly by the quantity and distribution of rainfall during the summer. The Sahel is covered with a thin carpet of vegetation, which has real food value. Are not the leaves of certain Capparidaceae a veritable concentration of proteins and mineral salts? This carpet of vegetation is a source of natural nourishment renewed each year as a function of the water which it receives. It is regularly eaten by such wild hoofed animals as dorcas and damas gazelles or oryx and addax antelopes; it is also liked by camels, zebu cattle (in particular, the peul, bororo, azaouak, and targui) and also sheep and goats.

1.   Nomadism

The life of the nomadic cattleman is governed by the presence of pastures. In the Sahel zone, and more particularly in the northern part, the rainstorms are both capricious and localized. An area of a few square kilometres might well receive 50 mm of rain in less than an hour, while neighbouring areas only a few kilometres away will have little or none. The distribution of rainfall does not obey a single fixed rule. An area well irrigated in one year will not necessarily be so in the following year, and so the location of pastures varies in time and space exactly in tune with the rains. Finding and exploiting these locations are the determining factors in the life of the nomadic shepherd. He heads his animals wherever the pasture is abundant and of good quality. His movements are not dictated by any law or code and are disconcerting for modern man, being responsive to only one thing — the whim of rainfall. The nomadic shepherd wanders from pasture to pasture across his immense domain and the duration of his stay in each locality depends on the state of his beasts and also a little on his mood. The nomad is content in his vast kingdom and knows neither frontiers nor administrative formalities. For him this vastness is twofold: that of time and that of space. He owns nothing except his flock which constitutes all his capital.

The camel, which is really the dromedary, is the nomad's animal *par excellence*. The sobriety of the camel is legendary and provided that it has decent grazing it can easily go for eight days without drinking and not be unduly worried. As a camel can easily cover 50 km a day, it can perfectly well exploit grazing situated at this distance from a well. Once a week the nomad, or rather his children, will be charged to lead the flock to the watering-hole. The journey there and back will last two days (Fig.6.6).

In certain cases the camel can get by without drinking at all if it is grazed on sufficiently moist pasture. This is so in certain years in northern Niger (Temesna or Azauak region) when the pastures grow green with the *gargir*, *Schouwia purpurea*, var. *schimperi*. This crucifer, which has pretty violet flowers, bears rather thick crassulaceous leaves containing sufficient water to satisfy the camels' need for moist food. Sometimes the nomads leave their livestock in the *Schouwia* pastures for months without worrying because they know that their beasts will be full and satisfied throughout the long period for which the *Schouwia* survives (Fig.6.7).

Apart from these special cases, however, the nomad roams ceaselessly in search of new pastures. His main pastime is to straddle his mount and provided with small supplies of tea, sugar, dried dates, and water, to journey according to the dictates of his will in search of the place where he will make his next camp.

Fig. 6.6.   A camel caravan (Ennedi, Chad).

2.   Transhumance

In the southern Sahel zone, benefiting from a more regular rainfall, there are locations renowned for the quality of their pastures. These are composed of particularly appetizing species, most often annuals, reappearing every year, more or less abundantly in accordance with the heaviness of the rainfall. It

Fig. 6.7.   Aone Ennedi 'Guelta' with *Adenia microcephala* (Chad).

follows that such pastures are especially sought after. Moreover, over the centuries a sort of customary code has been established according to which each pasture is reserved for such-and-such a tribe. Thus each year at precise times, may be seen major movements of men and beasts as they journey in a manner established by tradition to fixed localities. These regular migrations follow known routes and are planned ahead, as opposed to nomadic migrations which obey no such calendar or code. They are carried out according to fixed rules which are respected each year, family by family, tribe by tribe, as against those of the nomads that are dictated essentially by meteorological events.

This is why in the first half of July, for example, the Bororos, who specialize in sheep-husbandry, cross the Batha at Ati (Chad) to lead their flocks to the tender pastures of *Dactyloctenium aegyptium* watered by the first rains north of the Rimé Wadi. It is for the same reason that Arab cattlemen travel to the north of Chad in August, where their beasts can thrive on *Zornia glochiata* on the banks of the Haddad Wadi. Others journey still further up north to the Kharma Wadi or the Chili Wadi so that their cattle, sheep, and goats can enjoy the young and succulent *Echinochloa colona*, which proliferates in the shade of the acacias in the wake of flash floods. One could easily add to these examples of seasonal migration, which have been studied with a wealth of detail by the ethnologist, E. Bernus (1979, 1980) in northern Niger.

These regular seasonal movements of pasture exploitation profit in general from the support and favour of local authorities, for as renewable natural resources they can be exploited in a planned and systematic way. At the present time when desertification is linked to growth in the human and animal populations, it is urgent to plan a rational exploitation of natural pastures. Thus, one can conceive of a pasture allocation for each ethnic group; in other words the establishment of ecological and socio-ecological units. This leads on to the idea of integrated management of the resources of the Sahel; to the creation of 'pastoral units'. It would be easy to increase the productivity of certain pastures simply by building little dams and by planting such forage trees as *Acacia senegal*, which would protect the soil and allow the development of a herbaceous cover. The ethnic groups concerned, being very conscious of the areas reserved for them would strive to avoid their over-exploitation.

It should be noted that these seasonal migrations by Sahel beasts from south to north are necessary for their good health. These animals weakened by the enforced rationing of the long dry season, rapidly regain their strength on eating their fill of the fresh, tender herbiage nurtured by the rains. This grassy vegetation found on fixed Sahel dunes has a quality found nowhere else. Thanks to it the animals become reinvigorated. Deficiencies of certain nutrients, such as boron and carotene, are remedied. The animals develop glossy coats. The consumption of certain plants (*Polygala erioptera* and *Polygala irregularis*) or of the pinnules of *Acacia tortilis* subsp. *raddiana* has positive effects on milk production. The cattlemen are happy despite the mosquitoes, which disturb their sleep. The milk flows. It is the season for marriages!

In certain cases, the expected early development of young leaves on trees sensitive to the humid air which provides browsing before the grass shoots up, encourages small-scale local migrations. This is exactly what happens at Kanem, to the north of Lake Chad, in the region of Liloa, Manga, and Chitati, where, at the end of May or in June, according to the year, the Goranes drive their camels to the *digui*, i.e. the depressions under *Commiphora africana*. In the inter-dunal basins the young camels are so completely occupied in searching out the young leaves, one by one, that all their time is spent in doing this and they remain in one place. The early foliation of *Commiphora* and its food value (14 per cent protein) are a blessing for the Goranes, who thus avoid migrating to the south at a time when the oppressive heat could not be borne by the animals. In addition, a small but not insignificant advantage, is that the animals do not venture very far from the trees, which greatly assists their surveillance.

VII.   The Climatic Constraints

The Sahel is one of the regions of the world where the climatic constraints are the hardest, partly because of their extreme nature and partly because of the great variation in precipitation from one year to the next.

1.   *Accentuated dryness — desertification*

Since the recent drought, which for seven consecutive years (1967–73) ravaged this area and which caused the death of 50 per cent of the livestock, the attention of the whole world has been drawn to the unfortunate Sahel.

The truth is that the drought occurs in the Sahel at intervals and the dry and rainy periods succeed each other with a rhythm that is not fully understood. Already in the present century, the Sahel had known periods of great drought. All the old people remember the notorious summer of 1913, when there was not a single drop of water in the northern Sahel, so that there was unprecedented mortality in the course of spring 1914. Nearer the present, the drought of 1940 passed unnoticed because in the summer of 1940 the West was preoccupied with matters other than the weather in the southern Sahel. Then followed up to 1965 a long period during which fairly regular rainfall occurred with favourable effects on the Sahelian pastures. Gradually the subdesert zones were covered with annual plants, and in certain zones hydraulic pasture services covered some areas with regularly spaced wells. These wells allowed the exploitation of pastures previously exploited only in the rainy season. Gradually human and animal populations expanded. It is estimated that the number of cattle doubled between 1955 and 1970. To feed all this livestock the stock-raisers ventured increasingly towards the northern limits where the pastoral biomass is highly vulnerable to the vagaries of the rainfall.

These pastures became more and more over-exploited by livestock, whose numbers exceeded the pastures' carrying capacity, but as the rains came each year they were replenished — survived might be more correct. The ecological balance was already in jeopardy. Certain particularly attractive species (*Schoenefeldia*) became very rare while others more weed-like (*Cymbopogon schoenanthus*) proliferated in an alarming fashion. Experts did not devote sufficient attention to this change in the composition of the flora. Desertification crept up insidiously, but while there were plants for the animals to feed on it did not seem to matter. Then came the catastrophe. Drought afflicted pastures already weakened by overgrazing. This drought really corresponds to a series of low-rainfall years, succeeding an earlier series of higher than average rainfall years which are, in fact, being mistakenly considered as years of average rainfall. Thus, desertification, which up to then had been advancing surreptitiously took a leap forward. The hungry animals threw themselves on anything, even on the *Cymbopogon*; intensified aridity with attendant famine and misery made a sudden entry.

This process is well known. The vegetation cover diminishes. Herds concentrate on the scarce remaining pasture. The close herding of the beasts causes heavy trampling underfoot of the plants which, thus, cannot regenerate. They are flattened to the ground, crushed, and then chopped up by the hooves. The soil is constantly churned up so that the tufts themselves are loosened and end up being carried away by the wind. Then the naked soil receives the rains, which by an unfortunate combination of circumstances, become increasingly intense even while remaining of short duration. There is less rain, but the little that actually falls does so in fierce storms. The fine soil particles are carried off in surface runoff; erosion does its work. The wind dries out these small soil particles and blows them far away. Then the soil, stripped and deprived of its humus, is exposed to the unimpeded scorching of the sun. Its capacity for radiation reflection increases. The albedo, the fraction of the sun's rays that are reflected, increases. The Sahel, viewed from a satellite, becomes a brilliant, overheated surface. The soil like a sheet of ice instead of absorbing reflects the heat on to the surviving plants. It is the well-known phenomenon of positive feedback. The dryness acts upon itself to produce a still drier state. Even when the rains return, they fall on a surface like ice and flow away instead of sinking into the soil.

The thinning of the vegetation cover remains the main cause of desertification, as Le Houerou and his team have shown (1980). By examining aerial photographs taken over a period of 23 years (1952–75) they discerned that the percentage of surface soil not stabilized by vegetation rose from 4 to 26 per cent over that period.

With the overgrazing of the pastures and the impoverishment of vegetation communities one witnesses a transformation of the landscape. Thus, zones receiving between 400 and 500 mm of rainfall now support a carpet of vegetation the same as one would have seen 30 years ago in areas of 250–300 mm rainfall. The vegetation is no longer in balance with the rains. In other words, the rains bring less benefit to the vegetation. It is one of the unfair causes of increased aridity.

One of the main causes of the 'vegetation poverty' of the Sahel is the irreversible disappearance of good plant species, particularly of perennial plants or annual plants with a long vegetative cycle, to the benefit, if one can call it that, of annual plants with a short vegetative cycle running their course in a few weeks and thereby evading the animals who need these plants in the period when they are actually most dispersed in their habitat. One sees therefore, a natural selection of ecotypes with a very short life-cycle. Thus, the good species, such as *Digitaria gayana* become scarce and are replaced by the little-eaten *Aristida stipoides*. The prostrate Euphorbiaceae with little leaves, such as *Euphorbia granulata* and *Euphorbia aegyptiaca*, or the erect Euphorbiaceae such as *Phyllanthus pentandrus* which are all unappetizing, are now free to invade the vacant places. Recently, in September 1981, in the Guilimouni region (50 km to the east of Zinder), the author observed *Aristida longiflora*, a perennial grass unfortunately becoming scarce, being gnawed savagely at the base throughout the dry season, whereas one could see fine stands of this same plant in protected areas.

Thus, because of over-exploitation the Sahel is becoming an ecosystem particularly vulnerable to desertification due to:
(a) the reduction of the vegetation cover;
(b) the stunting of perennial plants;
(c) the elimination of organic matter from the soil due to leaching;
(d) the disappearance of vegetation species capable of lasting for a few months and their replacement by useless, inedible tropical annuals which last for only a few weeks.

2. Great variations in precipitation

In the Sahel zone no rainy season is like the one before, or the one after it, and no climatologist, in the present state of knowledge, is able to predict the course of the next rainy season. Anyone who appreciates the importance of the rain for all economic activiity will see the implications of our ignorance. At present, the climatic laws, if there are such laws, are just as hidden from scientists as they are from the insight of men closely acquainted with nature.

The best way of illustrating annual variations is to take an example and for this purpose the Zinder Station (13° 49′N) in the east-central part of Niger where the rainfall has been recorded since 1922 may be chosen (see Table 6.1).

One can see (Table 6.2) that for 20 consecutive years between 1922 and 1941 the rainfall never fell below the 400 mm level, as it did twice in the next 20 years (1942–61) and eight times in the 20 years following (1962–81). Thus, in the last half-century, the number of years when the rainfall was low and insufficient to produce a worthwhile amount of pasture, has grown in an exponential fashion which is very disturbing.

It seems that in recent years, it has become worse. For Zinder, the average rainfall which was 479 mm for the period 1960–69, fell to 404 mm between 1970 and 1979. It is true that the distribution of rains during the rainy season assumes an importance worthy of consideration. For instance, in 1981 in Zinder region, despite a very feeble annual rainfall of less than 300 mm, the harvest of millet was reasonable, thanks to well-spaced rains and regular weeding, but the level of the ponds was so low that certain of them which should have been overflowing by September 1981 contained no more water than in the previous April.

Table 6.1. Annual rainfall at Zinder

Year	Rainfall (mm)	Year	Rainfall (mm)	Year	Rainfall (mm)
1922	490.1	1942	339.5	1962	467.8
1923	425.0	1943	750.0	1963	362.5
1924	460.6	1944	547.2	1964	658.5
1925	580.0	1945	542.8	1965	434.0
1926	421.9	1946	800.3	1966	590.1
1927	559.0	1947	469.5	1967	404.5
1928	658.9	1948	371.4	1968	375.7
1929	548.8	1949	256.0	1969	436.0
1930	517.5	1950	609.0	1970	354.7
1931	566.2	1951	500.2	1971	352.4
1932	576.1	1952	661.9	1972	302.8
1933	563.4	1953	584.4	1973	297.5
1934	442.4	1954	699.7	1974	480.3
1935	524.5	1955	500.4	1975	470.7
1936	677.3	1956	610.2	1976	474.7
1937	435.2	1957	599.7	1977	256.9
1938	462.5	1958	526.0	1978	607.1
1939	690.5	1959	481.0	1979	442.8
1940	439.3	1960	583.3	1980	526.1
1941	426.0	1961	577.2	1981	c. 300

Table 6.2. Zinder Station. Number
of years with rainfall below 400
mm

1922–41	0
1942–61	2
1962–81	8

The Sahel's ecosystem, increasingly overloaded and diminishingly watered, now finds itself in a more fragile state than ever before. The degrading of its natural vegetation has reached the point of no return. The reconquest of the Sahel must take the form of a reconstitution of its tree cover.

VIII.   Biological forms

1.   Dominance of therophytes

It is hardly possible for any plant to get through a dry season of 10 months with temperatures which in April and May oscillate between 38 and 40 °C and a relative humidity (of the air) of the order of 10–12 per cent. The dormant buds have to be effectively protected to remain alive.

Thus, it is not surprising that annual plant or therophytes predominate in the Sahel zone. For them, there is no problem. They flourish during the only period when the Sahel is not dry and they drop their seeds on the ground when the drought returns. But seeds are a form of life which can live at a very slow pace: their respiratory exchanges are very slight and the tiny embryo encased in food reserves can survive for years awaiting favourable conditions. Their slight hydration (15 per cent maximum) and their double integument (impermeable exterior tegument, testa) is adequate protection. Herbarium seeds have been known to conserve their power of germination for three centuries, but in natural conditions under which seeds are submitted to great variations in temperature and humidity, it is probable that except in exceptional cases, their life expectancy is not more than 10 years or 20 at the most.

2.   Seed dormancy

It is obvious that the seeds of therophytes are endowed with a physiological regulation which prevents their immediate germination. If this were not so, late rains would germinate all the years' seeds, yielding plantlets which would abort through lack of water and threaten the survival of the species. But happily the seeds go dormant and they have to be awakened to enter active life.

It appears as if the seeds contain an inhibiting hormone varying in quantity from one type of seed to another which has to be soaked by the rains to rouse them from dormancy. This hypothesis at least, has the merit of being confirmed by precise observations. Thus, in the Massif de l'Ennedi (North Chad), and more precisely on the Basso Plateau, circumstances were such that the first important rain of 1962 took place on 20 July and the next rains did not fall until 4 and 6 August. Then by 10 August it was easy to see under the *Acacia tortilis* subsp. *raddiana* trees two lots of *Achyranthes aspera* (Amaranthaceae annual, sciaphile) one 12 cm high produced by the July rain and the other younger with a height of 3–4 cm from the August rain (personal observations). It appears clearly enough, therefore, that after the first rain some seeds remain in the soil ungerminated but active. In general, there is not such a long lapse between the first rains and the next and the seeds germinate continuously as their dormancy ends.

On the other hand, in the northern Sahel zone precise observations show that for a given species the delay between the first important rains of the year (of the order of at least 10 mm) and the appearance of plantlets on the surface of the soil is constant. This delay is three days for the seed of *Tribulus terrestris*, which is the quickest to germinate, four days for *Aristida adscensionis*, *Aristida mutabilis*, *Cenchrus biflorus*, and five days for *Aristida stipoides*, etc. The leguminous plants come later.

3. Production of seeds

With the return of the dry continental wind from the north-east all vegetative growth ceases among the annual plants and one witnesses a greater production of fruit. The seeds fall to the ground, where some are taken by the ants while most are borne away by the wind. Thus, they are scattered over greater or lesser distances until they meet an obstacle such as a rock or the base of a clump. Often they even accumulate to a considerable depth in depressions sheltered from the wind. The seeds of *Aristida funiculata* end in three long awns which gather together into a sort of ball, the hard parts inwards and the beards outwards. Propelled by the wind, they roll along the ground, just like balls, sowing, dropping, and picking up seeds throughout their course, thus ensuring excellent dispersal. All of which mean that the number of seeds produced by the Sahel annuals is unbelievably high in comparison with features of their vegetative development.

Therophytes have only one priority and that is to produce seeds in the minimum possible time. At Fété Olé (northern Senegal), which has under 350 mm of rainfall, Bille and Poupon (1974) have evaluated the number of seeds produced as 45 million ha^{-1}, being a total of 36.6 kg ha^{-1}. But careful counting allows one to discover that the soil contains more seeds than are produced each year. These seeds accumulate in the soil where they are stored. Not all germinate because some will be eaten by animals, while others will rot away. According to Bille and Poupon (1974) only one-tenth of the seeds actually germinate, to ensure the continuation of the species, thus indicating a tremendous wastage. The production of seeds is highly concentrated in time and then this source of food is suddenly put at the disposal of consumers.

A rational use of starchy grains, such as *Brachiaria, Panicum, Echinochloa, Dactyloctenium,* etc. all gathered incidentally by nomads and eaten as hungry-season food, could well be attempted using animal consumers such as the blue-bearded guinea-fowl or Clapperton's francolin, both of which are well adapted to such an experiment because of their Sahel origin. One is tempted to ask by such extensive animal-raising methods have not yet been tried.

4. The biological spectrum

Though the Sahel flora is on the whole well known, it includes a large element from neighbouring regions — complete census returns covering areas extensive enough to be significant are few in number. Thus, in Table 6.3 we are forced to use those of Boudet (1972), for the Hodh region of Mauritania, and our own for the Wadi Rimé Ranch covering 70 000 ha in Chad (Gillet, 1960, 1961).

It is readily apparent that the annual plants very largely dominate. After the therophytes and occupying second place are the tree species. The tree plays a major role in the Sahel. It not only follows faithfully the drainage system but also encourages directly the development of annual plants, and acts effectively in increasing their diversity. Certain therophytes never venture beyond the immediate vicinity of trees.

The difference noticed in the percentage of phanerophytes between Hodh and Wadi Rimé can be explained by pedological differences, namely the

Table 6.3. Life form spectra of Sahel vegetation

	Hodh (Mauritania) North Sahel	Wadi Rimé Ranch (Chad) South Sahel
Total number of species listed	161	207
Therophytes	61.4	67.1
Geophytes	1.8	4.8
Hemicryptophytes	5.6	6.7
Chamaephytes	6.8	7.9
Phanerophytes	23.6	13.5
	100.0%	100.0%

varied nature of soil in the Hodh as opposed to only sandy soil in the Wadi Rimé.

5.   Presence of geophytes

The presence of plants with bulbs or rhizomes is characteristic of the Sahel flora, especially on sandy substrata. They are too often neglected because they do not last for long, representing an important element in the countryside's vegetation only at the very beginning of the rains. But the geophytes have the advantage of taking off first and surpassing other plants by the vigour of their growth, at least in the early stages. The bulb becomes active as soon as it is humidified: it possesses greater reserves than seed, which enables it to produce a flowered stalk while taking in only small quantities of external nutrition.

The bulb of *Pancratium trianthum*, clothed in several layers can remain for years in the soil without suffering. It can even be covered with a thick layer of sand without inconvenience.

Certain geophytes (*Cyperus bulbosus*) with little bulbs positioned near the surface of the soil can respond to a small shower giving only a few millimetres of rain which is insufficient to prompt germination of seeds. In certain shallow depressions, during chronically dry years, the only sign of greenery is displayed by these humble bulbous (*Cyperus bulbosus*) or rhizomatous (*Cyperus rotundus*) members of the Cyperaceae.

It is justifiable to speak of early facies of geophytes, for example, the Dipcadi facies, which appear before the massive development of the grasses.

The rhizomatous geophytes constitute one of the most climatically resistant biological forms. The rhizome can remain in a dormant state, better protected than buried seeds. It is a perennial organ unaffected by trampling or by the wind. An annual plant, which is grazed and close-cropped, loses its reproductive capacity, but a grazed geophyte though weakened, still preserves intact its ability to put out new flowers.

IX.   Biomass and nutritive value of Sahel pasturage

As the principal use of Sahel vegetation, both herbaceous and woody, is to serve as animal forage, there is obvious need to ascertain accurately the biomass and nutritional value of the pasture. On the quantity and quality of the forage will depend the meat yield and, beyond that, the Sahel economy, who sole biological resource is stock-raising. If, on the whole, the amount of vegetation present on a given surface is a direct function of the amount of water (in absolute value and daily distribution) and the infiltration capacity of water, it is a completely different thing from the forage value of these areas which is fixed. The palatability is also an important factor.

1. Herbaceous biomass

What is understood by this is the mass of annuals evaluated as dried material and measured at the beginning of the dry season (October). It is the amount of forage which will actually be available to livestock during the long dry season (Fig. 6.8).

(a) Pastures situated to the north of the 14th parallel (Kanem, Chad)

Pastures of *Eragrostis tremula*. This type of pasture occurs on stable land with little rainfall (250–300 mm). The two main species are *Eragrostis tremula* and *Aristida mutabilis*. In this case, the biomass is of the order of 600 kg to the hectare when *Eragrostis* and *Aristida* are taken together. In the same context but in more favoured localities, hollows or channels between the dunes, the herbaceous carpet is more dense and the number of leafy stems put out by each individual plant is greater (six on average). The predominant species is *Schoenefeldia gracilis*, which is abundantly fruitful. In this case, the fresh stalk biomass is more than 1800 kg ha^{-1}. As it is generally acknowledged that the equivalence of dry Sahel forage is 0.33 FU (Forageable Units) kg^{-1}, it can be seen that in the first case the small amount of 180 FU ha^{-1} is at the disposal of the livestock, while in the second case it is 540 FU ha^{-1}.

Fig. 6.8. A grass (*Chrysopogon aucheri*) growing on a high plateau (Ennedi, Chad).

(b) Pastures situated on the 14th parallel (Kanem, Chad)

Here greater rainfall (about 400mm) promotes a richer pasture, not because the composition of the flora is really modified, but simply because the individual biomass of each individual plant increases. Each one develops further units more vigorous than the previous ones. In a pasture of *Aristida mutabilis* comprising about 200 stems to the square metre, the fresh weight biomass (October) is 2000 kg ha^{-1}. In mixed pasture of *Aristida mutabilis* and *Schoenefeldia gracilis* of 0.45 m in height, there is the same fresh weight biomass of 2000 kg ha^{-1}. But if large-leaved grasses (*Brachiaria deflexa*) are mixed with such narrow-leaved ones, the biomass rises to 2500 kg ha^{-1} (of which 1000 kg are *Aristida* and 1500 kg are *Brachiaria*). As these pastures on the 14th parallel benefit from higher rainfall, their water content can be estimated at 50 per cent which brings down their respective dry biomasses to 1000 and 1250 kg ha^{-1}.

(c) General rule

Following measures effected under different conditions (Senegal) the agrostologist Boudel (1972) estimates that the annual vegetation biomass of a Sahel pasture considered in November, after vegetative growth has ceased, can be estimated at 2 kg of dry matter per hectare per millimetre of rain received. This average does not take into account the soil retention capacity. The average value of 2 kg can be modulated in the following manner as a function of different soil types:

0.25 kg for shallow gravel soil, which is not retentive;
1.5   kg on dune undulations with oblique drainage.
3.5   kg on sandy peneplain with good water retention.

(d) Grazing load per hectare

These biomass calculations are useful for determining the number of hectares needed to support an average cow of 250 kg (= 1 tropical bovine unit = 1 UBT). This load can be light if, for example, the cow has 16 ha available or, on the contrary, heavy if the cow has only 8 ha. The most recent studies for determining the optimum load in the Sahel zone have been made by the agrostologist Klein (1980) on Ekrafane Ranch, north-east of Niamey for 400 kg of grass per hectare. His observations are summarized in Table 6.4. If the animals are herded at one head per 8 ha the gain in live weight is higher per hectare but less per head of livestock. There are also the risks of damaging the pasture and jeopardizing its renewal. Klein estimates that the optimum yield for the 1979–80 season was 1 UBT for 15 ha on the basis, as indicated above, of a biomass of 400 kg ha^{-1}.

(e) Productivity

It is important to distinguish the biomass, which is the net amount of vegetation present on the terrain at the moment when measurement is carried out, and the

Table 6.4. Grazing load and growth of cattle

Number of hectares per UBT	Coefficient of load	Individual gain of live weight (kg)	Gain obtained in kilograms of live weight per hectare
16	Light	88	3.8
8	Heavy	71	6.1

amount of vegetation mass yielded in a given time period, in general a growing season for annual vegetation. Productivity measurement is not easy to carry out because it involves a prolonged stay at a given station. In general, it is estimated that the maximum yield in the Sahel zone is 20 per cent above the maximum biomass. Bille and Poupon (1974) have published some measurements registered for the Senegalese Ferlo (200 mm) for various groupings:

For dune top stands of *Aristida mutabilis* the productivity is 35 g m^{-2} for August, 39 g m^{-2} for September, and 8 g m^{-2} for October, — being 82 g m^{-2} or 820 kg ha^{-1} (cf. our 600 kg ha^{-1} biomass observed in North Kanem).

For a stand of *Chloris prieurii*: 176 g m^{-2} for August; 65 g m^{-2} for September; 14 g m^{-2} for October. Total 255 g m^{-2} or 2550 kg ha^{-1}.

For a stand of *Panicum humile* on a moist depression growing up to November, the productivity would be 4800 kg ha^{-1}.

(f) Nutritional value

As the grass contains a lot of water its food value is correspondingly low, but as it is abundant, at least during the rainy season, the animals can compensate for that by eating large quantities. Its gross nitrogen content is, generally, between 5 and 9 per cent. The values of some of the most common and appetizing species are given in Table 6.5.

If a value of 0.5 FU kg^{-1} for this fresh grass is taken, an animal consuming 6 kg of dry matter would benefit from an energy input of 3 FU, which would correspond to the maintenance ration (2.3 FU), plus a reserve (0.4 FU), and allows for a gain in live weight. As straw the nutritional value falls to 0.3 FU, so it is obvious that value of the grass varies in accordance with its growth from a plantlet to the end of the fruiting period.

If the dry matter content is considered, that can only increase as the plant becomes straw. The content of digestible nitrogenous matter (m.a.d.) increases slowly at first, then more rapidly, reaching its maximum just when the flowering shoots are ready to extend vertically, after which it diminishes and becomes much reduced as the plant becomes straw-like. Thus, the straw of the Sahel is poorer than has been thought in its content of digestible nitrogenous

Table 6.5.   Nutritional value of some common grasses

Species	Percentage crude nitrogenous matter	FU kg^{-1}	Cellulose content
Brachiaria deflexa and other *Brachiaria* spp.	5.3	0.48	33.6
Dactyloctenium aegyptium	7.7	0.63	32.1
Echinochloa colona	7.9	0.47	35.3
Panicum laetum	8.1	0.61	33.7
Echinochloa stagnina (= borgou)	14.4	0.63	—

matter. In the dry season the only way cattle can find assimilable protein is to have recourse to the aerial leafy pasturage of trees and shrubs.

2.   Tree biomass

The consumption of aerial forage is an absolute necessity for cattle in the Sahel zone.

The determination of the tree biomass is, for different reasons, even more difficult to establish than that of a few square metres of herbs. It requires the employment of labour and machines such as lifting tackle, weighing machines, and even excavators if roots are to be weighed. For such reasons proper measurements of tree biomass are very rare.

But measurements have been made at the International Biological Programme's (IBP) plot at Fété-Olé (North Senegal, 343 mm). They have the disadvantage of being limited in extent but the advantage of being precise. The values recorded reflect great differences according to the locality, because the density of tree vegetation undergoes much more important variations than herbaceous vegetation. Schematically, it is convenient to distinguish three types of stations following a topographic gradient:
(a)   the upper levels of dunes where the sand is well drained;
(b)   the well-drained moderately inclined gradients;
(c)   the hollows constituted of accumulated horizons of lime and clay materials.

(a) Total biomass
This total, fresh biomass average is 5.5 t ha^{-1}, of which 3.3 is above the surface and 2.2 underground, or 60 per cent in the air and 40 per cent below ground. In actual fact, this biomass varies from 2 t ha^{-1} on high ground to 24 t ha^{-1} in the hollows. In the aerial part the bulk of material is mainly in the

trunk. In young trees, unlike those that are some years old, the underground biomass is more important than that above the surface.

The aerial biomass has a direct relation to the percentage cover of the woody layer. This cover value is the total projected area covered by the crowns as a percentage of the land surface. As a general average 5.5 t ha^{-1} corresponds to an 8 per cent cover value divided between a minimum of 2.8 per cent for the summits and 40 per cent for the hollows, which appear as a semi-closed canopy. Under these conditions at Fété-Olé, the total vegetation biomass expressed as an average in fresh weight is 5.5 plus 4.5 t ha^{-1} for the herbaceous strata, being 10 t ha^{-1} in all.

(b) Foliage biomass
It requires great patience to take all the leaves off a tree in order to weigh them. To avoid this laborious task mathematicians have proposed equations giving leaf weight as a function of the diameter of the base (cf. *Acacia tortilis* subsp. *raddiana, Acacia seyal, Balanites aegyptiaca*, etc.), but these formulae are valid only for a given station, because they depend on the size of the tree, which is not a constant. For trees of equal trunk diameter those with fine leaflets such as *Acacia tortilis* var. *raddiana* have less foliage biomass than acacias with bigger leaflets.

The dry foliage mass in grams as a function of trunk diameters for three major species are given in Table 6.6.

(c) General rule
The Niono Station (550 mm) researchers who are in close touch with actual conditions, have established a formula that dry above-ground consumable biomass is 1 kg mm^{-1} of rainfall per hectare per year. It would be, therefore, 500 kg in the south Sahel zone if the consumption of 50 per cent of the biomass by animals is assumed. But Hieurnaux (1980) at Niono has carefully measured 1000 kg of dried leaves per hectare, an average figure ranging from 21 kg ha^{-1} for an open area of *Combretum* and *Sclerocarya birrea* to 3536 kg for a dense

Table 6.6. Comparison of relation between stem diameter and dry leaf weight (g)

	20 cm dia.	30 cm dia.
Acacia tortilis subsp. *raddiana*	1 100	1 600
Acacia seyal	3 800	8 000
Balanites aegyptiaca	6 000	10 500

These figures are from Oursi Lake Station (400 mm) in northern Burkina Faso.

stand of *Pterocarpus lucens*. It is clear that this formula cannot be applied to the average western Saharan zone where the tree cover is much more open. In Fété-Olé's 25 ha plot (343 mm) Poupon (1980) reports 120 kg of leaves and fruits per hectare (although it should be 680 kg there using the formula of 2 kg of leaves for each millimetre of rainfall per hectare per year).

Generally, the woody layer is the only source of digestible protein during the dry season, so that in the big stock-raising regions it is overgrazed, when it is not actually massacred, for the shepherds have no compunction in chopping down branches to bring them within the reach of smaller animals. It is very rare during the dry season that Sahel cattle are assured of their 2 kg daily ration of forage. For another example at Oursi Lake (450 mm), Nebout (1978) calculates the above-ground leaf mass at 2260 t for a surface area of 37,000 ha: that is, 61 kg of tree leaves per hectare, which is far from our theoretical 900 kg! It is noted that animals consume 49 per cent of the total biomass which represents in effect only 2 per cent of their ration. This demonstrates the degree to which this fodder is utilized. Practically everything which is accessible is consumed and that is far from being sufficient.

(d) Productivity
If it is already so difficult to measure the productivity of the herb layer what can be said about that of the woody biomass? For Fété-Olé Poupon (1980) estimates 0.42 t ha^{-1} yr^{-1}, which is slightly under one-tenth of the total biomass (5.5 t ha^{-1}). This means that the annual increase of the woody biomass is 8 per cent both above and below ground.

(e) Nutritional value
It is now amply demonstrated that the aerial foliage supplies indispensible nutritional elements to support animals throughout the long dry season. The protein content of *Acacia* leaves is almost three times that of grass and it is about the same for the commonest Sahel trees with persistent leaves. The Capparidaceae family merits particular attention, for the leaves of all of them (shrubs, trees, and climbers) have both high protein and mineral salt contents.

X.   Conclusion

Everywhere throughout the Sahel zone the tree stratum, stricken by drought and ravaged by intense over-exploitation, is regressing. From north to south in the Sahel the most beautiful trees (*Acacia tortilis* ssp. *radiana*, *Commiphora africana*) are rapidly disappearing. In most regions the young plants, arising from germination, have no chance of surviving beyond one year and regeneration is practically nil. If it is seriously desired to save the Sahel from complete desertification the only way is to protect the trees and allow them to multiply.

References

Bernus, E. (1979). L'arbre et le nomade, in *Journal d'Agric. Trad. et de Bot. Appliq. (Paris)*, **36**, 103–128.

Bernus, E. (1980). Tourregs nigériens. Unité Culturelle et diversite régionale d'un peuple pasteur. *Mémoires ORSTOM*, No.94, Paris, 508 pp., 30 cartes et figures, 8 Pl. photo, 5 cartes bors texte Bibliographie (20 pp.)

Bille, J. C., and Poupon, H. (1974). Recherches écologiques sur une savane sahélienne du Ferlo septentrional, Sénégal. La régénération de la strate herbacée. *La Terre et la Vie*, **28**(1), 21–48.

Boudet, G. (1972). Désertification de l'Afrique tropicale sèche. *Adansonia*, sér. 2, **12**(4), 505–524.

Delwaulle, J. C. (1978). Plantations forestières en Afrique tropical sèche. *Rev. Boils et Forêts des Trop.*, 181–184. CTFT, Nogent-sur-Marne.

Gillet, H. (1960). Etude des pâturages de l'O. Rimé. *Journal d'Agric. Trop. et de Bot. Appliq.*, **7**(2).

Gillet, H. (1961). Pâturages sahéliens. Le Ranch de l'Ouadi Rimé. *Journal d'Agric. Trop. et de Bot. Appliq.* Paris **8**(10–11).

Gillet, H. (1968). Le peuplement végétal du Massif de l'Ennedi (Tchad). *Memoires du Museum National d'Histoire Naturelle*, nouvelle série. Série B, Botanique, Vol.17. Paris, 206 pp., 23 pl.

Hiernaux, P. (1980). L'inventaire due potentiel des arbres et arbustes d'une région du Sahel malien. *Coll. Intern. sur fourrages ligneux en Afrique*. CIPEA, Addis Ababa.

Klein, (1980). Rapport d'activités. IEMVT (unpublished).

Le Houerou, H. N. (1980). Le rôle des ligneux fourragers dans les zones saheliennes et soudaniennes. *Colloque intern. sur les fourrages ligneux en Afrique*. Addis Ababa, 8–12 April 1980.

Monod, Th. (1973). Les déserts. *Coll. Horizons de France*. Paris.

Nebout, J. P. (1978). Etude sur les arbres fourragers dans la zone sahelienne. Centre Technique Forestier Tropical, Nogent-sur-Marne et Inst. d'Elev. et de Medecine Vétér. des ayx Tropicaux. Maisons Alfort (miméogr.).

Poupon, H. (1980). Structure dynamique de la strate ligneuse d'une steppe sahelienne au nord du Sénégal. *Travaux et documents ORSTOM*, No. 115, 1–351, Paris.

Trochain, J. L. (1980). *Ecologie Végétale de la zone intertropicale non desertique*. Université Paul Sabatier. Toulouse.

Plant Ecology in West Africa
Edited by G. W. Lawson
© 1986 John Wiley & Sons Ltd

CHAPTER *7*

Aquatic vegetation

David M. John
Department of Botany, British Museum (Natural
History), Cromwell Road, London SW7 5BD, UK

I. Introduction

Past geological events in tropical West Africa did not produce any features equivalent in extent to the Rift Valleys of East Africa, or to the relatively flat and largely swamp-filled basins lying between them. Of the seven major swamp complexes mentioned by Thompson and Hamilton (1983) for the continent, the only one in tropical West Africa is in the Lake Chad basin. Naturally, there are smaller permanent and seasonal swamps in the region and many of these are associated with the floodplains and delta plains of river systems. They include the floodplains of the major rivers (Gambia, Senegal, Benue, Shari, and Niger) and the so-called internal delta of the Niger which is situated in Mali. Such areas owe their existence to the deposits of river-borne material laid down wherever the surrounding countryside is relatively flat. Often the swamp vegetation in large lakes is best developed in the deltas of rivers draining into their basins. Marginal swamp is little developed in many man-made ponds and lakes where the seasonal rise and fall in water-level is significantly greater than in many naturally occurring lacustrine habitats.

The areas of swamp vegetation undergo seasonal expansion and contraction in response to shorter- or longer-term changes in water-level due to the amount and distribution of rainfall. Often the nature of the swamp vegetation depends upon short-term changes in the water balance, with seasonal or ephemeral swamps containing a mixture of truly aquatic species and flood-tolerant terrestrial species. In the region covered by this chapter, the so-called 'reed swamp' has a relatively restricted distribution as compared to some of the areas covered by swamp forest such as, for example, that lying to the west of the coastal delta of the Niger (see Adejuwon, 1973).

II. Aquatic Associations

The more permanent reed swamp and other truly aquatic associations of macrophytes, are widespread and remarkably similar throughout tropical West

Africa. Various authors have described aquatic and semi-aquatic associations in the region. Some have adopted the concepts of the Braun–Blanquet School of Phytosociology and have attempted to place the associations into a hierarchy. Associations of aquatic plants have been described for Senegal (Adam, 1958, 1964; Trochain, 1940), Mali (Chevalier, 1932; Duong-Huu-Thoï, 1950; Roberty, 1946), Sierra Leone (Cole, 1968), Nigeria (Cook, 1965, 1968), and Lake Chad (Iltis and Lemoalle, 1983). Some of these accounts are concerned primarily with terrestrial associations and only brief mention is given to those occurring in aquatic habitats. Useful keys, descriptions, and illustrations of some of the more important aquatic and semi-aquatic plants found in the Sahelo-Sudanian region (includes much of tropical West Africa) are given by Raynal-Roques (1980) and in Lake Volta by Hall *et al.* (1971). Various other accounts of West African aquatics exist such as that by Horn af Rantzien (1951) on the aquatic plants of Liberia and present-day Ghana.

The aquatic and semi-aquatic associations recognized in different parts of West Africa are often remarkably similar in structure. Only two accounts of these associations are considered in some detail below. These are those by Cook (1965, 1968) on the associations in the lower Niger floodplain (Kainji area) and by Iltis and Lemoalle (1983) on those found in Lake Chad. The following associations are those recognized by Cook (1968) in Nigeria:

1. *Tristicha trifara* association (*Tristichietum trifarae* C. Cook 1968). This grows on rocks in the swift-flowing parts of larger rivers (River Niger, River Temo, River Doro) and consists of a single species (*Tristicha trifara*) with a few attached algae. The Podostemaceae are extremely specialized and are the only higher plants that occupy this type of lotic habitat. Similar associations are to be recognized in other river systems in West Africa. For instance, Hall (1971) describes other members of the family Podostemaceae growing in the River Ankasa in the moist semi-deciduous forest zone of western Ghana.

2. *Lemna* and *Pistia* association (*Lemneto-Pistietum* Lebrun 1947). This is found in the Niger valley in still or very slowly flowing water in aquatic habitats which are flooded during the wet season. The association is characterized by *Lemna paucicostata* and *Pistia stratiotes*, and is occasionally accompanied by *Ceratopteris cornuta*; all free-floating plants. According to Cook (1968), this association is able to develop in turbid water and so is limited by the mineral status of the water, competition with rooted emergents, water movement, and an available surface of water. It is widespread in the West African region and these two genera form a similar association in Lake Chad (see Iltis and Lemoalle, 1983).

3. *Ceratopteris cornuta* association (*Ceratopteretum cornutae* C. Cook 1968). According to Cook (1968), this is closely related to the former association when it is in the free-floating state. It differs from the *Lemna–Pistia* association in that it manifests itself in the wet season where it occupies very shallow water (<10 cm deep) and may be rooted in the substrate.

4. *Salvinia nymphellula* association (*Salvinietum nymphellulae* C. Cook 1968). This forms single-species, free-floating mats on still and permanent bodies of water that are usually occupied by other vascular plants. The association is widely reported in West Africa (see Hall *et al.*, 1971) and is often the first stage in a succession on small ponds leading to eventual filling and terrestrialization (see Aké Assi, 1977).

5. *Nymphaea lotus* association (*Nymphaeetum loti* Lebrun 1947). This was found in varying sized bodies of still or gently flowing water (permanent or nearly so) between 50 and 150 cm deep and is even tolerant of greater depths during periods of flash floods. Other species mentioned from this probably very widespread association in West Africa include *Nymphaea micrantha*, *Aeschynomene crassicaulis*, *Polygonum salicifolium*, and *Utricularia inflexa*. In Lake Chad, an association of *Nymphaea* spp. and *Utricularia* spp. (Nymphaeidae) is recognized by Iltis and Lemoalle (1983).

6. *Eichhornia* and *Ranalisma* association (*Eichhornieto-Ranalismetum* Léonard 1951). This is a short-lived association which dies when the shallow ponds or pools (<5 cm deep) disappear in the dry season. It is characterized by *Eichhornia natans* and *Ranalisma humile*.

7. *Ottelia* and *Nymphaea* association (*Ottelia-Nymphaeetum maculatae* C. Cook 1968). This develops in shall water (<50 cm) shaded by phanerogams and does not dry out over the dry season. Occasionally in the rainy season the habitat may be subject to fast-flowing water in times of flood. The characteristic species are *Ottelia ulvifolia* and *Nymphaea maculata*, though the charophyte *Nitella furcata* may occasionally be present.

8. *Najas* and *Nitella furcata* association (*Najado-Nitelletum furcatae* C. Cook 1968). This occurs in shallow (<1 m deep) clear water in the dry season, but in considerably deeper and faster-flowing water on occasion over the rainy season. It is characterized by *Najas affine* and *Nitella furcata* and *Chara* spp. in the rainy season. These aquatics are accompanied by *Aponogeton subconjugatus* and by much *Utricularia* spp., *Lemna*, and *Pistia* especially in the dry season.

9. *Ceratophyllum demersum* association (*Ceratophyllum demersi* Den Hartog & Segal 1964). This submerged aquatic was found in shallow backwaters of the River Niger and in the mouths of streams discharging into the main river. It is a very common association in West Africa and is present in all man-made lake systems and in Lake Chad.

10. *Echinochloa pyramidalis* association (*Echinochloetum pyramidalis* Léonard 1951). This and the next association react to flooding by floating on the surface when the water is deep and root on the bottom as the water-level drops and is very shallow. It is mentioned by Cook (1965) as common along the edges of rivers with steep banks, though this species has possibly been confused with *E. scabra* (=*stagnina*), a truly aquatic species (see Hall *et al.*, 1971). During the flood period floating portions of *Echinochloa* may become

detached and form free-floating islands. In the Volta and Kaingi lakes and other lacustrine habitats in West Africa, this grass is common in the drawdown area and is also a component of the sudd where it is associated with various aquatics including *Vossia cuspidata, Ipomoea aquatica, Alternanthera nodiflora,* and *Cyperus pustulatus.*

11. *Aeschynomene nilotica* association (*Aeschyonemeetum niloticae* C. Cook 1968). In the area of the Kaingi dam, this association was regularly flooded (<2 m) in the wet season but was without standing water in the dry season. Plants of this species are much smaller than *Aeschynomene elaphroxylon* which forms an associaton recognized along the border of the reed islands in Lake Chad (see Iltis and Lemoalle, 1983). It increased considerably in extent on the vast areas of lake floor exposed during the 'Little Chad' stage.

12. *Polygonum senegalense* association (*Polygonetum senegalense* C. Cook 1968). This is found along the edges of the river on sloping banks and sand-bars providing that there is no strong current. It is often the most common association along the banks of large rivers along with the *Echinochloa* association. These two associations are also to be found along the margin of such man-made lakes as Lake Volta and Lake Kainji. Common plants of this association are *Ipomoea asarifolia, I. aquatica,* and *Cyperus maculatus.*

13. *Cyperus procerus* association (*Cyperetum proceri* C. Cook 1983). This is found in relatively flat areas in the Niger floodplain that are often flooded in the rainy season and still remain wet over the dry season. It is perhaps similar to the *Cyperus laevigatus* association that characteristically fringes the saline ponds and lakes in the Kanem region of Chad. Such an association also occurs very sporadically in the extreme north of Lake Chad where the salinity of the lake water is greatest (see Iltis and Lemoalle, 1983).

All these associations are well defined and appear to be widespread in tropical West Africa. They depend for their existence on hydrological features that are themselves dependent on relatively small-scale climatic and geographical changes. Such azonal associations are best regarded as habitat opportunist vegetation types.

Some of the associations recognized by Cook (1965, 1968) in aquatic habitats in the area of the Niger to be flooded by Lake Kainji (see above) are similar to those in Lake Chad that lies in the Sahelian zone. However there are to be found several other associations mentioned by Iltis and Lemoalle (1983) that seem to have no equivalents in this lower Niger floodplain. These include the *Phragmites australis* subsp. *altissimus* (Phragmitidae) association (frequent throughout the lake), the *Typha domingensis* (=*australis*) (Typhidae) association (forms reed belts; absent in south), and the one formed by the semi-aquatic meadows of *Vossia cuspidata* (dominant in the Shari and El Beïd deltas). The aquatic macrophytes showed readily discernible changes away from the Shari delta to the northernmost part of the lake during the 'Normal Chad' stage. For instance, the meadows of *Vossia* became less common and eventually disappeared, papyrus progressively disappeared to the north of the Grande Barrière, *Typha* assumed along with *Phragmites* increasing dominance, *Cyperus laevigatus* appeared, and a number of genera such as *Nyphaea* and *Utricularia* gradually disappeared. This impoverishment of the aquatic flora further north is probably accounted for by the increased salinity and by exposure to stronger and drier winds. The aquatic vegetation changed very dramatically in the northern basin when the lake bed almost completely dried out in 1975. Most of the vegetation that existed before the drought period of 1972–73 had almost completely disappeared by this time. *Aeschynomene elaphroxylon* was still present and *Typha domingensis* persisted along some leeward shores in the eastern side of the basin and around some temporary ponds. Another association not mentioned by Cook is the *Potamogeton schweinfurthii* association which occurs, for example, at Lake Tiga (Nigeria) in slowly moving deeper water (G. W. Lawson, pers. comm.).

III. General features of lentic habitats

The dominant aquatics forming the emergent reed swamp are normally rooted in waterlogged soil and rarely grow in water deeper than about 1.5 m. Characteristic species of these swamps in West Africa are such widely distributed grasses as *Phragmites australis* and *P. mauritanica*. Papyrus (*Cyperus papyrus*) is another distinctive swamp plant that is found throughout central and eastern Africa, but its distribution in West Africa is very much restricted with records existing for Benin, Guinea, Ivory Coast, Niger, and Senegal. It is not mentioned by Iltis and Lemoalle (1983) as forming an association in Lake Chad though extensive papyrus stands occur in the southern basin. Its limited distribution in West Africa has been accounted for by the intolerance of the plant to the often considerable difference between high and low water levels during the annual flood cycle (Thompson and Hamilton, 1983). Further data are needed before this explanation can be accepted (see remarks in Newton, 1986a). The stout rhizomes of papyrus and other sedges and the long stems of the grasses *Vossia cuspidata* and species of *Echinochloa*, often extend out from the edge of the swamp as floating mats. These mats frequently become detached to form floating islands commonly referred to as 'sudd'. Such floating islands are especially common in Lake Chad, but are also reported from several man-made lakes including Lake Volta, Lake Kainji, and the Weija (=Densu) lake or reservoir in Ghana.

The two grasses that characteristically fringe the outer edge of swamp vegetation (*Vossia cuspidata*, *Echinochloa scabra*/pyramidalis) are capable of keeping pace with the rapid rise in water-level associated with the flood season. They are the only common aquatics fringing the shore of man-made lakes in which there is a considerable annual drawdown. The increase in cover of *Echinochloa* between 1976 and 1977 in Lake Kainji was viewed with some alarm by Chachu and Chaudhry (1978) who considered it a threat to both fishing and navigation. Morton and Obot (1984) believe that between 28 per cent and 40 per cent of the lake's surface was covered by the grass during the years 1972 to 1983. They point out that the power generation authority consider its growth will reduce the life expectancy of Lake Kainji by water displacement and increasing salinity. On the positive side, Morton and Obot consider that 75 per cent of the standing crop of the grass could be cut for harvest each year on a sustainable basis and much of this used as a livestock fodder. It has long been harvested between Gao and Niamey as a building material and for fodder, and so is regarded as an asset by the local people (see Rzóska, 1985). There are two types of plants with floating leaves to be found in somewhat deeper water: rooted in the bottom mud (*Potamogeton* spp., water-lilies such as *Nymphaea* spp. and *Nymphoides indica*) and completely free-floating (e.g. the ferns, *Azolla africana* and *Salvinia nymphellula*; the duckweeds, *Lemna* spp. and *Wolffia arrhiza*; the Nile cabbage or water-lettuce, *Pistia stratiotes*; the bladderwort, *Utricularia* spp.). Beyond the emer-gent or floating-leaved aquatics in lakes or ponds, there is often a zone of completely submerged plants including *Ceratophyllum demersum*, *Najas* spp., *Vallisneria aethiopica*, *V. sporoides*, and various stoneworts or charophytes (principally *Nitella* spp.).

The explosive development of the free-floating *Pistia* has been reported immediately following the impoundment of West African rivers to form man-made lakes. This plant does not build up in this region the enormous populations found in India and Central America. Hall and Okali (1974) recognized four growth phases of *Pistia* in the Pawmpawm arm of Lake Volta which are accounted for by seasonal fluctuations in nutrient conditions. It grew most vigorously in this arm of the lake over the rainy season and they estimated a rate of production of 14.2 g dry wt m^{-2} day^{-1}, if a correction is made for losses and death. The main difference in behaviour of *Pistia* in western Nigeria (see Pettet and Pettet, 1970) was that the drastic die-back of the population occurred within the main dry season rather than in the wet season. Though this and other floating aquatics (e.g. *Wolffia arrhiza*, *Salvinia nymphellula*) are sometimes troublesome, they generally decline in numbers as a pond or lake stabilizes and sometimes almost completely disappear. Fortunately the two most notorious species, the water-hyacinth (*Eichhornia crassipes*) and the water-fern *Salvinia molesta* (formerly incorrectly known as *auriculata*), are not known from tropical West Africa north of the equator apart from a minor occurrence of the former in Senegal. These species are present in the Congo and every effort should be made to ensure that they do not become more widely distributed. They have created serious management problems in certain East Africa lakes such as Lake Kariba due to their explosive development (Boughey, 1963). Indigenous to tropical West Africa is *Salvinia nymphellula*, but this small water-fern is not regarded as being especially troublesome. More of a problem are the floating islands of sudd as they hinder navigation when they cover vast areas. Sudd formation is encouraged when floating mats are trapped by trees flooded during the rise in water-level following the impoundment of a river to form a lake. This happened initially in Lake Volta, though it was not encountered to any great extent in Lake Kainji and Lake Asejire where much of the area to be flooded was cleared of woody vegetation beforehand.

IV.   General features of lotic habitats

Swiftly flowing streams and larger rivers often have members of the family Podostemaceae growing attached to rocks. Commonly accompanying these phanerogams are various filamentous green and red algae. Such aquatic macrophytes are confined to the shallows and to temporary pools or backwaters in major rivers. Along the river-bank there is sometimes found a gradual transition from aquatic to semi-aquatic associations tolerant of brief exposure to the air, through to terrestrial plants tolerant of only brief periods of submergence. Over the dry season the river levels drop and terrestrial species colonize the newly exposed river margin, only to be drowned by the return of the floods. The mat-forming *Echinochloa* is commonly found in places fringing the river. This aquatic grass sometimes becomes free-floating and detaches as the current velocity increases following the rise in water-level in the wet season. It is usually to be found on steep eroding banks, whereas almost pure stands of *Polygonum senegalense* occur on sloping banks where there is active accretion. In shallow backwaters and in the mouths of streams draining to main rivers, there are occasional submerged beds of *Ceratophyllum demersum*. This plant easily becomes dislodged from the river-bed and so is commonly found free-floating in the flood season. The River Volta has become transformed downstream of Lake Volta from a flood- into a reservoir-type of river. Such a transformation has led to the stabilization of its flow and to an increase in transparency of the water which has favoured the year-long development of *Potamogeton octandra* and *Vallisneria aethiopica* whose stands have consequently increased greatly in extent (see Hall and Pople, 1968). Another effect of this transformation has been to encourage the colonization of some low-lying sand-banks and the shallow river margin by *Cyperus articulatus* and *Typha domingensis*.

V.   Succession

Various successional pathways have been suggested for aquatic and semi-aquatic vegetation in different water bodies in West Africa. One of the recently suggested successional sequences for Lake Chad is shown in Fig. 7.1. The direction of succession is indicated by arrows and suggests, for example, that an assemblage of species dominated by *Vossia cuspidata* is succeeded by

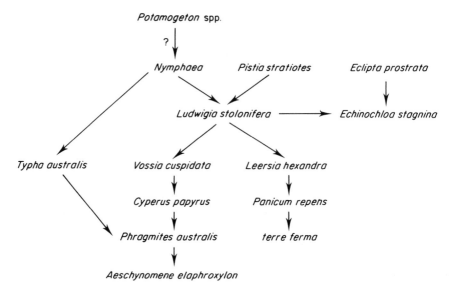

Fig. 7.1.   A successional scheme proposed by Léonard (1969) for the aquatic and semi-aquatic phanerogams of Lake Chad.

another dominated by papyrus. The final stage in the sequence is an assemblage or association dominated by the ambach *Aeschynomene elaphroxylon* that covered extensive areas of the lake bed as it dried out during the 'Little Chad' stage. The course of succession is similar in most other West African lakes, though the floristic composition of the assemblages often shows some minor variation from those found in Lake Chad.

Floating plants are commonly regarded as pioneers in newly formed open areas of standing water. Mats of such floating hydrophytes as *Pistia stratiotes* and *Vossia cuspidata* often become colonized by other species which eventually replace them to form a more persistent sudd. In Lake Volta, and many smaller water bodies, the first colonizer of *Pistia* mats is often *Scirpus cubensis*. This sedge produces horizontal stolons which grow just below the water surface in among the *Pistia* roots. They give rise at intervals to erect leafy shoots or tillers between the *Pistia* rosettes. The ability of *Scirpus* to colonize open water is apparently very limited, possibly it requires the support given to it by such plants as *Pistia* in order for it to become established. *Pistia* declines markedly in density in mats invaded by *Scirpus cubensis*, whereas there was no such decrease found by Okali and Hall (1974) in those consisting almost wholly of *Pistia*. They suggest that surface and subsurface interactions are probably more important than mere shading of the *Pistia* by *Scirpus*. There seems to be no evidence to support Pierce's (1971) contention that the disappearance of *Pistia* leads to the sinking of the floating sudd except perhaps where it is very extensive and well protected. The areas of sudd studied by Okali and Hall (1974) in the Pawmpawm arm of Lake Volta actually enlarged as the sedge colonized new *Pistia* mats blown against it.

Large plants such as the cat's-tails or bulrush *Typha domingensis* may colonize floating *Scirpus*-dominated sudd. The sudd in some aquatic habitats sometimes becomes colonized by trees that form floating thickets 3–4 m high. For instance, in the Avu lagoon near the lower Volta River (Lawson *et al.*, 1969) and in the Weija (=Densu) lake or reservoir near Accra (Okali and Hall, 1974), there are trees such as *Ficus congensis* and *Aeschynoneme elaphroxylon* growing on the sudd. In another body of standing water in Ghana, Lake Barekese near Kumasi, the most important colonizer of the *Pistia* mat was a large sedge known as *Rhynchospora corymbosa*. According to Lawson *et al.* (1969), the submerged aquatic *Ceratophyllum demersum* often provided the opportunity for sedges such as *Scirpus cubensis* to root and grow in some of the freshwater lagoons in the lowermost parts of the River Volta. Other floating aquatics, quite often forming a scum on the surface of sheltered bays, are the duckweeds (*Lemna, Spirodela, Wolffia*). It is sometimes possible to trace a successional sequence in small ponds from the floating fern *Salvinia nymphellula* gradually through to complete terrestrialization (see Aké Assi, 1977).

A seasonal succession of plants occurs in the floodplains of major river systems and on the area of land covered and uncovered by water along the edge of lakes. Terrestrial forms commonly colonize areas exposed when the water-level drops, and most of these are destroyed with the return of the flood (Fig. 7.2). The rooting of the stranded sudd is believed to be of little or no significance in the establishment of littoral vegetation (see Hall, 1975). Surveys concluded by Hall (1970) in Lake Volta showed a continuing succession in the drawdown with new plants coming in as colonizers while others were dying out. He was able to recognize three broad zones of vegetation (annual forb zone, perennial grass/*Polygonum* zone, sedge zone), the details of which varied from place to place. The continuing alternation of flooding and exposure does not permit the vegetation in such areas to achieve any great

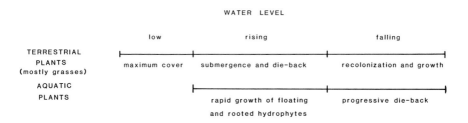

Fig. 7.2.   A summary of the changes that occur seasonally in the aquatic and terrestrial plants growing in a drawdown area or on a river floodplain.

stability. Nutrients are released from the newly inundated soil and from the flooded or 'drowned' plants that eventually decay. These probably stimulate the growth of truly aquatic macrophytes especially those that form mats covering the rising water (e.g. *Echinochloa scabra*, *Polygonum senegalense*, *Pistia stratiotes*). Such plants are likely to interfere with phytoplankton production by reducing the amount of light entering the water and by competing with the algae for available nutrients. It is interesting to note the growth of planktonic algae and macrophytes is generally out of phase, viz. the primary production of phytoplankton declines in the wet season as that of emergent and floating aquatics is increasing.

VI. Conclusion

Long-term and quantitative studies on the plants found in aquatic habitats in West Africa are either lacking or else are far from complete. Often such data have been collected over a limited period of time (rarely more than a year), and so are unlikely to represent conditions over the long term in what is sometimes a large and heterogeneous aquatic system. No doubt remote sensing has an important role to play in the future for giving a synoptic picture of the distribution of useful parameters over a large area. For instance, it is now possible using satellite imagery to estimate the area covered and the distribution pattern in a lake (see Chachu, 1979, for Lake Kainji). The general lack of basic information on many aquatic habitats makes it almost impossible for botanists to present realistic solutions to the many problems of water management encountered in the region.

Some of the less arid areas of West Africa have high population densities that are leading to increasing pressure being imposed upon freshwater resources. In such populous areas there is increasing pollution and an actual loss of aquatic habitats as industrialization and urbanization gather pace. Though water bodies are becoming lost by the filling-in of ponds and lakes, yet others are being created by the construction of irrigation systems and ponds of various kinds (ornamental, aquaculture, watering of livestock) together with the impoundment of streams and rivers to conserve water that otherwise would drain away. Sometimes the creation of new water bodies has led to problems not fully anticipated in their planning. Often these relate to the explosive development of aquatic vegetation and of water-borne diseases. Though some attention has been directed to these problems (see Newton, 1986b), little or none has been given to the need to conserve aquatic habitats or to undertake baseline surveillance or monitoring studies so necessary if future generations of researchers are to follow the changes as man's activities increasingly modify the West African environment.

References

Adam, J.-G. (1958). Floristique des pâturages salés et végétation des rizières du Sine-Saloum (Sénégal). *J. Agric. trop. Bot. appl.,* **5**, 505–541, 638–664.

Adam, J.-G. (1964). Contribution à l'étude de la végétation du lac de Guiers. *Bull. Inst. fr. Afr. noire,* sér. A, **26**, 1–72.

Adejuwon, J. O. (1973). The ecological status of freshwater swamp savannas in the forest zone of Nigeria. *J. W. Afr. Sci. Ass.,* **16**, 133–154 (1971).

Aké Assi, L. (1977). *Salvinia nymphellula* Desv. (Salviniacées) fléau en extension vers l'ouest de l'Afrique intertropicale. *Bull. Inst. fond. Afr. noire,* sér. A, **39**, 555–562.

Beadle, L. C. (1981). *The Inland Waters of Tropical Africa. An introduction to Tropical Limnology.* Second Edition. Longmans, London and New York.

Boughey, A. S. (1963). The explosive development of a floating weed vegetation on Lake Kariba. *Adamsonia,* **3**, 49–61.

Chachu, R. E. O. (1979). The vascular flora of Lake Kainji. In Anon., *Proceedings of the International Conference on Kainji Lake and River Basins Development in Africa.* Vol. II. pp. 479–487. Lake Kainji Research Institute.

Chachu, R. E. O., and Chaudhry, A. B. (1978). Vascular plants on Lake Kainji. *Annual Report of the Kainji Lake Research Institute.* **1977–78**, 48–51.

Chevalier, A. (1932). Les associations végétales du lit du Moyen-Niger. *C.r. somm. Séanc. Soc. Biogéogr.,* **9**, 73–77.

Cole, N. H. A. (1968). *The Vegetation of Sierra Leone (incorporating a Field Guide to Common Plants).* Njala University College Press, Sierra Leone.

Cook, C. D. K. (1965). The aquatic and marsh plant communities of the reservoir site. In E. White (Ed.), *The First Scientific Report of the Kainji Biological Research Team,* pp. 21–42. University of Liverpool, England.

Cook, C. D. K. (1968). The vegetation of the Kainji Reservoir site in Northern Nigeria. *Vegetatio,* **15**, 225–243.

Duong-Huu-Thoï (1950). Étude préliminaire de la végétation du delta central Nigerien. In *Conferência Internacional dos Africanistas ocidentais,* 2.ᴬ. *Conferência Bissau,* 1947. Vol. II, *Trabalhos apresentados à 2.ᴬ Secção (Meio Biologico),* 1.ª Parte, pp. 53–156, Lisbon.

Hall, J. B. (1970). Observations on the vegetation of the Lake Volta drawdown area. Volta Basin Research Project Technical Report No. X35, 21 pp. and 22 figs, Accra (cyclostyled).

Hall, J. B. (1971). New Podostemaceae from Ghana with notes on related species. *Kew Bulletin,* **26**, 125–136.

Hall, J. B. (1975). The vascular flora of Lake Kainji and its shores. In A. M. A. Imevbore and O. S. Adegoke (Eds.), *The Ecology of Lake Kainji: The Transition from River to Lake,* pp. 149–156. University of Ife Press, Ife.

Hall, J. B., and Okali, D. U. U. (1974). Phenology and productivity of *Pistia stratiotes* L. on the Volta Lake, Ghana. *J. appl. Ecol.,* **11**, 709–726.

Hall, J. B., Pierce, P. C., and Lawson, G. W. (1971). *Common Plants of the Volta Lake.* Ghana Universities Press, Accra.

Hall, J. B., and Pople, W. (1968). Recent vegetation changes in the lower Volta River. *Ghana J. Sci.,* **8**, 24–29.

Horn af Rantzien, H. (1951). Certain aquatic plants collected by Dr. J. T. Baldwin Jr. in Liberia and the Gold Coast. *Bot. Notiser,* **1951**, 368–398.

Iltis, A., and Lemoalle, J. (1983). The main types of communities and their evolution during a drought period. 5. The aquatic vegetation of Lake Chad. In J. - P. Carmouze, J. - R.Durand, and C. Lévêque (Eds.), *Lake Chad: Ecology and Productivity of a*

Shallow Tropical Ecosystem, pp. 125–143. W. Junk, The Hague, Boston, Lancaster.

Lawson, G. W., Petr, T., Biswas, S., Biswas, E. R. I., and Reynolds, J. D. (1969). Hydrobiological work of the Volta Basin Research Project 1963-1968. *Bull. Inst. fond. Afr. noire.* sér. A, **31**, 965–1003.

Morton, A. J., and Obot, E. A. (1984). The control of *Echinochloa stagnina* (Retz.) P. Beav. by harvesting for dry season livestock fodder in Lake Kainji, Nigeria — a modelling approach. *J. appl. Ecol.,* **21**, 687—694.

Newton, L. E. (1986a) The aquatic macrophytes. In D. M. John, *The Inland Waters of Tropical West Africa: An Introduction and Botanical Review.* Advances in Limnology, no. 23. E. Schweizerbartsche, in press.

Newton, L. E. (1986b). Economic aspects of aquatic plants. In D. M. John, *The Inland Waters of Tropical West Africa: An Introduction and Botanical Review.* Advances in Limnology, no. 23. E. Schweizerbartsche, in press.

Okali, D. U. U., and Hall, J. B. (1974). Colonization of *Pistia stratiotes* L. mats by *Scirpus cubensis* Poeppig & Kunth on the Volta Lake, Ghana. *J. Agric. Sci.,* **7**, 31–36.

Pettet, A., and Pettet, S. J. (1970). Biological control of *Pistia stratiotes* L. in Western State, Nigeria. *Nature, Lond.,* **226**, 282.

Pierce, P. C. (1971). *Aquatic Weed Development, Impact and Control at Volta Lake 1967–71.* USAID, Accra, Ghana (cyclostyled).

Raynal-Roques, A. (1980). Les plantes aquatiques (Plantes à fleurs et Fougères) 2. In J. - R. Durand and C. Lévêque (Eds.), *Flore et Faune Aquatiques de l'Afrique Sahelo-Soudanienne,* Off. Rech. Sci. Tech., Outre-Mer, Documentations Techniques No. 44, pp. 63–152, Paris.

Roberty, G. (1946). Les associations végétales de la vallée moyenne du Niger. *Veroff. geobot. Inst. Zurich,* **22**, 1–168.

Rzóska, J. (1985). The water quality and hydrobiology of the Niger. In A. T. Grove (Ed.), *The Niger and its Neighbours,* pp. 79–99. A. A. Balkema, Rotterdam and Boston.

Thompson, K., and Hamilton, A. C. (1983). Peatlands and swamps of the African Continent. In A. J. P.Gore (Ed.), *Ecosystems of the World. 4B. Mires: Swamp, Bog, Fen and Moor. Regional Studies,* pp. 331–373. Elsevier, Amsterdam.

Trochain, J. (1940). Contribution à l'étude de la végétation du Sénégal. *Mem. Inst. fr. Afr. noire,* **2**, 1–433.

Plant Ecology in West Africa
Edited by G. W. Lawson
© 1986 John Wiley & Sons Ltd

CHAPTER *8*

Coastal vegetation

G. W. Lawson
Department of Biological Sciences,
Bayero University, Kano, Nigeria

I. Introduction

The exact boundary between land and water on the coast is not an easy one to determine bearing in mind the continued influences of seasonal and diurnal changes in tides, wave action, wind, and possible flooding by fresh water. It is usually feasible, however, to separate the coastal plant communities into those that are predominantly aquatic and those that are predominantly terrestrial. The marine aquatic communities are almost exclusively algal in West Africa except for some marine monocots that occur towards the outer limits of the tropics, i.e. in Mauritania in the north and Angola in the south. Such communities are dealt with in this book in Chapter 9 on the littoral and sublittoral marine vegetation and therefore will not be treated here except incidentally. There are, however, predominantly aquatic communities of brackish water in lagoons, estuaries, and swamps that are not dealt with in Chapter 9 but which will therefore be included in the present chapter. Such communities usually grade through a number of zones into purely terrestrial systems. Also included in the present chapter will be those mainly angiosperm communities occurring on sandy shores that are occasionally flooded by the sea but which are undoubtedly predominantly terrestrial in nature.

II. Formation of the coastline

The present coastline of West Africa consists mainly of sandy beaches with occasional rocky outcrops and is often backed by lagoons running parallel to it. One has to go as far north as Cape Blanc to encounter really massive cliffs but the Cape Verde Peninsula much further south consists of extensive volcanic rocks. Further south again, rocky beaches are fairly common in Sierra Leone, Liberia, and Ghana but in other areas such as Ivory Coast, Togo, and Benin, they are virtually absent. Nigeria too, with the massive Niger delta forming a considerable part of its coastline, is free from rocky shores though immediately east of it in Cameroun and further east and south in Gabon, they are often fairly extensive.

In Chapter 11, Dr Sowunmi points out that the West African coastline in the remote past has at times been both further inland and further seaward than it is at present, and suggests that the lagoons of the Guinea coast were formed about 40 000 to 30 000 years BP. Webb (1958) believes they were formed less than 25,000 years ago. On either estimate they were apparently formed, geologically speaking, in relatively recent times from an initially indented coastline by the building up of barrier beaches of sand resulting in a shorter, straighter coastline.

From the shape of sand spits formed, it is clear that on most West African shores the movement of sand is from west to east and a figure often quoted is that it moves at the rate of 1–1.5 million t yr^{-1}. Though the direction of the Guinea current is normally from west to east, it is, according to Webb (1958), more likely that the sand movement is caused by longshore drift. When a wave hits a beach at an angle, it tends to wash sand in the direction in which it is moving but when spent, the water runs off at right angles to the beach, dropping suspended sand or other particles as it does so. This pattern repeating itself again and again gives rise to the gradual movement of sand along the shore — a phenomenon known as longshore drift. The angle at which waves hit the beach is associated with the direction of the wind and as Dr Sowunmi indicates in Chapter 11, it is believed that the south-westerly monsoon wind system has been in operation in West Africa from the Palaeocene.

Webb (1958) has pointed out that the main lagoon systems of the Guinea coast run roughly east and west, i.e. at approximately 45° to the prevailing wind. There are, however, many small lagoons of the 'closed' type (see Chapter 9) on the coast of Ghana, for example, whose orientation does not exactly bear out this generalization, though they do run parallel to the beaches.

Where man-made structures such as piers or moles have been constructed roughly at right angles to the beaches, the effect of longshore drift is often exhibited in a spectacular way with regard to coastline formation. The classic case is that of the two moles built to protect Lagos harbour. The west mole which was built in 1907 (see Usoroh and Sonuga, 1972) has obstructed the sand movement from west to east with the result that a massive build-up of sand has occurred on its western side. On the other hand, sand has continued to move eastwards from the mole lying east of the harbour entrance and this sand has not of course been replaced by the sand trapped by the west mole. The result has been that considerable erosion has taken place from the bar beach with a continued threat to property at the south of Victoria Island. Another area which has been under threat for many years is the town of Keta on Keta lagoon just east of the Volta estuary in Ghana. A different type of man-made factor is fairly recently in operation here though whether it will act to preserve Keta or bring about further erosion is not yet entirely clear. This is the damming of the Volta River at Akosombo. Formerly, flow from the mouth of the Volta was very seasonal because of flooding during and after the rainy season. Since the building of the dam, however, this whole region has been altered as the dammed water is now released at a steady rate which gives a constant flow through the year. Such examples are evidence that formative factors are still operative on the West African coastline and emphasize the need for studies on the shore to take into account the dynamic nature of the processes involved.

Webb proposed the view that parts of the coastline have been advancing steadily at an average rate of 1.6 km per 1000 years and cites as evidence of this the fact that from the air it is possible to distinguish alternate bands of vegetation running parallel to the shore. The more luxuriant vegetation is on the silty soils of former lagoon beds; the poorer vegetation is growing on the sandy soil of former sand-bars that separated the lagoons from the sea. He postulates a continuous process whereby sand-bars are built up in the shallow offshore water till they reach a height where they cut off a new lagoon. This lagoon gradually becomes filled in by silt and debris until it reaches a high enough level to become dry land. In the meantime further sand-bars will be cutting off new lagoons to seaward and repeating the process.

III. Factors affecting coastal vegetation

As there are a number of factors that affect to some extent all types of coastal vegetation, it will be useful to begin with a brief account of these before turning to the communities in detail. It should be noted that some of these factors are also treated from a slightly different point of view in Chapter 9.

The build-up of a sand-bar is, as already indicated, due mainly to the continued action of wind and waves and it continues to be affected by these factors on its exposed southerly side after formation. The strong winds that blow from the sea may affect plants mechanically by breaking them or blowing them down, and it should be remembered that the very loose sandy substratum will make them particularly vulnerable to uprooting. Wind also exerts an effect on plants and soil by depositing salt spray especially near the edge of the sea (Jeník and Lawson, 1968, Fig. 1). More importantly, wind may exert its effect by increasing the rate of transpiration. Jeník and Lawson (1967) found that highest wind speeds tended to occur on the shore at Nungua (Ghana) at midday and early afternoon coinciding with the period of high air temperatures and lowest relative humidity, thus giving rise to the maximum rate of evaporation (Jeník and Lawson, 1967, Figs. 2 and 3). Loss of water becomes especially important where the plants are growing in sand which has little capacity to retain moisture. Furthermore, such water as is present is mostly sea-water which can only become available to plants whose cell sap has a sufficiently high osmotic pressure to overcome that of the strongly saline environment.

The combined effect of all the factors — wind, highly saline water, loose substratum with little water retention capacity — is to produce an extreme type of environment hostile to most forms of plant life.

On the other hand, the area behind the barrier beach, namely the lagoon, is a much more sheltered one. Here the effect of the onshore winds is relatively moderated and the waves are eliminated so that the water surface is quiet and placid. Fine particles of silt that enter from streams and rivers are allowed to settle on the bottom which becomes muddy rather than sandy though it may nevertheless also remain relatively unstable.

The question of salinity depends a lot on the location of the lagoon. In areas of high rainfall or where the lagoon is fed by large rivers, the lagoon retains a connection with the sea through which fresh water is debouched during and after the rainy season. Through this same connection, there is entry of sea-water into the lagoon which may be tidal at any time of the year or more especially during the dry season when the flow of fresh water is much reduced. Salinity conditions within the lagoon may therefore vary from nil to full sea-water, and the static organisms such as rooted plants have to be able to withstand this range of variability. In a very large lagoon such as Lagos lagoon, the salinity changes throughout the year are considerable though the innermost parts may remain fresh.

IV. The vegetation of sandy shores

As already explained, sandy beaches form an extreme environment for plant life. Indeed, on the intertidal parts of such beaches in West Africa there is an entire absence of any form of plant life except for some microscopic forms. Evidence of these is sometimes seen at low water when variously coloured patches on the surface of the sand betray the presence of millions of individuals of such organisms as species of diatoms or dinoflagellates, or even *Euglena* (Taylor, 1967).

Above the high-water level, however, or on parts that receive only the high spring tides, occurs a very characteristic community of plants usually referred to as the strand vegetation. Boughey (1957) has described this vegetation as it occurs in Ghana and the features he delineates apply to most of the West African coast and beyond.

1.   The pioneer zone

On the seaward side of the beach can be found, in the most extreme conditions of all, what may be called the 'pioneer' community. This is made up of a small number of plants that are highly adapted to these extreme conditions. In the first place, they are all either rhizomatous or stoloniferous, with horizontal spreading underground stems, or stems straggling over the surface and rooted at intervals. Both of these features apparently enable plants to persist in areas of very mobile sand and also enable them to bind it to some extent and thereby render it less mobile. The most prevalent and striking of the stoloniferous plants is *Ipomoea pes-caprae* (Convolvulaceae) whose trailing stems can often be of considerable length. Of very similar vegetative appearance though less common and of a very different family is *Canavalia rosea* (Papilionaceae). Both of these species have soft and thick semi-succulent leaves.

By way of contrast, the commonest rhizomatous plant of this zone, the grass *Sporobolus virginicus*, and the sedges *Remeiria maritima* and *Cyperus maritimus* have narrow and thinner though otherwise xeromorphic leaves occurring in tufts on the erect shoots that arise from the underground stems. Other plants of a more distinctive succulent type include *Sesuvium portulacastrum* and *Philoxerus vermicularis*, two plants of remarkably similar appearance though from quite dissimilar families (Ficoidaceae and Amarantaceae respectively). *Alternanthera maritima*, another amarantaceous plant, and *Diodea maritima* are also fairly common in this zone, and a second *Ipomoea, I. stolonifera*, often occurs near the west of the slope.

All of the plants so far mentioned have very wide distribution in the tropics and all are tolerant of salt water. Many have seeds, fruits, or propagules that are resistant to high salinities and capable of spreading for considerable distances by sea by remaining viable after prolonged immersion in sea-water. Though the term 'pioneer community' perhaps implies that continuous colonization of bare sand is taking place, this is not always true and the community occurs also on the edges of beaches where the sand is being eroded.

At the uppermost limits of the pioneer zone may be found a number of species that are not creeping but upward-growing perennials. These include *Euphorbia glaucophylla, Portulaca foliosa, Phyllanthus amarus*, and *Sansevieria* sp. Though the composition of the pioneer community is very constant there are some differences associated with rainfall. Thus *Scaevola plumieri* is found in the drier parts of the coast, while *Stenotaphrum secundatum*, a grass, occurs in the wetter areas.

2. The main strand zone

The width of the pioneeer community is not normally very great, being usually only a few metres. Immediately landward lies a much more stable community of considerably greater width on a flat stretch of ground. This is what has been referred to as the main strand vegetation zone. Unlike the pioneer community, which is an open one with much bare sand lying between individual plants, the main strand community is much denser with the plants closer together and showing little bare earth. The sand is here mixed with a good deal of plant humus making it darker and less loose, and also allowing it a much greater capacity to hold water rather than let it flow straight through as happens in sand. Again, being somewhat further distanced from the sea, it is less liable to waves and therefore less directly saline though it receives a good deal of salt spray (see Jeník and Lawson, 1968, Fig. 1). The result is that the environment is a much less extreme one for plant life than obtains in the pioneer zone. This allows a greater variety of plants to grow in it and furthermore ones that are less adapted to strictly marine conditions, and which are found in other types of habitat. It may be noted, however, that some plants of the pioneer zone such as *Ipomoea pes-caprae*, *Canavalia rosea*, *Sporobolus virginicus*, and *Remeirea maritima* may also be found here. Many weedy species are present and the main strand zone is a place where exotic species such as weeds from South America often become established initially.

Though many of the soil factors on the shore are thus moderated in this zone, the wind is still strong here and exerts an appreciable effect (Jeník and Lawson, 1967, Fig. 2). Many of the species are low-lying and it is interesting to see that in some cases of plants with wide distribution, those present on the shore have adopted a prostrate or semi-prostrate habit. This applies for example, to such grasses as *Heteropogon contortus* and *Andropogon gayanus*. Such coastal ecotypes, when transplanted to places well away from the sea, retain their prostrate characteristic indicating that it is genetic rather than merely caused directly by the environment. Along many parts of the coast the main strand zone has been much disturbed by human activity and especially by the planting of coconut palms.

3.   The evergreen shrub zone

On the landward side behind the main strand zone, Boughey (1957) has described what he terms an evergreen shrub zone. The shrubs tend to be rather low-lying and on the seaward side are frequently beautifully wind-trimmed into a wedge shape with the thin end pointing towards the sea. Such trimming is probably due to the combined effects of desiccation and salt deposition on any emergent twigs or leaves so that the whole shrub acquires a plane-like surface. The shrubs represented are such Pan-African species as *Eugenia coronatus*, *Chrysobalanus orbicularis*, and *Sophora occidentalis*, but also present are the pan-tropical *Thespesia populnea*, originally cultivated but now naturalized, and the small savanna date-palm *Phoenix reclinata*.

As with the main strand zone, the evergreen shrub zone is often much disturbed by agriculture. Cassava, okro, groundnuts (*Voandzeia* and *Arachis*) are commonly planted, as well as the tiger-nut (*Cyperus esculentus*) in the drier parts. Prickly pear (*Opuntia*) and sisal (*Agave*), originally planted as hedges to keep pigs out of the crops, have become naturalized and spread accordingly.

According to Boughey, the zones that have been described above never constitute a succession as far as Ghana is concerned. However, in areas of great sand accumulation, such as Lighthouse beach to the west of Lagos west mole, it is apparent that some kind of succession has taken place. The pioneer species occupy the outermost and more recently formed parts while the older and inner areas are much more stabilized by increasing numbers of plants with the exotics *Casuarina* and sometimes *Terminalia catappa* well established.

Sand-dunes are usually absent from tropical coasts (though large dunes occur, for example, near Walvis Bay in Namibia) and according to Morton (1957) this is due to the prevention of wind-generated sand movement by the binding effect of salt which is deposited in the surface sand by the rapid evaporation of sea-water caused by high temperatures. The thin crust of sand and salt so formed is usually sufficient to prevent the movement of appreciable quantities of sand thay may occur in temperate regions. He has reported, however, that genuine dunes occur on the west side of Old Ningo lagoon in Ghana running parallel to the direction of the prevailing wind. The plant mainly responsible for the formation of mobile dunes is *Scaevola plumieri*. This is a broad-leaved succulent plant with woody stems and short rhizomes which produces large clumps of erect shoots in which the sand lodges. Its ability to continue slow upward growth prevents it from being submerged under the sand. Behind the mobile dunes are fixed dunes that have a larger number of plants, mostly species already mentioned as forming the pioneer zone, growing on them. According to Morton, the reason why dunes have developed at Old Ningo is because the sand there is very coarse and light, being mostly made up from broken fragments of thin shells which present a large surface to the wind. The binding action of salt is therefore not sufficient to prevent their movement by wind action.

Sourie indicates in his map (1957, Fig. 1) that dunes occur behind the beaches to the north of the Cape Verde Peninsula in Senegal.

V. Mangrove vegetation

The word 'mangrove' is used collectively to describe woody plants occupying tidal land in the tropics but, as will be seen later in this chapter, mangroves may be found in places lacking regular tides and may penetrate far inland to occur in virtually freshwater situations. Though related ecologically, and frequently of superficially very similar appearance, they are often from widely diverse taxonomic groups. The West African mangrove vegetation or mangal (see Chapman, 1976a for definitions) is relatively poor in genera and species compared with its counterpart on the East African coast but includes *Rhizophora* (Rhizophoraceae), *Avicennia* (Avicenniaceae), and *Laguncularia* (Combretaceae), the same grouping that occurs, in fact, on the western side of the Atlantic. There are also a number of associated plants with some mangrove features that may be referred to as semi-mangroves, such as *Conocarpus* (Combretaceae), *Pandanus* (Pandanaceae), and *Nypa* (Palmae).

1. Distribution of mangroves

In West Africa, mangroves have a rather limited distribution with a northern limit at about the Senegal River, but *Avicennia* has been reported as far north as Cape Timiris by Monod as cited by Schnell (1952). Apparently mangroves once extended further north than they do now and Trochain (1940) mentions that subfossil pneumatophores of *Avicennia* with the roots covered by ferrugineous material have been found 50 km north of St Louis. They also occur as far south as northern Angola, but reach their maximum expression in the Niger delta area. Chapman (1977) gives a map showing the distribution of mangal in Africa, as do also Aubreville *et al.* (1958) and White (1983).

2. Common features of mangroves

As mentioned above, mangroves are linked ecologically by having a number of features in common with each other. They are all evergreen shrubs and trees, with leaves that are thick and often of simple shape. All tend to have aerial roots of one sort or another. These are especially evident in *Rhizophora* which has very prominent arching prop roots, liberally supplied with lenticels, that arise from quite high up the stem and branch into rootlets when they reach the soil. Again, many have a tendency towards pneumatophores that arise from underground horizontal roots and grow upwards till clear of the soil. These are covered by large lenticels which no doubt help to aerate the underground roots submerged in waterlogged conditions, hence their alternative name of breathing roots. The lenticels allow the passage of gases but are impermeable to water. At high tides when the pneumatophores are covered by water, oxygen is used up in the roots. The carbon dioxide produced simultaneously, being very soluble in water, is rapidly dissolved leading to reduced gas pressure in the roots. At low water new air is drawn into the lenticels equalizing the pressure with that of the outside air and providing the roots with fresh oxygen (Walter, 1979).

Jeník (1970) has described a rather different type of aerial root for *Laguncularia racemosa* where the erect peg roots, corresponding to the pneumatophores of *Avicennia*, are normally underground except where excavated by erosion and bear very small short-lived roots above the surface. These latter, which live for only a few months and are constantly being replaced, he terms pneumathodes.

Mangroves in addition have a tendency towards vivipary. This is best seen in *Rhizophora* where the seed germinates while still attached to the parent plant, and the enlarged hypocotyls are seen handing for some time before dropping to the ground. If they do so at low water, they may fall dart-like into the uncovered mud and rapidly produce roots according to Jackson (1964), but if they fall into water they may be washed around for some time before settling and during this period the emergence of roots is suppressed. This appears to be a useful adaptation to prevent premature development of roots before a suitable substratum is reached. With *Avicennia*, on the other hand, the plantlets develop a tangle of long thin roots while still afloat and Jackson believes that this plant does not achieve such a wide distribution as *Rhizophora* for that reason. In *Avicennia* the seed is in a very advanced state before it leaves the parent tree with its cotyledons already green and ready to function. The reason why mangroves have evolved a tendency toward vivipary must be because in the very unstable environment of the swamp they have to establish themselves very quickly in order to survive. Chapman (1976a) estimates that after landing on the mud, root growth can be sufficient after two days to retain the seedling.

In addition to morphological (and partly physiological) adaptations already mentioned, mangroves have other more strictly physiological features that adapt them to their environment. Prominent among these are the wide tolerance they show to salinity changes. In open lagoons the salinity may vary from that of full sea-water in the dry season to fresh water in the rainy season and *Rhizophora* appears to tolerate such variation. But in closed lagoons the spectrum of salinity may be much greater as sea-water trapped in the lagoon when it closes may be concentrated by evaporation. Thus Pople (cited by Ewer and Hall, 1972) provides figures showing the salinity of a closed lagoon as twice that of sea-water. In actual fact, salinities much higher than this must occur since some closed lagoons dry out completely until salt is crystallized on the surface of the sand. In such areas there may be no vegetation at all except sometimes for crusts of blue-green algae, but the salinities of adjacent areas bearing vegetation must be colossal just before drying out occurs. The fact that *Avicennia* is found in such places indicates that its range of tolerance to salinities must be very much higher than that of *Rhizophora*, though Hiernaux (*in litt.*) has reported that *Rhizophora* grows in salinities of as much as 50 parts per 1000 on the coast of the Republic of Guinea. Another factor Hiernaux mentions that may influence the distribution of *Rhizophora* and *Avicennia* is that the former is much more tolerant of high acidity than the latter.

3. Types of mangrove habitat

There are three main types of habitat in which mangroves occur: around lagoons — both open and closed as already mentioned, in estuaries and deltas, and more rarely on the open shore. These habitats are not always entirely distinct from one another.

Lagoon vegetation
Two zones are discernible around lagoons in West Africa, an outer herbaceous zone of low halophytic herbs, and an inner zone at the water's edge of mangroves proper.

(i) The herbaceous zone   The upper limit of this zone is often represented by the main strand vegetation on the seaward side and by the evergreen shrub community on the offshore side. Relatively few species of plants make up the herbaceous zone and they usually occur as virtually pure stands forming extensive patches or more often as subzones within the herbaceous zone. Boughey (1957) for example, describes three main subzones, an upper subzone of *Sporobolus virginicus*, a middle subzone of *Fimbristylis obtusifolia*, and a lower subzone of *Paspalum vaginatum*. He points out, however, that these belts are often broken by patches of other species, notably the sedges *Mariscus ligulatus* and *Cyperus articulatus*, and the grass *Imperata cylindrica*. At the lowest levels are *Sesuvium portulucastrum* mixed with, or replaced by, *Phyloxerus vermicularis*.

To illustrate the variability of arrangement of these subzones, attention may be drawn to the zonation at Botianaw, Ghana, where the following occur from above downwards. A belt of *Andropogon* sp. and other grasses gives way to *Imperata cylindrica*, then *Panicum repens* and finally *Sporobolus robustus* or *Paspalum vaginatum* before the *Avicennia* is reached (see Swaine *et al.*, 1979, for a full description of this area). In other places, islands of the large tufted fern *Acrostichum aureum* are a distinctive feature of the herbaceous zone, for example around Lagos lagoon.

(ii) The mangrove zone   This zone starts with *Sesuvium* and *Phylloxerus* and goes down to the level of permanent water. Flooding is either seasonal (closed lagoons) or diurnal (open lagoons).
Open lagoons: Open lagoons are characterized by the presence of *Rhizophora*, though, as Lawson (1966a) has pointed out, an algal mat consisting of *Bostrychia* species (*B. binderi, B. radicans, B. tenella*) accompanied by other red algae such as *Caloglossa* (*C. leprieurii, C. ogasawaraensis*) and *Catenella* (*C. impudica, C. caespitosa*) is more indicative of genuine diurnal tidal conditions as these algae cannot exist without regular immersion in sea-water or brackish water. Such algal mats occur both on the prop roots of *Rhizophora* and on the pneumatophores of *Avicennia* under appropriate conditions.

Three species of *Rhizophora* are present in West Africa — the same three indeed that occur on the western side of the Atlantic (Keay, 1953b; Savory, 1953). *Rhizophora racemosa* is the commonest pioneer species and is found on the outer side of the zone, *R. harrisonii* dominates the middle parts of the zone where wetter conditions prevail, and *R. mangle* occupies the drier inner areas in West Africa though it acts as a pioneer in America. These differences in position appear to be related to differences in salt tolerance of the three species and it is also interesting to note that Breteler (1969) claims that *R. harrisonii* is a hybrid between *R. mangle* and *R. racemosa* and therefore might be expected to have intermediate characters. *Rhizophora racemosa* can grow in fresh water whereas *R. mangle* is confined to salt swamp. The fact that the latter occurs on the higher ground means that it is less liable to be flooded by fresh water in the rainy season: in the dry season it is flooded only by the high tides and subsequent evaporation raises the salinity to higher concentrations than those found in the areas occupied by *Rhizophora racemosa*.

The Rhizophoras tend to be confined to those parts of the lagoons where their roots are regularly flooded by high water. The area behind them, which may be flooded only by the spring tides at fortnightly intervals, on the other hand, is often dominated by *Avicennia germinans*.

Closed lagoons: In closed lagoons the species of *Rhizophora* are usually quite absent but the highest seasonal flooding levels are occupied by a belt of *Avicennia germinans* (Fig. 8.1) accompanied occasionally by *Laguncularia racemosa*. In this mangrove zone, but at the outer edges and at a slightly higher level, may be found *Conocarpus erectus* and *Dodonia viscosa*. Other plants

Fig. 8.1   Closed lagoon near Teshie, Ghana. The lagoon has almost dried out except for some pools such as that at the extreme left in the middle of the picture. *Avicennia germinans* forms a zone at the outer edge on the dried-up lagoon and its pneumatophores are clearly seen in the foreground protruding upwards through the sand.

present especially on the landward side, include *Drepanocarpus lunatus* and *Phoenix reclinata* though these two are not infrequently met with also around open lagoons. Parts of closed lagoons may become completely dried out during the dry season and in these parts *Sesuvium* and *Philoxerus* may be found straggling over the surface. In addition, a blackish encrusting layer of dried blue-green algae is very common in such areas.

Estuaries and deltas

The deposition of material at the mouths of large rivers, whether this is caused by a decrease in the speed of the flow or by the actual flocculation of particles by the electrolytes in sea-water, may result in the formation of large expanses of tidal mud-flats which increase the land surface available for colonization by plants. Such expansion seaward is taking place to the east and west of the Niger River according to Keay (1953a). These areas provide a particularly good oportunity for the study of the process of succession in mangrove swamps. Care must be taken, however, not to assume that any spatial zonation of species necessarily represents a temporal succession series. Gledhill (1963) has pointed out, for example, that in Sierra Leone the sequential spatial zones *Avicennia germinans*–*Rhizophora racemosa* (sensu lato)–*Drepanocarpus lunatus*–*Raphia* with *Pandanus*–*Raphia* as reported in Department of Agriculture, Sierra Leone Sessional Paper No. 7, 1938, is, in fact, not a temporal succession but a transition up a river from brackish water to fresh. The real succession on mud-flats in West Africa appears to be as given in the following account.

The soil, which is very soft, is first colonized by a pioneer community of *Rhizophora racemosa*. The many prop roots that are produced tend to slow down the flow of water as it passes through them causing it to drop the suspended particles and debris it is carrying. The prop roots also break up into numerous rootlets on entering the soil and these eventually form a mat that holds the soil together. The root raft that is thus formed together with the detritus it collects tends to form a peat that raises the general surface gradually until it is only just flooded at high waters. Crabs dig their burrows in the soil thus produced and this activity, no doubt, helps to mix the soil and keep it aerated. Further growth of *Rhizophora* leads to the formation of a mature *Rhizophora* community and it is presumably during this process that the three species mentioned earlier sort themselves out into three subzones with *R. racemosa* outermost, *R. harrisonii* next, and *R. mangle* innermost, though the exact details of how this happens have not been worked out in West Africa. Again, if the spatial zonation of mangroves as observed corresponds to a temporal zonation, it may be assumed that the raising of the soil level accomplished by *Rhizophora* alters the environment to one in which *Rhizophora* is less well able to compete. Walter (1979) has pointed out that for East Africa the zonation of mangroves is due to competition and that this is related to their salinity tolerances. *Avicennia* (*A. marina*) though a weak competitor, has a high salinity tolerance and can grow where the osmotic pressure of the water is 36.4–65.0 atm, whereas *Rhizophora* is restricted to where it is less than 37.5 atm. Thus, where the soil level is raised sufficiently to allow less frequent immersion in sea-water, it is much more strongly concentrated by evaporation during the periods of relative drought. Under such conditions, the environment is no longer suitable for *Rhizophora* and it must cede to *Avicennia*. It should be borne in mind, however, that the picture is not always as simple as this and in some places *Rhizophora* and *Avicennia* can grow alongside one another (Fig. 8.2). Zonation may also be interfered with by the type of substratum and *Avicennia* grows well on very sandy beaches, whereas *Rhizophora* always appears to be restricted to muddy areas. This feature also correlates well with the ability of *Avicennia* to withstand drier conditions since sand does not retain water. One feature possessed by *Avicennia* that enables it to grow in areas of very high salinity is its ability to secrete excess salt by means of special glands in its leaves. According to Walter the secretion has a higher concentration of salt than sea-water and it is common to see crystals of salt on the leaves of *Avicennia* caused by evaporation of this brine.

According to Gledhill (1963) the correct sequence in the succession for West Africa is as described in Sierra Leone Department of Agriculture Sessional Paper No. 1, 1951, namely, *Rhizophora racemosa–Avicennia–Laguncularia racemosa–Conocarpus erectus*, the last-named being rarely reached by the tides. This corresponds to the succession on the Atlantic coast of tropical America as given, for example, by Hedgpeth (1957) except that *Laguncularia* is not mentioned. It should be noted that in the drier parts where a more 'normal' soil occurs, other species including *Dalbergia ecastophyllum*, *Drepanocarpus lunatus*, and *Hibiscus tiliaceous* may be found, as is also the introduced species *Terminalia catappa*.

The mangrove zone frequently, as in the Niger delta area, passes over into freshwater swamp with *Raphia* palms, and *Pandanas candelabrum* which may itself be regarded as a precursor of rain forest. It is interesting to note that the prostrate stemmed *Nypa* palm which, as Dr Sowunmi points out in Chapter 11, was an important constituent of prehistoric vegetation of the Niger delta, was reintroduced into Calabar in 1906 from Singapore and now appears to have established itself and is growing well, for example, near Port Harcourt.

Open shores

For the most part mangroves are absent from open shores on the West African coast except for the odd stunted *Avicennia*, for example, that may be found in the meagre soil of a high pool on an exposed rocky shore. This may be due to the considerable wave action along most of its length and to the mobility of the

Fig. 8.2 Contrary to the generally accepted view that *Avicennia* takes over from *Rhizophora* in mangrove succession, this picture shows young plants of *Rhizophora* growing among the breathing roots of *Avicennia* and apparently colonizing an area where *Avicennia* has reached maturity.

sand (cf. the East African coast where a fringing coral reef shelters the sandy beaches and where mangroves are therefore fairly commonly found). In Guinea, however, Schnell (1950a, b 1952), and Hiernaux (*in litt.*) have described the coast as one in which the wide coastal plain dips gently towards the sea and then continues for a considerable distance as a submarine shelf.

This lateritic carapace is covered by a thick layer of grey mud up to several metres deep for the most part or by shelly and sandy deposits in places. Presumably, the very gentle slope continuing out to sea minimizes wave action and this allows for the development of mangrove vegetation. The numerous rivers of this coast form large meandering estuaries and lead finally to a maze of tidal creeks and it is in these that the mangroves mostly occur. In an area of land estimated at 300 000 ha about one-third is covered by mangroves. The tidal rise and fall in Guinea is greater than in most other parts of West Africa and is the reason why such vast areas are uncovered. Near Conakry, it is 3–4.5 m and Hiernaux even mentions a figure of 7 m. *Rhizophora* tends to dominate the parts visited by daily tides and *Avicennia* is more characteristic of places reached by spring tides, but sometimes the two grow together. On more solid ground are grasslands of *Paspalum vaginatum* interrupted by stands of *Sesuvium* and *Philoxerus*.

4. Productivity and exploitation

The combination of constant water availability, warm temperatures, and continuing nutrient replenishment means that all the factors needed for high organic productivity are abundantly present in mangrove swamps. Though data on primary productivity of mangrove swamp from other parts of the world confirm this, few actual figures have been obtained for West Africa though, no doubt, they would be comparable. Pauly (1975) gives a modest figure of about 350 mg C m^{-2} day^{-1} for Sakumo lagoon in Ghana, though it seems that this figure refers to the phytoplankton and bottom algae only.

Mangrove swamp appears to have all the features of what Margalef (1963) has called an 'immature' ecosystem, namely a constantly disturbed environment, few species occurring in large numbers, and high but 'wasteful' primary productivity. Though Savory (1953) mentions *Rhizophora racemosa* growing to 40 m high in the Niger delta most mangrove growth is very much less than this (Fig. 8.3) and the question must be asked what happens to all the production if it does not manifest itself as wood? The answer seems to be that a great deal of energy is used up in reproductive activity and in producing vast numbers of heavy propagules few of which survive. Fagade and Olaniyan (1974) have pointed out that the young of 31 marine species of fishes enter Lagos lagoon during the dry season when relatively high salinity conditions prevail but return to the sea to reproduce. It would seem that the reason they come into the lagoon is to exploit the excess food supply there. If this is so, then it probably means that there is a net export of the surplus energy of the lagoon into the sea, a 'mature' ecosystem where there are more species making much more efficient use of the energy available.

Compared to East Africa, where there is considerable exploitation of mangrove poles for building and other purposes, the West African mangroves

Fig. 8.3   Lagoon near Princes Town, Ghana, showing *Rhizophora* growing densely
with abundant prop roots. *Paspalum vaginatum* in right foreground.

Fig. 8.4   Aerial view of Panbros salt-works occupying an area of the lagoon system
near the mouth of the Densu River.

have been relatively little disturbed by human activity. Paradis and Adjano-houn (1974), however, describe disturbance of mangrove vegetation in Benin in connection with the local salt industry, and some of the lagoons on the coast of Ghana have been utilized for this purpose (Fig. 8.4). Hiernaux mentions the use of mangrove areas in rice production. For the latter *Avicennia* areas are preferred by farmers as the soil in *Rhizophora* areas often has a very low pH value. In addition, there are various oyster-growing and fish-pond projects along several parts of the coast. In some localities draining of lagoons or filling them with sand to reclaim land for building purposes in proceeding and it is to be expected that with the high concentration of population near the coast, more and more human interference with mangrove and lagoons will take place in the future especially with regard to pollution such as occurs now, for example at Korle lagoon, Ghana. It is to be hoped that all those planning such development will take into account fully the possible ecological effects of their actions (see Kinako, 1977).

References and Bibliography

Anon. (1981). The coastal ecosystems of West Africa: coastal lagoons, estuaries and mangroves. *Unesco Reports in Marine Science*, **17**, iii + 60 pp.

Anon. (1981). Coastal lagoon research, present and future. *Unesco Technical Papers in Marine Science*, **33**, 1–348.

Aubreville, A., Duvigneaud, P., Hoyle, A. C., Keay, R. W. J., Mendonça, F. A., and Pichi-Sermolli, R. E. G. (1958). *Vegetation Map of Africa South of the Tropic of Cancer*. Clarendon Press, Oxford.

Boughey, A. S. (1957). Ecological studies of tropical coastlines. I. The Gold Coast, West Africa. *J. Ecol.*, **45**, 665–687.

Breteler, F. J. (1969). The atlantic species of *Rhizophora*. *Acta Bot. Neerl.*, **18**(3), 434–441.

Chapman, V. J. (1976a). *Mangrove Vegetation*. Cramer, Vaduz, viii + 449 pp.

Chapman, V. J. (1976b). *Coastal Vegetation*. Pergamon, Oxford, viii + 292 pp.

Chapman, V. J. (1977). Africa B. the remainder of Africa. In V. J. Chapman (Ed.), *Wet Coastal Ecosystems* (Ecosystems of the world, Vol. I (D. W. Goodall, Editor in Chief), pp. 233–240. Elsevier, Amsterdam, xi + 428 pp.

Ewer, D. W., and Hall, J. B. (Eds.) (1972). *Ecological Biology I*. Longman, London, X + 334 pp.

Fagade, S. O., and Olaniyan, C. I. O. (1974). Seasonal distribution of the fish fauna of the Lagos lagoon. *Bull. inst. fond. Afr. noire*, sér. A, **36**(1), 244–252.

Gaston, J. (1911). Le littoral et les lagunes de la Côte d'Ivoire. *Bull. Com. Afr. Fr. Rens. Col.*, No. 6, 154–158.

Gledhill, D. (1963). The ecology of the Aberdeen Creek mangrove swamp. *J. Ecol.*, **51**, 693–703.

Hedgpeth, J. W. (1957). II. Biological aspects, in chap. 23 *Estuaries and Lagoons* of *Geol. Soc. Amer. Memoirs*, No. 67, Vol. I, 693–729.

Hedin, L. (1932). Esquisse rapide de la végétation des bords languinaires dans la région de Grand Bassam et de Bingerville (Côte d'Ivoire). *Bull. Mens. Ag. Econ A.O.F.*, No. 139, 211–215, 3 Figs; No. 140, 251–255, 3 Figs. (reproduced in *La Terre et la Vie* 1933, pp. 596–609, 7 Figs).

Hiernaux, C. R. (unpublished). Aspects géographiques des zones de Mangroves. Paper read at West Afr. Sci. Assn. Meeting Kumasi, 1963. Privately circulated.

Jackson, G. (1964). Notes on West African Vegetation. I. Mangrove Vegetation at Ikorodu, Western Nigeria. *J. West Afr. Sci. Assn.*, **9**, 98–110.

Jacques-Felix, H. (1957). Les *Rhizophora* de la mangrove atlantique d'Afrique. *Jé. Agric. trop. Bot. appl.*, **4** (7–8), 343–347.

Jeník, J. (1970). Root system of tropical trees. 5. The peg roots and the pneumathodes of *Laguncularia racemosa* Gaetn. *Preslia (Praha)*, **42**, 105–113.

Jeník, J., and Lawson, G. W. (1967). Observations on water loss of seaweed in relation to microclimate on a tropical shore (Ghana). *J. Phycol.*, **3**(3), 113–116.

Jeník, J., and Lawson, G. W. (1968). Zonation of microclimate and vegetation on a tropical shore in Ghana. *Oikos*, **19**, 198–205.

Keay, R. W. J. (1953a). *An Outline of Nigerian Vegetation*, 2nd edn. Govt. Printer, Lagos, 55 pp.

Keay, R. W. J. (1953b). *Rhizophora* in West Africa. *Kew. Bull.*, **8**(1), 121–127.

Kimpe, P. de (1967). Les facteurs de production piscicole des lagunes de l'est Dahomey et leur évolution récente. *Bois et Forêts Trop.*, No. 111, 53–62.

Kinako, P. P. S. (1977). Conserving the mangrove forest of the Niger. *Delta. Biol. Cons.*, **11**(1), 35–39.

Kunkel, G. (1969). The structure and succession of the mangroves of Liberia and their fringe formations (über die Struktur und Sukzession der Mangrove Liberias und deren Rondformation). *Berichte der Schweizerischen Botanischen Gesellschaft*, **75**, 20–48 (Berne).

Kwei, E. A. (1977). Biological, chemical and hydrological characters of coastal lagoons of Ghana, West Africa. *Hydrobiologia*, **56**(2), 157–174.

Lang, J., and Paradis, G. (1977). Un exemple d'environnement sédimentaire bio-dendritique non carbonate marni et continental holocene en climat intertropical; le domaine margino–littoral au Benin méridional (Ex-Dahomey). *Rev. Geog. Physique et Geol. Dynamique (2)*, **19**(3), 295–312.

Lawson, G. W. (1966a). The littoral ecology of West Africa. *Oceanogr. Mar. Biol. Ann. Rev.*, **4**, 405–448.

Lawson, G. W. (1966b). *Plant Life in West Africa*. Oxford University Press, London, ix + 150 pp.

Lawson, G. W., and John, D. M. (1982). *The Marine Algae and Coastal Environment of Tropical West Africa*. Cramer, Vaduz, 450 pp.

Margalef, R. (1963). On certain unifying principles in ecology. *Amer. Naturalist*, **97**, 357–374.

Mornet, J. (1907). Les lagunes de la côte d'Ivoire. *Bull. Com. Afr. Franc. Rens. Col.*, No. 7, 157–163.

Morton, J. K. (1957). Sand dune formation on a tropical shore. *J. Ecol.*, **45**, 495–497.

Nicou, R. (1956). Présence du *Laguncularia racemosa* dans la mangrove du pseudo-delta du Senegal. *Notes Africaines*, No. 71, 67–68, 3 figs.

Paradis, G., and Adjanohoun, E. (1974). L'impact de la fabrication du sel sur la végétation de mangrove et la géomorphologie dans le Bas-Dahomey. *Ann. Univ. Abidjan*, ser. E (Ecologie), **7**(1), 599–612.

Pauly, D. (1975). On the ecology of a small West African lagoon. *Ber. dt. wiss. Komm. Meeresf.*, **24**, 46–62.

Savory, H. J. (1953). A note on the ecology of *Rhizophora* in Nigeria. *Kew. Bull.*, **8**(1), 127–128.

Schnell, R. (1950a). Contribution préliminaire à l'étude botanique de la basse Guinée Française. *Etudes Guinéennes IFAN*, No. 6, 29–76.

Schnell, R. (1950b). Esquisse de la végétation cotière de la basse Guinée Française. *Conf. Internat. Africanistas Ocidentais em Bissau 1947*, **2**(1), 203–214 (Lisbon).

Schnell, R. (1952). Contribution à une étude phytosociologique et phytogéographique de l'Afrique occidentale. *Mém. Inst. fr. Afr. noire*, **18**, 43.

Schunck de Goldfiem, J. (1936). La mangrove guinéenne. *Rev. Gen. Sc. (pur et appl.)*, **47** (17, 18), 477–482.

Sourie, R. (1957). Étude écologique des plages de la côte Sénégalaise aux environs de Dakar (Macrofaune). *Ann. L'Ecole Supérieure des Sciences*, **3**, 1–110.

Swaine, M. D., Okali, D. U. U., Hall, J. B., and Lock, J. M. (1979). Zonation of a typical grassland in Ghana, West Africa. *Folia Geobot. Phytotax.*, Prague, **14**, 11–27.

Taylor, F. J. (1967). The occurrence of *Euglena* on the sands of the Sierra Leone peninsula. *J. Ecol.*, **55**(2), 345–359.

Tessier, F. (1956). Formes mineures d'erosion: cannelures horizontales des lagunes d'Afrique occidentale. *Congreso Geologico Internacional XX session, Mexico.*

Tricart, J. (1957). Aspects et problèmes géomorphologiques du littoral occidental de la côte d'Ivoire. *Bull. Inst. franc. Afr. noire*, sér. A**19**(1), 1–20.

Trochain, J. (1940). Contribution à l'étude de la végétation de Sénégal. *Mém. Inst. fr. Afr. noire*, No. 2, 1–433 + 30 Pl.

Usoroh, E. J., and Sonuga, J. O. (1972). Human activity as a cause of shoreline retreat: an example of Victoria Beach, Lagos. *Lagos Notes and Records*, **3**(2), 19–30.

Walter, H. (1979). *Vegetation of the Earth and Ecological Systems of the Geo-biosphere.* Springer-Verlag, New York, XX + 274 pp.

Webb, J. E. (1958). The ecology of Lagos lagoon. *Phil. Trans. Roy. Soc.*, **B**, **681**(241), 307–419.

Webb, J. E. (1960). *The Erosion of Victoria Beach: Its Cause and Cure.* Ibadan University Press, Ibadan.

White, F. (1983). *UNESCO/AETFAT/UNSO Vegetation Map of Africa.* Descriptive memoir and map. UNESCO, Paris.

Plant Ecology in West Africa
Edited by G. W. Lawson
© 1986 John Wiley & Sons Ltd

CHAPTER *9*

Littoral and sub-littoral marine vegetation

David M. John*
Department of Botany, University of Ghana, Legon, Ghana

I. Introduction

This is an account of the littoral and sublittoral vegetation of the stretch of West African coast from Cameroun to the Cape Verde Peninsula in Senegal, and includes those islands lying close offshore (Fig. 8.1). It deals with the marine algae and commonly associated animals occurring along this mostly unprotected and almost constantly surf-pounded coast. Also considered are those living in the sheltered and less strictly marine conditions of lagoons, deltas, and estuaries. The mangrove vegetation commonly bordering lagoons and the lower courses of rivers is dealt with in a separate chapter (Chapter 8). Only brief mention is given to sea-grasses as they are comparatively rare in Senegal which is their southernmost limit along this coast.

Our knowledge of the distribution and ecology of benthic organisms has progressed considerably since the publication of Lawson's (1966) review of the littoral ecology of West Africa. The shores of Ghana and Senegal are still those most intensively studied in the region, but most others have now received variable amounts of attention. Only the islands of the Archipel de Bijagos and the mainland of Equatorial Guinea (formerly Rio Muni, Mbini) remain virtually unknown. For the first time some consideration has been given to functional as well as structural aspects of shallow sublittoral algal vegetation (see John et al., 1977; Lieberman et al., 1979).

II. General environmental features

1. The West African coast

Much of the Atlantic seaboard of tropical North Africa consists of palm-fringed sand beaches backed by lagoon systems, low wave-cut platforms, and rocky

---

* Present address: Department of Botany, British Museum (Natural History), Cromwell Road, London SW7 5BD.

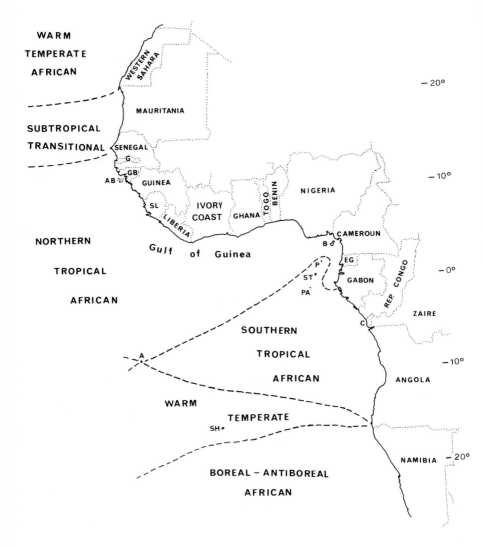

Fig. 9.1. Map of tropical West Africa showing the marine phytogeographical regions and including the stretch of coastline (northern Senegal to Cameroun) and islands covered by this review. Key to letters: mainland countries: (G) Gambia; (GN) Guinea-Bissau; (SL) Sierra Leone; (EG) Equatorial Guinea; (C) Cabinda; islands: (AB) Archipel de Bijagos; (B) Bioko; (ST) São Tomé; (P) Príncipe; (PA) Pagalu; (A) Ascension; (SH) St Helena.

headlands. These inshore rocky areas vary in structure from very hard igneous rocks (e.g. granites, basalt) through to conglomerates, limestones, shales, and soft sandstones. Rocky shores are long and fairly continuous in northern Senegal, Sierra Leone, Liberia, Ghana, Cameroun, and are much less extensive in Gambia and the western part of Ivory Coast. In places the coast is interrupted by large rivers flowing into the Atlantic Ocean through wide estuaries or extensive deltas consisting of systems of anastomosing creeks often edged by mangroves (e.g. Niger delta). Mud-flats are commonly extensive though there are occasional rocky or lateritic banks. Much of the Nigerian coast is characterized by mangrove vegetation as is the coastal region from Gambia southwards to the Sierra Leone-Liberia border. Along some stretches man-made structures including harbour breakwaters or moles, bridge or pier supports, outfall pipes and oil-rigs, afford the only suitable surfaces for the attachment of benthic organisms.

The lagoons in the central part of the West African coast running almost parallel to the equator have been classified by Boughey (1957) into two types — 'open' and 'closed' lagoons, depending on whether they retain a permanent connection with the sea (open) or an annual or less frequent connection (closed). Recently Pauly (1975) has recognized a third type, the 'semi-closed' lagoon, which has an artificially restricted connection to the sea. Those coastal lagoons fed by such permanently flowing rivers as the River Volta in Ghana and River Niger in Nigeria are examples of the open type, while there are numerous examples of the closed type which are fed mostly by small seasonal rivers and streams. In open lagoons the salinity never reaches above that of the sea itself (c. 35 ‰), and the diurnal rise and fall of the tides allows for the development of belts of marine organisms on suitable surfaces. This contrasts with lagoons that become cut off from the sea for one to several years (closed lagoons) or for several times in the same year (semi-closed), and where salinity may range from almost fresh water in the wet season to more than twice that of sea-water in the dry season. Owing to the widely fluctuating conditions these lagoons do not normally contain a distinctive assemblage of benthic organisms (including algae) and so are only briefly mentioned here.

The sea-bed off the West African coast is largely of sand, gravel, or mud that in places is overlain with shells or shell fragments. Most rocky headlands have shallow underwater extensions and off the coast between Ivory Coast and Nigeria there is a series of narrow rocky banks which stretch almost unbroken for considerable distances (Allen, 1965; Martin, 1971). In many places the sandy sea-bed is strewn with cobbles or nodules known as rhodoliths which are formed by the long-term accretion of calcareous red algae. Such cobble areas, sometimes misleadingly called 'coral banks', have been reported off Sierra Leone (Longhurst, 1958; McMaster *et al.*, 1970), Ivory Coast (Barbey, 1968; Martin, 1971), Ghana (Buchanan, 1958; John and Lawson, 1972a, John *et al.*, 1977; Lieberman *et al.*, 1979), and Togo (Barbey, 1968). Some of the deeper water beds, such as those reported at 100 m off Ivory Coast consist of non-living cobbles believed to have been formed under 20 ± 10 m of water when the sea was about 80 m below its present level (Martin, 1971).

Coral reefs do not exist along the West African coast though some coral species are abundant and form large colonies in just a few shallow-water coves (e.g. western side of Cape Three Points in Ghana). Such concretionary development is believed to be a consequence of the water temperature remaining high throughout the year in these protected coves where there is solar heating and restricted mixing with colder upwelled water (Laborel, 1974). The upwelling of colder water, the lack of suitable bedrock close inshore, and high turbidity during the rainy season, are all factors that have been suggested to account for the generally low numbers and poor development of true reef-building corals in tropical West Africa (see Buchanan, 1958; Ekman, 1953; Laborel, 1974).

2.   The coastal climate

The main features of the coastal climate of West Africa are summarized in various atlases and other publications dealing specifically with the climate of Africa (see Chi-Bonnardel, 1973; Griffiths, 1972; Jackson, 1961; Thompson, 1965; Toupet, 1968). Relevant climatic data are included in many papers on shore ecology (John, 1972; John and Lawson, 1977b; Longhurst, 1964; Sourie, 1954a, among others).

The coastal region of West Africa stretching southwards from the southern border of Sierra Leone to Nigeria has a four-season climatic regime: major dry season (November–March/April), when air temperatures are generally highest, and rainfall and relative humidity minimal; a two-peak wet or rainy season (March/April–August; September–October/November) associated with lower annual air temperatures and higher rainfall and relative humidity; and a short dry season (August) between the two rainfall peaks. Places north of about 8–10° N have a typical tropical or savanna climatic regime of one rainy season and only one dry season. In northern Senegal the dry season is particularly long (November–June) and the mean monthly air temperature is generally lower (often 20°C) over this period than elsewhere in the region.

The actual amount of rainfall and the exact period of the rains not only relates to the position of the intertropical boundary or front lying between an equatorial low-pressure air mass and a subtropical continental high-pressure air mass, but also to differences in the winds that cross the coast, the presence of coastal currents, and to coastal orientation and relief. The relatively dry climate from south-eastern Ghana eastwards through Togo and Benin is believed to be due to the prevailing wind (south-south-westerly) blowing almost parallel to the coast, and to a cool current of water immediately offshore. In contrast, the rainfall from the southern border of Liberia to the

Casamance region of southern Senegal is exceptionally high ($\geqslant$ 3000 mm yr^{-1}) as it receives the full force of the moisture-laden south-westerlies during the northern hemisphere summer. Rainfall is also particularly heavy where there is high internal relief such as along the Liberian–Guinean mountain range and around Mount Cameroun in the eastern corner of the Bight of Benin. Winds are generally from the south-west, but for the period from about November to February northerlies or north-westerlies are dominant and sometimes the very dry harmattan wind reaches the southern coast of the Gulf of Guinea.

3. Oceanographic features

The oceanography of West Africa has been the subject of a number of review articles and papers (see Berrit, 1969; Ingham, 1970; Longhurst, 1962), and these form the basis of the account that follows. Even at the present time many of the details of the circulation of surface waters in the region are uncertain and the mechanism of upwelling has still to find a satisfactory explanation (Houghton, 1976).

A number of surface currents influence the West African coast and often extend cold-water conditions far into the tropics. The north to south flowing cold Canary current is part of a large northern gyral system influencing the North-west African coast. This Canary current diverges westward between Cape Blanc and Cape Verde and flows across the Atlantic as the lower limb of the North Atlantic gyre. Another branch of this current continues southwards into the Gulf of Guinea region of West Africa and contributes to the so-called Guinea current. Also contributing to this Guinea current is the easterly flowing Equatorial counter current lying between the large gyrals of the North and South Atlantic. From time to time the Guinea current reverses its easterly flow along the West African coast. This relatively warm (>20°C) current is of lowered salinity (<35‰) and shows some seasonal variation, with a minimum temperature between July and October when the ambient air temperature in the region is low, rainfall and cloud cover are at a maximum, and solar radiation is minimal. Furthermore, the boundary of this warm water is not constant but moves northward during the northern hemisphere summer and southward during the winter months. This results in an alternating regime of colder and warmer surface waters between Cape Blanc and the region just south of Cape Verde. The importance of sea-water temperature on the phytogeography of West African marine algae has been discussed by Lawson (1978) and Lawson and John (1977).

The overall hydrological conditions are complicated by the periodic upwelling of colder and more saline water (< 24°C, > 35‰) taking place from Cape Palmas in southern Liberia to just east of Lagos in Nigeria. There are usually two or more periods in the year when the temperature of the surface water suddenly drops due to the upwelling of colder water. The principal period of upwelling occurs from late June to October while weaker and more limited periods take place between December and March. A consequence of the upwelling of this deeper and also nutrient-rich water is a sudden increase in the production of phytoplankton (Anang, 1979; Longhurst, 1964; Reyssac, 1970) and fish such as *Sardinella aurita* (see Houghton and Mensah, 1978).

Changes in the salinity of the inshore water may be due to causes other than upwelling. The surface salinity may show an appreciable drop during the rainy season in those parts of the region where there is a considerable discharge of fresh water from rivers and lagoons. Off Sierra Leona and Guinea, surface-water salinities of less than 20‰ have been reported by Watts (1958) and Marchal (1960) respectively, while in the Bight of Benin it may fall below 28‰ during the rains (Berrit, 1969). The diminutive size and apparent low production of shore algae in some West African countries has been attributed to such lowered inshore salinities (see Lawson, 1955; John and Lawson, 1977b).

The relatively unindented coastline of West Africa is subject to a more or less continuous pounding by surf, with the largest swells most commonly occurring during the rainy season. Wave amplitude is greatest in July and August at Tema in Ghana, and this is also the roughest period of the year along the Nigerian coast (Lawson, 1966). Sourie (1954a) gives a detailed account of the patterns of wave action found on the coast of Senegal. During the rainy season the discharge of silt-laden water and heavy seas contribute to a local inshore band of turbid water about 1–6 km wide in the central part of the Gulf of Guinea (Longhurst, 1964). This is the period of the year considered least favourable for the growth and development of subtidal algae (John et al., 1977).

In common with other tropical shores the theoretical tidal range is small (Table 9.1) although the sphere of marine influence may be extended vertically well above tidal levels by up-carry due to wave action. Wave action is an important factor influencing the extent of shore-inhabiting organisms in West Africa since the wave fronts commonly exceed the small tidal range. This range may be increased in river estuaries where the water is channelled between converging shores, for example at Freetown in the estuary of the Sierra Leone River where it is almost twice that of most other parts of the region. The tides are of the semi-diurnal type with the two low and two high tides each day differing in height by as much as 0.3 m. The ecological implications of this phenomenon for shore algae have been considered by Lawson (1957b, 1966).

III. Marine phytogeography

According to Lawson (1978) the marine algal flora of West Africa can be divided into four groupings: warm temperate African, tropical African, tropical transitional African, and boreal-antiboreal African. Recently Lawson (pers. comm.) has suggested a modification to this terminology as 'tropical transitional' implies a transition from the southern African flora, the tropical African now

Table 9.1. The heights of the tides above chart datum at various places along the West African coast. (From *Admiralty Tide Tables*, Vol. 2, 1980.). Key to abbreviations: Mean High Water Spring, MHWS; Mean High Water Neap, MHWN; Mean Low Water Spring, MLWS; Mean Low Water Neap, MLWN.

Place	Lat. N	Long. W	MHWS	MHWN	MLWN	MLWS
				(Metres)		
Banjul, The Gambia	13 35	16 46	1.8	1.4	0.5	0.2
Freetown, Sierra Leone[a]	8 30	13 14	3.0	2.3	1.0	0.4
Tabu River, Ivory Coast	4 25	7 21	1.2	0.9	0.5	0.2
Takoradi, Ghana	4 52 N	2 15 E	1.5	1.2	0.6	0.2
Bonny Town, Nigeria	4 20	7 10	2.3	1.9	0.9	0.4
Libreville, Gabon	0 23	9 26	2.1	1.7	0.8	0.4

[a] Amplitude greatly influenced by configuration of river estuary.

becoming the northern tropical African group and the tropical transitional African now the southern tropical African group. Much of the coastline covered by this present review falls into the very coherent tropical African flora group (see Fig. 9.1). This comprises the floras of all the mainland countries stretching from Cameroun in the south to as far north as Gambia. Over 30 per cent of the total algal flora (excluding blue-green algae) of Gambia is absent from Cape Verde which is a relatively short distance to the north (see John and Lawson, 1977b). Sourie (1954a) has pointed out the transitional nature of the marine flora and fauna of the Cape Verde Peninsula, with warm and colder water species occurring side by side. This transitional flora of Senegal has been found to extend through Mauritania to the southern side of the Cape Blanc Peninsula. Many truly tropical species are able to penetrate as far north as the Baie du Levrier side of this peninsula, while a great many colder-water species do not extend any further south than its tip (see Lawson and John, 1977). The flora of Senegal is included by Lawson (1978) in his warm temperate grouping, but he recognizes that it could with perhaps equal justification be linked to the tropical transitional grouping (=southern tropical African). On present evidence it would seem reasonable to regard Cape Blanc as the true boundary between the warm temperate flora and the subtropical transitional flora of Mauritania and northern Senegal (see Fig. 9.1). These floristic changes appear to correlate well with water temperature, with the coast to the north of Cape Blanc subject to cold water throughout the year and the stretch extending to just south of Cape Verde characterized by alternating warmer and colder water conditions. Another transitional, or possibly warm temperate flora (see Lawson, 1978), may exist around the Angola–Namibia (=South West Africa) border where a fairly abrupt change takes place from tropical to boreal-antiboreal species.

IV.   Shore habitats

The three principal groups of shore organisms can be recognized on suitable stable surfaces along the West African coast: macroalgae and suspension-feeding animals (barnacles, mussels, oysters, tube worms, anemones), the respective grazers and predators of these sessile forms (littorinids, limpets, echinoids, thaid gastropods, crabs), and secondary species of small size that are physically dependent on the microhabitats provided by the ground-cover species (epifauna, epiflora, nematodes, polychaetes). These organisms are not haphazardly arranged, but different species occupy wide or narrow, overlapping or more or less exclusive, belts or zones one above the other. The broad pattern of this zonation is determined by the gross physical gradient of emersion/submersion related to the tidal cycle, and superimposed upon this is one caused by biological interactions.

The West African tidal range is small (commonly <2 m) and so in many wave-sheltered habitats (e.g. tidal lagoons, inlets or coves, harbour basins, deltas, some estuaries), the belts or zones maybe telescoped into a very narrow strip. In such wave-sheltered habitats the biota may undergo striking changes over a very small vertical range. On the unprotected open coast the most important factor determining the distribution of shore organisms is the almost continuous and often heavy surf. Increasing wave action widens and vertically extends the belts of sessile organisms, and also influences their species composition. Marine influence may extend over 10 m or more above low-water level where wave action is exceptionally severe. It is a complex factor often varying considerably over a short stretch of shoreline according to local differences in topography, slope, and aspect.

The early recolonizers of shores experimentally cleared of benthic organisms are most commonly opportunistic species which are absent or not normally found in abundance at some of the levels they now come to occupy. In West Africa such observations have been made on experimentally denuded areas of shore in Ghana (Lawson, 1957b, 1966) and on surfaces newly available for colonization along the coasts of Togo and Benin (John and Lawson, 1972b). The findings suggest that many species have wider tolerance ranges than would be indicated from their observed distribution on undisturbed shores where the normal complement of species is present. The distribution and abundance of many shore organisms seems therefore to relate less to purely physical conditions than to the dispersal of their reproductive stages and the biological interactions undergone by the settlement, juvenile, and adult stages.

1. The description of zonation

A convenient basis for the description and classification of West African shores, or indeed shores elsewhere in the world, is the pattern of zonation of the most common plants and animals. One of the earliest schemes for describing shore zonation, and to claim widespread applicability, was proposed by T. A. and A. Stephenson (1949). This scheme was adopted for West African shores by Gauld and Buchanan (1959), Lawson (1955, 1956, 1957a) and Longhurst (1958), while Sourie (1954a, 1954b) followed a still earlier one proposed by Feldmann (1937). The Stephensons' terminology has been modified and improved upon by Lewis (1961, 1964), and most recent publications have used his scheme as a framework for summarizing shore zonation in West Africa (De May et al.; 1977, John, 1972; John and Lawson, 1972b, 1974, 1977a, b; John and Pople, 1973; Lawson, 1966).

According to Lewis (1961, 1964) the shore can be divided into a littoral zone and a sublittoral fringe. The littoral zone is that part of the shore where the organisms are adapted to, or have a requirement for, alternate emersion from and submersion in sea-water, wave splash, or spray. The disappearance of marine organisms marks the upper limit of the littoral while at the junction of this zone with the sublittoral zone there is a rather ill-defined sublittoral fringe. This is the upper part of the sublittoral zone and is uncovered only at extreme low water of spring tides (ELWS). The littoral fringe is that part of the shore beyond the tidal range and so is only influenced by sea-water splash and spray. Sandwiched between the two fringes is the eulittoral zone which is subject to a more or less regular daily tidal emersion and submersion. On open rocky shores the eulittoral zone may be often subdivided into an upper and lower subzone. In Senegal Sourie (1954a, b) has recognized five 'horizons' or belts of zone-forming organisms. The upper and lower of these correspond to the fringes of the littoral and sublittoral respectively, and the middle ones to the eulittoral zone.

2. Open rocky shores

The littoral ecology of nearly all West African countries having rocky shorelines has now received some attention. Descriptions of shore zonation have been mainly confined to more or less continuous areas of rock and little attention has been paid to the ill-defined patterns found on boulders. Where the boulders are unstable the littoral organisms are poorly represented, though even when they are relatively stable the algal diversity may be low and animals tend to predominate (see Bassindale, 1961; Gauld and Buchanan, 1959; Sourie, 1954a). The zonation of organisms encountered on rocky shores (Fig. 9.2) under conditions of moderate to strong wave action may be regarded as

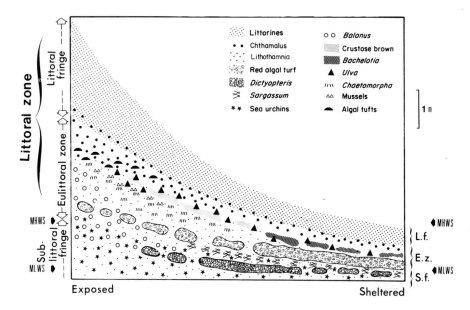

Fig. 9.2. A diagrammatic representation of the zonation of the dominant littoral organisms on rocky shores in tropical West Africa in relation to exposure to wave action. MHWS = mean high water spring tides; MLWS = mean low water spring tides. (After Lawson and John, 1982.) Reproduced by permission of A. R. Gantner Verlag K.G..

the most typical for this region where few strictly marine habitats are truly sheltered.

(a) Littoral fringe

This zone shows the most constant composition and appearance in spite of receiving very different amounts of sea spray and splash. It extends upwards from the limit of barnacles in quantity to the disappearance of small snails known as littorines. The two snails characterizing this level on the shore are *Littorina punctata* and *Nodolittorina granosa* (=*Tectarius granosus*). In excessively wave-exposed places these snails may reach over 10 m above low water, with *Nodolittorina* occupying the highermost levels when the two occur together. *Littorina cingulifera* may occasionally be present in low numbers, particularly under conditions of reduced salinity.

The fringe is comparatively barren of algae except for cushion-like forms or low mats which, like the snails, are mostly confined to cracks, crevices, or fissures, and are only found on open rocks when these are shaded by trees or rock overhangs. The most common are somewhat bleached red algae (principally *Bostrychia* spp., *Lophosiphonia reptabunda*) which are occasionally entangled with the green hair-like strands of *Rhizoclonium*, and heavily epiphytized by tufts of filamentous blue-green algae. In such places the small isopod *Ligia* may be present in some abundance. On occasion the rock surfaces at this level may be covered by a blackish, powdery, penetrating layer of the blue-green alga *Entophysalis deusta*.

(b) Eulittoral zone

This zone corresponds to the intertidal and may conveniently be divided into two subzones: an upper one, which becomes more or less completely dried out during low water and is characterized by the barnacle *Chthamalus*, and a lower one, which rarely dries out completely even during low water spring tides, and is characterized by a crust of calcareous red algae collectively termed lithothamnia. On wave-exposed shores the barnacle subzone appears from a distance as a greyish band immediately above the well-developed, bright pinkish-red band which is the lithothamnia subzone.

(i) Barnacle subzone   This is normally dominated by barnacles but on some very wave-sheltered shores they may be almost completely replaced by the small limpet *Siphonaria pectinata*. Where wave action is negligible *Ostrea cucullata* may sometimes form a belt in the lower part of this subzone. These oysters may also be present on the horizontal rocks of more wave-exposed shores, where they are often confined to water-filled depressions and so cannot be regarded as a distinctive feature of 'open' rocks at this level on the shore (see Lawson, 1956). The limpet *Siphonaria* is often accompanied by *Nerita atrata* (=*N. senegalensis*) though this snail is usually absent from wave-sheltered rocks, and both these molluscs sometimes extend into the subzone below.

Algae are not well represented in this upper subzone though a belt of crustose brown algae (*Ralfsia expansa, Basispora africana*) is a feature common to many West African rocky shores. This brown algal belt is best developed in the lower part of the subzone where the barnacles are often present in low numbers. Sourie (1954a, b) considers his middle 'horizon' or belt in Senegal to be characterized by the brown alga *Ralfsia* and the red alga *Caulacanthus ustulatus*. He also mentions the presence in this belt of *Bangia atropurpurea* (as *B. fuscopurpurea*) and *Porphyra*. This latter red alga and *Caulacanthus* disappear to the south of Senegal though a *Porphyra* (*P. ledermannii*) is known from Cameroun. On somewhat sand-scoured and gently sloping beach rock there may be a dense covering of green algae such as *Enteromorpha flexuosa* or, less commonly, *Chaetomorpha linum* (=*C. aerea*). On more wave-sheltered shores this green covering to the rocks may be replaced by *Bachelotia antillarum* which is often associated with a good deal of sand.

(ii) Lithothamnia subzone   This subzone is normally kept moist even at low water by wave wash, splash, or spray, and often by a springy 'algal turf' which

is normally well developed on wave-sheltered shores. Algae dominate this subzone though on wave-exposed rocks many animals may be associated with the characteristic pinkish-red layer of calcareous red algae which are collectively termed lithothamnia. The most constant of the sessile animals include the mussel *Perna perna* (=*Mytilus perna*), the limpets *Patella safiana* and *Fissurella nubecula*, and the carnivorous snails *Thais nodosa* and *T. haemostoma*. Where wave action is considerable the large barnacle *Balanus tintinnabulum* may be present in abundance, while the shore crab *Grapsus grapsus* ranges in large numbers over the entire littoral zone.

The greatest development and diversity of shore algae are in this subzone and the sublittoral fringe immediately below. On wave-exposed rocks crustose algae predominate while other algae are diminutive and form cushion-like tufts, felty mats, or small ring-like patches. Many of the algae have erect branches arising from a prostrate system and belong to such genera as *Gelidium*, *Herposiphonia*, *Laurencia*, *Taenioma*, and *Polysiphonia* (all red algae). One of the most conspicuous features of these areas are the coarse hair-like tufts of *Chaetomorpha antennina* and, more rarely, the membranaceous fronds of another green alga, *Ulva*. With increase in shelter many of the low mats or cushions become denser, more extensive, and often coalesce to form a variegated or sometimes almost continuous algal turf. The crustose lithothamnia are now overgrown by this turf and other larger algae including species of *Bryopsis*, *Caulerpa*, *Gracilaria*, *Hypnea*, *Padina*, *Colpomenia*, and articulated corallines (e.g. *Amphiroa*, *Corallina*, *Jania*, *Haliptilon*). Steeply sloping and moderately wave-exposed rocks sometimes have a slimy greyish to brownish-green covering of the zoanthid *Palythoa*. The small mussel *Brachydontes puniceus* is present on some sheltered shores and the tubes of the worm *Vermetes* may sometimes form a zone partly overgrown by lithothamnia and blue-green algae.

(c) Sublittoral fringe

In West Africa there has always been great difficulty in defining the junction between the littoral and sublittoral zones because of the seeming lack of a reliable indicator (see Lawson, 1966). *Dictyopteris delicatula* and *Sargassum vulgare* have in the past been used to define this fringe, but these brown algae are not a constant feature of it. Almost invariably they are absent from very wave-exposed shores and on occasion are found in the sublittoral proper. On moderately wave-exposed rocks along the Cape Verde Peninsula *Sargassum vulgare* may be accompanied by two large brown algae (*Cystoseira*, *Ecklonia*) at the southernmost limit of their North Atlantic range (Sourie, 1954a). The black sea-urchin *Echinometra lucunter* appears to be the most reliable indicator of the sublittoral fringe (see Lawson and John, 1982). This sea-urchin occupies a relatively narrow depth range and a wide range of wave-exposure not only in West Africa, but also on the western side of the tropical Atlantic

(John and Price, 1979) and around the mid-Atlantic island of Ascension (Price and John, 1980).

3.   Harbour breakwaters

Increasing commercial activity in West Africa has led to the construction of many new harbours consisting mainly of rough stone breakwater systems. Such man-made constructions are soon colonized by shore organisms and a zonation pattern develops which in time begins to approach that found on nearby rocky shores. None the less, some significant differences remain between the distribution of littoral organisms observed along sheltered to moderately wave-exposed parts of breakwaters in Benin, Ivory Coast, Ghana, and Togo, as compared to similarly exposed rocky shores near by (John and Pople, 1973).

In the very sheltered innermost parts of many harbours there is often gross pollution and all hard surfaces in the lower eulittoral zone are usually covered to a variable depth by a layer of oil and sediment. Over this mixture occurs a mat of filamentous blue-green algae consisting most commonly of *Microcoleus lyngbyaceus*. The upper level of the littoral zone differs from similar natural rocky areas in the paucity or sometimes complete absence of littorines and barnacles. Occasionally the barnacle *Balanus amphitrite* is present, and this is the only organism characterizing the upper subzone of the eulittoral. Macro-algae are absent and there is usually no organism to define the lower subzone of the eulittoral or the sublittoral fringe.

Water movement increases towards the entrances of the harbours, and sessile shore animals become abundant while the most common algae are crustose forms. The littoral zonation pattern now shows some resemblance to that found on nearby moderately wave-exposed rocky shores (Fig. 9.3), though erect filamentous or fleshy algae are absent or inconspicuous. In the sublittoral fringe and the shallow sublittoral there is a patchy covering of cushion- or mat-forming species, articulated corallines, and the occasional clump of *Hypnea cervicornis* growing over the lithothamnia. Blue-green algae are common and are often associated with much silt and diatoms. Oysters are common around the low-water level though are not conspicuous due to the covering of other sessile organisms. Reef-fishes known to be algivorous or omnivorous (see Sanusi, 1980) congregate in large numbers close to the rough stone surface of the breakwaters. The most common of these small fishes are the largely territorial damselfishes (*Stegastes imbricata, Microspathodon fronta-tus*), the sergeant-major fish (*Abudefduf saxatilis*), the surgeon fish (*Acanthurus monroviae*), and the redlip blenny (*Rupescartes atlanticus*). In the immediate sublittoral the obvious grazers commonly include the short-spined black sea-urchin *Echinometra lucunter*, the very long-spined *Diadema antillarum*, and the slate pencil urchin *Eucidaris*. Heavy swells break almost continuously

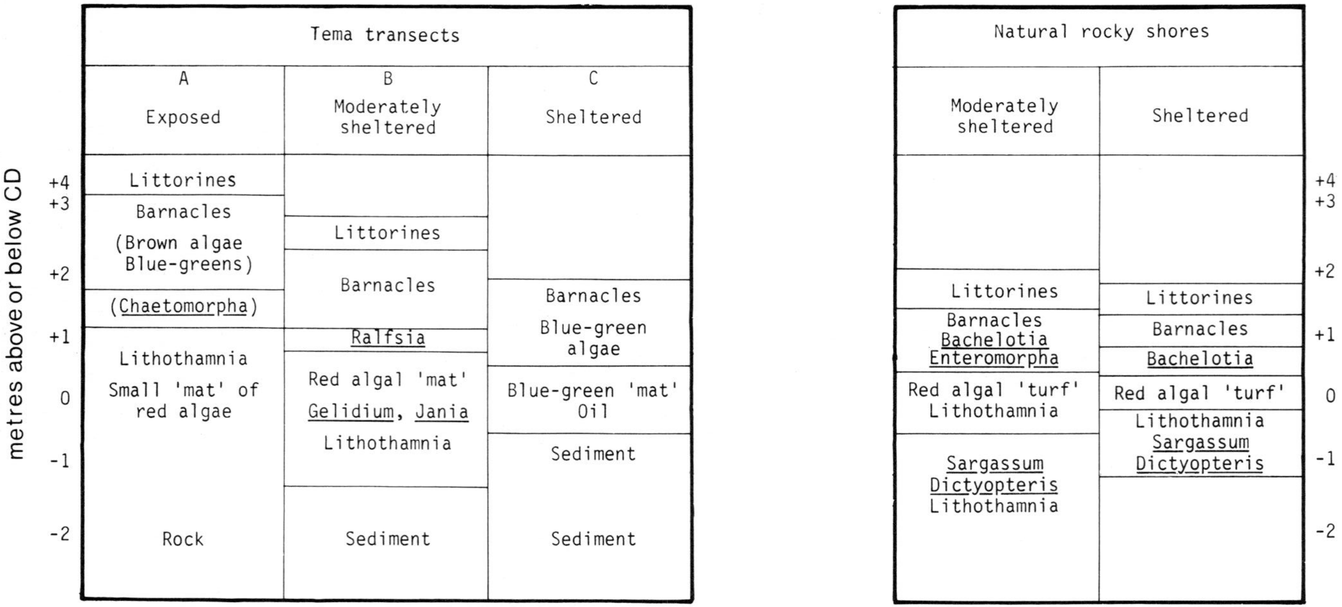

Natural rocky shores

metres above or below CD	Moderately sheltered	Sheltered
+4		
+3		
+2	Littorines	Littorines
+1	Barnacles / Bachelotia / Enteromorpha	Barnacles / Bachelotia
0	Red algal 'turf' / Lithothamnia	Red algal 'turf' / Lithothamnia / Sargassum / Dictyopteris
-1	Sargassum / Dictyopteris / Lithothamnia	
-2		

Tema transects

metres above or below CD	A Exposed	B Moderately sheltered	C Sheltered
+4			
+3	Littorines		
+2	Barnacles / (Brown algae Blue-greens)	Littorines	Barnacles
+1	(Chaetomorpha)	Barnacles / Ralfsia	Blue-green algae
0	Lithothamnia / Small 'mat' of red algae	Red algal 'mat' Gelidium, Jania	Blue-green 'mat' Oil
-1		Lithothamnia	Sediment
-2	Rock	Sediment	Sediment

Fig. 9.3. The distribution of the principal organisms on the breakwaters at Tema harbour in Ghana compared with the pattern on local rocky shores. Parentheses around names indicate that the organisms are only seasonally common. CD = chart datum. (After John and Pople, 1973.) Reproduced by permission of Elsevier Biomedical Press (= North-Holland Publishing Co.).

around the head and seaward side of the harbour breakwaters. The species composition and pattern of littoral zonation along these parts are almost identical to those found on rocky cliffs subject to severe wave action.

4. Tidepools

Irregularities in the surface of littoral rocks result in the formation of tidepools of varying size and topography. During the time they are out of contact with the sea, changes may take place in water temperature, salinity, oxygen tension, and hydrogen-ion concentration. The extent of these changs depends on a number of factors including the time of low water, the season of the year, the size of the pool, and its level on the shore. The following account is based on observations of tidepools in Senegal by Sourie (1954a), Ivory Coast and Liberia by John (1972), and in Ghana by Gauld and Buchanan (1959), Sanusi (1980), and John (unpublished).

Tidepools in the littoral fringe are not replenished with sea-water by the rise and fall of the tides, but depend for their existence on salt spray, wave splash, and rainwater. Unshaded pools are often lined by blue-green algal mats which are best developed in those polluted by bird droppings. Portions of this spongy lining may become detached and float on the surface as flock-like masses. Some of the polluted pools may be green and turbid due to the presence of large populations of planktonic algae.

Shallow tidepools in the upper eulittoral subzone may contain brown mats of *Bachelotia antillarum* which are often associated with a good deal of sand. Conditions in large and deep pools may be similar to those found in the open sea providing they are not too high on the shore. Such pools contain many of the algae that commonly grow in the sublittoral fringe and shallow sublittoral zone (e.g. *Sargassum vulgare*, *Dictyopteris delicatula*, several species of *Dictyota*, *Padina*, *Galaxaura*). The algae are normally confined to a narrow fringe along the often steep rocky sides if the pools are sand-lined. Dense mats of blue-green algae, sometimes accompanies by large clumps of *Hypnea cervicornis* and species of *Gelidium*, may develop in moderately wave-sheltered pools lined with boulders. A feature of these pools is the abundance of fish known to feed on algae such as the territorial damselfish *Microspathodon frontatus*, the redlip blenny *Rupescartes atlanticus*, and juveniles of the damselfish *Abudefduf hamyi*, the sergeant-major fish *Abudefduf saxatilis*, and the parrot fish *Pseudoscarus hoefleri*.

Tidepools in rocky areas away from beaches are usually free from sand and often lined by lithothamnia accompanied by brown crusts of *Ralfsia expansa* and *Lobophora variegata* (prostrate form). The sides of such pools may be elevated by the development of branched and anastomosing lithothamnia to form small shelf-like platforms level with the water surface. Some pools may be lined by greenish or greyish-brown zoanthids, or the hard and brownish colonies of the coral *Siderastrea radicans*. In wave-exposed places these pools may contain dense populations of the sea-urchin *Echinometra lucunter* and a number of small algal-feeding fishes. The only evident algae, other than crustose forms, are those forming the more or less persistent algal mat confined to the territories of the damselfish *Microspathodon frontatus*.

5. Lagoons, estuaries, and deltas

These commonly occurring West African habitats provide a rather special and somewhat inhospitable environment for most macroalgae. In such brackish-water situations the algae have to withstand widely fluctuating salinities, very little wave action, high turbidity, and smothering by sediment. There is a predominance of mud-covered surfaces which are unsuitable for most algae other than small blue-greens. Sessile littoral organisms are almost wholly confined to the rocky or lateritic banks of estuaries and to the roots of the fringing mangrove vegetation. The mangroves form extensive swamps in many deltas and lagoons and often cover coastal flats which become inundated only at high water. Lawson (1966) has reviewed littoral zonation in brackish-water habitats in several West African countries (principally Ghana and Sierra Leone), while more recently observations have also been made in Ivory Coast (John, 1972), Gabon (John and Lawson, 1974), Gambia (John and Lawson, 1977b), and Liberia (De May et al., 1977).

(a) Mangrove areas

In many brackish-water habitats the roots of the mangroves often afford the principal surfaces available for the attachment of littoral organisms. The two most common mangroves in West Africa are *Rhizophora* spp. (the red mangrove) and *Avicennia germinans* (the white mangrove). In *Avicennia* it is the upright breathing roots (pneumatophores) spreading some distance around the tree that are colonized. This mangrove is found in periodically closed and permanently open lagoons, though only in the latter is there an appreciable development of algae on its roots. *Rhizophora* is mainly confined to open lagoons and its arching stilt roots are most commonly covered with littoral organisms only when it grows in main channels where there are strong marine influences. On the muddy shores of many lagoons there remains an almost black crust of blue-green algae when the surface dries out and becomes cracked during the dry season. Shallow pools left when the water-level drops during the dry season are often floored by a thick spongy mat of blue-green algae. Portions of this mat become detached due to the entrapment of gas bubbles and float on the surface of the pool as flock-like masses. The prolific development of blue-green algae in these seasonal pools is similar to that observed along the open coast in some littoral fringe pools and in shallow salt pans where sea-water is evaporated to obtain salt.

The vertical height of the littoral zone may be very restricted because of the sheltered conditions that normally prevail in mangrove-characterized areas. This zone can be divided into a littoral fringe and a eulittoral zone, although the number of belts and their species composition are somewhat variable. The sublittoral zone and sublittoral fringe are generally absent as the mangroves are usually rooted in mud and so there is no suitable surface for the establishment of benthic organisms at and below low water.

The littoral fringe is generally characterized by the small snail *Littorina scabra* (=*L. angulifera*). In many estuaries and lagoons the eulittoral zone on the mangrove roots may be divided into three subzones (Fig. 9.4): an upper barnacle subzone dominated by *Chthamalus* (usually *C. rhizophorae*), or more rarely by *Balanus amphitrite*; a middle zone of the oyster *Ostrea tulipa*; and, on occasion, a lower algal subzone. This lowermost subzone of algae usually consists of an inconspicuous greyish furry mat of one or more species of *Bostrychia*, other small red algae (e.g. *Caloglossa*, *Catenella*, *Polysiphonia*), and green filaments of species of *Cladophora*, *Chaetomorpha*, and *Rhizoclonium*. These latter green algae are often especially common on the breathing roots of *Avicennia*. The zone-forming organisms are not always exclusive to a particular level, often when green algae are present by themselves they occupy a vertical height corresponding to the uppermost subzone or even the littoral fringe.

(b) Rocky banks

Some estuaries have rocky or stable lateritic banks which are often shaded by overhanging vegetation. On these surfaces there may be a distinct zonation of littoral organisms though the pattern gradually changes away from the mouth as tidal influences, wave action, and salinity become gradually more reduced.

In estuaries the littorines (*Littorina punctata*, *Nodolittorina granosa*) that characterize the littoral fringe along open shores are gradually replaced by *Littorina cingulifera* as marine influences diminish (see Fig. 9.4). These small snails are often accompanied in very shaded places by the green hair-like strands of *Rhizoclonium*, and in the lower part of the fringe by various blue-green algae. There is below this fringe a barnacle subzone which is the first of three subzones that form the eulittoral zone. The barnacles may on occasion be crowded out in the upper part of this subzone by such red algae as *Bostrychia* and *Caloglossa*, and sometimes become completely replaced by a dirty grey covering of *Bostrychia radicans* (Lawson, 1957a). The middle is characterized by *Ostrea tulipa*, and this oyster is often accompanied by the small limpet *Siphonaria pectinata*. Further up an estuary this limpet disappears and eventually the oysters become replaced by the small mussel *Brachydontes puniceus*. Characterizing the lowermost subzone of the eulittoral is a low and dirty grey turf of *Murrayella periclados*, other red algae (principally *Bostrychia*, *Caloglossa*), several green algae including *Cladophora*, *Enteromorpha*, and

		MANGROVE ROOTS	ESTUARINE ROCKS
LITTORAL FRINGE		**Littorinid Zone** – *Littorina* *scabra angulifera*	**Littorinid Zone** – *Littorina* *cingulifera*
	Upper	**Barnacle Subzone** – *Chthamalus* *rhizophorae*	**Barnacle Subzone** – *Chthamalus* spp., – sometimes an algal mat, often *Bostrychia* *radicans*
EULITTORAL ZONE	Middle	**Oyster Subzone** – *Ostrea tulipa*, *Thais callifera*	**Oyster Subzone** – *Ostrea tulipa*, *Siphonaria pectinata*
	Lower	**Algal Subzone** – algal mat of *Bostrychia* spp., *Caloglossa* spp., *Catenella impudica*	**Algal Subzone** – algal mat of *Bostrychia* spp., *Caloglossa* spp., ?*Siphonocladus* *brachyartus*, or algae replaced by sessile animals such as *Balanus amphitrite*, *Pomatoleis* sp.

Fig. 9.4. The distribution of the principal organisms on the roots of mangroves and on rocky banks in estuaries in West Africa where marine influences are still appreciable.

Rhizoclonium, and blue-greens. In the Sierra Leone River estuary this subzone is often replaced by a belt dominated by animals such as serpulid worms, the barnacle *Balanus amphitrite*, and sponges (see Longhurst, 1958).

The position of the sublittoral fringe is often occupied by mud-flats from which algae are usually absent. Rocks occur at this level on the shore at the mouth of the Ancobra River in Ghana and bear an almost pure stand of the red alga *Grateloupia filicina* (Lawson, 1966). This alga also forms a mat at the entrance to the Tabou River in Ivory Coast (John, 1972), but here it is accompanied by *Bryocladia thyrsigera* and both grow in the lower eulittoral subzone. Small and bushy *Gelidium* plants form a rather distinctive belt just below the barnacle zone at the entrances to the Tabou and Ancobra rivers.

V. Sublittoral habitats

Brief descriptons of shallow-water algal vegetation exist for Gabon (John and Lawson, 1974), Ghana (John and Lawson, 1972a), and Sierra Leone (John and Lawson, 1977a). Bodard and Mollion (1974) have described such vegetation along the northern part of the coast of Senegal though based solely on dredge samples. Much of the account that follows is of the algal vegetation growing on areas of calcareous cobbles and rocky banks lying off the Ghanaian coast. This is the most well-known section of West African coast as a result of the investigations by John *et al.* (1977, 1980) and Lieberman *et al.* (1979, 1984) using the aqualung (scuba).

There is no obvious zonation of the algal vegetation with a gradual increase in depth. The overall drop in species numbers with depth and the restriction to shallow water of certain algae may be accounted for by differences in their compensation points. Some algae are restricted to or are better developed in deeper water and this might be due to the longer growing season found further offshore where turbidity is less, to a reduction in turbulence associated with depth, or to an inability to compete successfully with the more abundant shallow-water plants (John *et al.*, 1977). The sublittoral algae show their greatest development over the calmer months of the dry season when shore algae are being adversely affected by severe dessication.

1. Inshore rocks

The brown algae *Sargassum vulgare* and *Dictyopteris delicatula* are absent from the sublittoral fringe where there is severe wave action.On such shores the fringe is covered by a continuous pinkish-red crust of lithothamnia and large numbers of *Echinometra lucunter*. During low water in the day these sea-urchins are commonly confined to almost spherical depressions or holes. The algae *Sargassum* and *Dictyopteris* are often restricted to the sublittoral fringe on more wave-sheltered shores, while the lithothamnia invariably continue down into the sublittoral zone proper. *Sargassum vulgare* extends into the shallow sublittoral on a few gently sloping and wave-sheltered shores (e.g. entrance of the Gabon River), but is soon replaced by *Sargassum filipendula*. Off the wave-exposed coast of Ghana this latter species is yet to be discovered growing in water shallower than 8 m. Often associated with these *Sargassum* beds are well-developed and diverse assemblages of epilithic and epiphytic algae.

The beds of *Sargassum* are very extensive and the individual plants well developed in sheltered bays where the rocky bottom has little surface relief (e.g. Man O' War Bay on the Sierra Leone Peninsula). Large algae including *Sargassum* are often absent from such wave-sheltered inlets when the rocky bottom is irregular due to the presence of holes, ledges, and gullies. Miemia Bay in the west of Ghana is an example of a protected bay showing much surface relief due to its boulder shore. The lower part of this shore is covered by lithothamnia which give way in the sublittoral to cushion- or mat-forming algae, articulated corallines, and larger clumps of just a few species (principally *Hypnea cervicornis*, *Dictyopteris delicatula*, *Dictyota* spp.). In such natural habitats showing much physical relief there are large populations of algivorous fishes similar to those found along moderately wave-exposed parts of the rough stone breakwaters which form many of the harbour systems in West Africa (see section IV.3). The territories of the damselfish *Microspathodon frontatus* are common in the immediate sublittoral (c. 0–0.5 m) and may be readily recognized in Miemia Bay by the presence of striking red mats of the filamentous alga *Polysiphonia*. In deeper water the low algal turf or mat is almost continuous but is often best developed in the territories of another black damselfish, *Stegastes imbricatus*.

Sea grasses normally root in sand or mud unlike the large majority of West African algae which require hard, stable surfaces for attachment. To the north of Senegal there are some very extensive areas of sheltered shallows covered by sea-grass vegetation. For example, the very wave-sheltered south-easterly shore of the Cape Blanc Peninsula is fringed in places by sea-grass meadows (see Lawson and John, 1977). The report by Sourie (1954a) of *Cymodocea* growing just south of the Cape Verde Peninsula in Senegal is one of the very few on sea-grass from the coastal stretch considered in this review.

2. Offshore rocky banks

A series of narrow rocky banks run almost parallel to the coast between Ivory Coast and Nigeria. Off the Ghana coast the most common plant on low and often thinly sand-buried parts of the shallowest of these banks, is the beautiful net-like red alga *Dictyurus fenestratus*. This alga is endemic to West Africa and commonly covers areas of several metres at depths ranging from 8 to 14 m. On low and little-structured rock abutting sand occur occasional beds of *Sargassum filipendula*. Often large individuals of this *Sargassum* become detached and are eventually cast up on the shore. Commonly growing on such periodically sand-buried rocks are other brown alga (principally *Dictyopteris delicatula*, *Spatoglossum schroederi*, *Dictyota* spp.) and various red algae including *Gelidiopsis variabilis*, *Corynomorpha prismatica*, and *Thamnoclonium claviferum*.

The algal vegetation is inconspicuous where the rocky banks are raised and form numerous low cliffs, ledges, and gullies. In these areas of much physical relief, the rock surfaces are covered with a crust of lithothamnia and the occasional mat or low turf of largely filamentous red and blue-green algae (Fig. 9.5). This turf or mat finds its greatest expression in the almost contiguous or overlapping territories of a small damselfish known as *Stegastes imbricatus*. Outside these damselfish territories are to be found the occasional large and often isolated clump or solitary individual of such algae as *Dictyopteris delicatula*, *Dictyota* spp., *Hypnea cervicornis*, *Laurencia majuscula*, and *Predaea feldmannii*.

In the vicinity of cliffs and ledges harbouring large populations of the nocturnal sea-urchin *Diadema antillarum*, there is an abundance of such sessile animals as oysters and barnacles while algae other than crustose forms

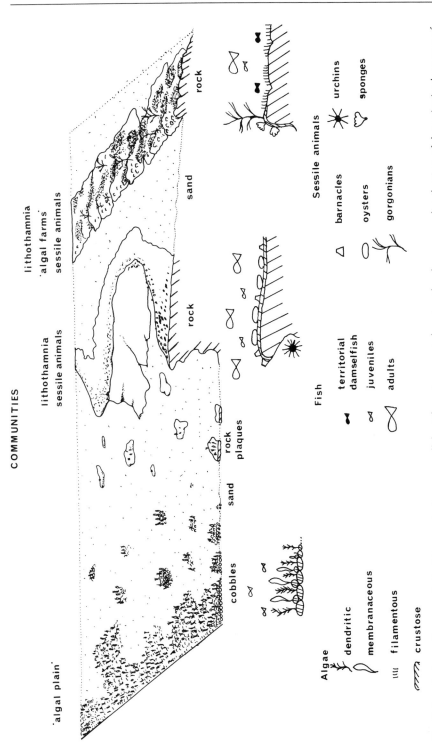

Fig. 9.5. A stereodiagram of a belt transect of an area off the Ghanaian coast showing the dry season distribution of the principal groups of plants and animals. (After Lieberman *et al.*, 1984.)

are rare and usually diminutive. Fish known to be algivorous or omnivorous (see Sanusi, 1980) congregate in large numbers around these raised rocky banks, and include *Acanthurus monroviae, Abudefduf saxatilis, Rupescartes atlanticus, Pseudoscarus hoefleri,* and *Stegastes imbricatus.*

3. Cobble and shell bottoms

In many places off the West African coast the only hard substrates for algal attachment are calcareous cobbles formed by the accretion of red algae and shells. The surface of these seasonally disturbed cobbles is often very irregular due to the morphology of the calcareous algae forming them, abrasion caused by tumbling and sand-scouring, and to the activities of various borers (principally bivalves). In shallow offshore areas a diversified algal community develops rapidly on these surfaces during the calmer months of the dry season (c. 6 months' duration off Ghana) when physical disturbance and sand burial are rare events. Within a few weeks after the return of stable conditions many algae have reached a height of 20 cm or more. Some of the largest and commonest plants forming this so-called 'algal plain' are red algae such as *Solieria filiformis, Laurencia majuscula, Rhodophyllis gracilarioides, Lophocladia trichoclados, Halymenia actinophysa, Gracilaria foliifera,* and *Anatheca montagnei.* Off Ghana there is a decline in algal diversity on the cobbles with increase in depth. Many of those growing over a wide depth range are larger and better developed in deeper water further offshore where the growing season is longer, turbidity is less, and competition from shallow-water species is removed. These plants include the green alga *Caulerpa taxifolia,* and various red algae including *Cryptonemia luxurians, Botryocladia guineense, Halymenia actinophysa, Rhodophyllis gracilarioides, Plocamium telfairiae,* and *Waldoia antillana.* Paradoxically the large and conspicuous algae present on the cobbles over the dry season are absent, or else small and diminutive on adjacent rocky banks where the extensive areas of hard substrate show much physical relief.

The shells of the snail *Turritella* also provide a suitable surface for the attachment of algae. Off the Ghanaian coast these conical shells are fairly common over areas of grey mud at depths in excess of 30 m, and are empty or else occupied by hermit crabs. The most common plants growing on these shells are red algae including *Spyridia filamentosa, Gracilaria foliifera,* and a foliaceous alga which probably belongs to the genus *Hymenena.* On rock and shell bottoms off northern Senegal there is an algal community characterized by *Halymenia senegalensis* and *Anatheca montagnei* (Bodard and Mollion, 1974). This community has been subdivided into a northern facies (Dakar to M'Bour) recognized by the abundance of such species as *Polyneura denticulata, Botryocladia senegalensis,* and *Hypnea cervicornis,* and a southern facies (M'Bour southwards) in which *Pseudobranchioglossum senegalense* and various species of *Hypnea* are dominant.

VI. Seasonality and succession

The belts or zones of shore-inhabiting organisms are dynamic and vary in position or level in response to changing physical and biological conditions. Sourie (1954a) has observed seasonal changes in the composition and abundance of shore organisms on the Cape Verde Peninsula in Senegal. During the wet season the barnacle belt was found to be invaded from below by algae including *Ralfsia expansa* and *Caulacanthus ustulatus*, and on more wave-exposed shores by crusts of lithothamnia. These algae persisted into the dry season, but became bleached and eventually disappeared during the months of November and December. Lawson (1957b, 1966) has similarly reported an upward migration of a number of Ghanaian shore algae during the wet season followed by a retreat from the upper levels in the dry season. He demonstrated that the regular seasonal movement of the belts of shore organisms was correlated with desiccation operating through the agency of the tides. After the equinox in September the lower low waters occur in the daytime and the higher low waters during the night, whereas when the sun is north of the equator (March–September) the position is reversed. In consequence, there is a period corresponding to the main dry season when shore organisms have to endure the obviously more intensive desiccation due to daytime rather than night-time exposure to the air (see Jeník and Lawson, 1967). This daytime exposure leads to the mass mortality of upper littoral algae and so to the apparent downward migration of their zones or belts. Desiccation will also be especially severe during the northern hemisphere winter at times when the very dry harmattan wind reaches the coast.

Seasonal changes in the ground cover of algae and zoanthids have been studied in lower eulittoral tidepools on Matrakni Point in Ghana. In such pools the cover of sessile organisms other than crustose algae increases in the rainy season (May/June–October/November) and shows a decline over the rest of the year (Fig. 9.6). Though higher solar insolation may be expected to promote the growth of mat- and clump-forming algae, it seems to be more than offset by the adverse affect of various physical and chemical changes taking place during low tide in the dry season. Indeed, the pools remain out of contact with the sea for especially long periods due to the lower wave fronts over this calmer period of the year.

Marine macroalgae and other benthic organisms cannot survive the agitation caused by surf and other types of water movement unless securely attached to stable surfaces such as rocky outcrops. Accordingly there is little or no development of marine vegetation along those unprotected stretches of coastline where sand, mud, pebbles, boulder, or shell substrates predominate. The distinctive assemblage of algae that develops on seabed cobbles (see section V.3) disappears over the rainy season when the cobbles become agitated and tumbled by the powerful back-and-forth water motion caused by

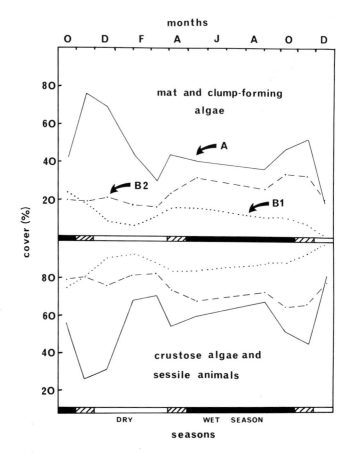

Fig. 9.6. The seasonal changes in the percentage cover of crustose algae/sessile animals and mat/clump-forming algae in lower eulittoral tidepools on Matrakni Point, Ghana. Pool A contains no sea-urchins or resident fish population; pool B has been subdivided into two sectors: sector B1 contains sea-urchins and non-territorial algivorous/omnivorous fish, and sector B2 has been cleared of urchins but contains non-territorial and territorial (damselfish *Microspathodon frontatus*) fish. (Data from Sanusi, 1980.)

the heavy swells which characterize this period of the year. Rapid recolonization of these cobbles takes place on the return of calm and surge-free conditions associated with the onset of the dry season. The fast-growing filamentous algae that develop initially are soon followed by slower-growing fleshy and dendritic forms which are commonly perennials and pseudoperennials. These algae probably survive the unfavourable season as spores, germlings, juveniles, persistent holdfasts, or stoloniferous bases. Some may arise from spores produced by plants that survive the unstable period of the year on littoral rocks, nearby rocky banks, or on deeper-water cobbles less affected by water movement. The conical shells of *Turritella* provide a surface for the development of a deep-water algal community where they lie below the surge zone and yet are shallow enough for there to be sufficient light for plant growth.

It is not always possible to find a ready explanation to account for some of the local departures from the typical pattern of shore zonation without an intimate knowledge of the biological history of an area. No doubt many of the anomalous distribution patterns represent no more than a stage or stages in a successional sequence following the creation of space caused by some past destructive event. If smaller algae such as diatoms and blue-green algae are ignored then three main phases of recolonization have been recognized on rocks at Christiansborg in Ghana (Lawson, 1957a, 1966). Within weeks of clearing an area of shore there was a complete covering of *Ulva* which had replaced an earlier colonizing green alga known as *Enteromorpha*. Small spots of lithothamnia appeared early in the succession and after several months the completion of the second phase was marked by the dominance of these crustose forms. The final phase involved invasion by a number of algae some of which were not normally found at the level in question. Over six years elapsed before the community became indistinguishable from that on surrounding rocks. Sourie (1954a) observed that the barnacle *Chthamalus stellatus* took over two years to achieve a 90 per cent recolonization on a denuded area of shore in Senegal where formerly it had a complete coverage. In general, recolonization seems to be much slower at higher than at lower levels on the shore.

The rate and pattern of succession on stable sublittoral surfaces have been investigated off Ghana using plates of Perspex (Plexiglass) fixed to frames placed on the sea-bed at depths of about 10–13 m. Irrespective of the time of year such plates were positioned in the sea, initial colonization was by bacteria, followed closely by diatoms and then a complete covering of filamentous brown algae (*Ectocarpus, Giffordia*). Spots of lithothamnia appeared within a few days, grew rapidly and coalesced after a month or so to cover almost 40 per cent of the primary surface (Fig. 9.7). In time these crustose algae became bleached and overgrown by a number of erect forms and sessile animals. The death and sloughing off of the sessile animals (particularly barnacles) resulted in the re-exposure of the primary plate surface. Opportunist species quickly recolonized the newly exposed surface, and so plates several months old often became covered by a mosaic of plant and sessile animal communities of different ages. Similar communities are likely to develop on re-exposed rock surfaces following a destructive event such as parrot-fish grazing. The algal vegetation on beach rocks and low submarine rock banks abutting sand often remains stable for only a short period, and simply represents a successional stage held in check by periodic sand burial and sand-scouring. One of the few plants able to grow successfully on such

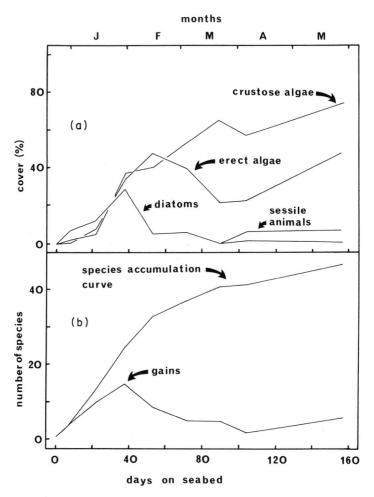

Fig. 9.7.   The changes in the cover of algae and sessile animals (a) and the number of algal species (b) developing on artificial surfaces fixed to a frame placed in a cobble area off Ghana during the dry season.

rocks is the red alga *Dictyurus fenestratus* whose net-like erect branches arise from a creeping stoloniferous base. These branches may actually assist in the entrapment of sand grains and so contribute to its accumulation around the base of the plant. In low-lying rocky areas off Ghana the climax community is probably one dominated by *Sargassum filipendula*. This community is normally short-lived because the constantly shifting sand leads to the loss of the *Sargassum* plants due to the decay of the holdfast when buried. It can only become established once again if the rock surface is re-exposed and so open to colonization by its oospores. This brown alga is apparently unable to tolerate grazing and therefore is only able to survive in intensively fish-grazed areas in the form of a low-growing basal rosette of foliar appendages. Similarly, many of the algae associated with the *Sargassum* beds, or growing on cobbles, are absent from raised rocky banks where large numbers of grazing animals congregate. Seasonal perturbations of the environment prevent the algal plain community ever stabilizing on the cobbles; this community rarely survives for more than six months off the Ghana coast.

VII. Plant-herbivore relations

Little is known of inter- and intraspecific interactions along the West African coast of primary ground-cover species, and of these species with their respective grazers and predators. The physical conditions and large numbers of known algal feeders (principally limpets and echinoids), may be responsible for the paucity and diminutive size of filamentous and fleshy algae at the most wave-exposed end of a shelter/wave-exposure gradient. On such shores crustose algae characterize the lower eulittoral subzone, while other forms are diminutive and inconspicuous unless they are growing directly on the large limpet *Patella safiana*. In the only study of limpet grazing in West Africa there followed a rapid and prolific growth of blue-green algae when *Siphonaria* was experimentally removed from an area of Ghanaian shore (Gauld and Buchanan, 1959).

The grazing activities of the sea-urchin *Echinometra lucunter* probably account in large part for the absence of algae other than crustose growth forms from many eulittoral tidepools and the sublittoral fringe of wave-exposed shores. Following the experimental removal of *Echinometra* from a sector (B2) of a lower eulittoral tidepool in Ghana there was a significant increase in the ground cover of filamentous and fleshy algae (see Fig. 9.6). This increase in algal cover was mostly due to the creation of new territories protected by the damselfish *Microspathodon frontatus* and to the expansion of previously existing ones (Table 9.2). Though grazing pressure was much reduced in this sector of the pool by the removal or urchins, there were still present large numbers of non-territorial algivorous or omnivorous fish (juveniles of *Abudefduf hamyi* and *Microspathodon frontatus*; *Rupescartes atlanticus*; various gobies, etc.). The aggressive behaviour of the adult damselfish discouraged other fish from actively grazing algae in their territories, and allowed the development within them of the so-called 'algal farms'. The non-territorial fishes and sea-urchins were the principal grazers in the other sector (B1) of the pool. In a small pool (A) with no evident population of macrograzers, the ground cover of algae (excluding crustose forms) was always high and any seasonal change was probably due to differences in the physico-chemical environment.

The paucity and diminutive size of filamentous and fleshy algae, the preponderance of lithothamnia, and the conspicuousness of small algal-

Table 9.2.   The percentage cover of mat- or clump-forming algae in lower eulittoral tidepools, or sectors of a tidepool, on Matrakni Point in Ghana: small pool with no resident macrograzers (A); large pool (B) divided into a control sector (B1) containing urchins and non-territorial algivorous/omnivorous fish, and another sector (B2) from which urchins have been cleared though containing algivorous/omnivorous fish (Data from Sanusi, 1980).

Pools/sectors	Damselfish terroritories	Outside territories	Totals
		Mean percentage cover (range)	
A	0	34.2 (17–63)	34.2 (17–63)
B1	0	12.1 (0–24.5)	12.1 (0–24.5)
B2	15.7 (5.6–22.2)	8.8 (0–13)	24.5 (1.62–35.1)

feeding fishes during the day (see Fig. 9.5), are all features of moderately wave-sheltered and shallow-water habitats showing much physical relief (e.g. rough stone breakwaters, boulder-lined bays, much structured offshore rocky banks). The important role played by these algivorous fish in determining the nature of the sublittoral algal vegetation has been demonstrated experimentally by John and Pople (1973) along the rough stone breakwaters of Tema harbour in Ghana. It appears that such so-called reef-fishes often congregate around rocky areas providing there are numerous holes, crevices, gullies, and ledges to afford them protection from predation by marauding carnivores. Open habitats lacking much surface relief (e.g. areas of sand, cobbles, or low and little-structured rocks) have a very different and less diversified fish community. For instance, the few demersal fishes observed over the cobble-strewn sea-bed off Ghana in the dry season were mostly juveniles (personal observations) who had apparently ventured away from the protection of the nearby rocky banks to feed in the seasonally devastated algal plain community. In this cobble area there appears to be no resident population of algal-feeding fish or sea-urchins. Non-territorial fishes have been observed to follow the rising tide in wave-sheltered areas along rough stone breakwaters and boulder-lined bays, and actively to feed on algae growing in the eulittoral zone.

The most abundant algae in heavily fish-grazed habitats appear to be those species having some defence against grazing. This may be structural (texture: calcareous, gelatinous, leathery; growth-form: crustose, peltate; growth region: protected meristem(s), ready reorganization of new meristem(s) if damaged or removed) or, if appearing palatable, then possibly biochemical (e.g. *Dictyota* spp., *Dictyopteris delicatula*, *Hypnea cervicornis*). Other potential defence mechanisms suggested by Ogden and Lobel (1978) include temporal escape (short-lived), or spatial escape (occupying cracks, crevices, wave-washed areas, etc.). The common occurrence of articulated coralline algae and crustose growth-forms in fish-grazed areas, may also be a consequence of the suppression of competition from faster growing but selectively grazed filamentous and fleshy forms which normally overgrow them. In some rocky areas the shallow-water territories of the damselfish *Microspathodon frontatus* are most conspicuous due to the presence of a reddish mat consisting almost solely of the filamentous alga *Polysiphonia*. This alga forms similar mats in the territories of the giant blue damselfish (*Microspathodon dorsalis*) in the Gulf of California, Mexico (Montgomery, 1980a, b). These mats are highly productive and are believed to represent an early successional stage held in check by grazing (see Brawley and Adey, 1977; Kaufman, 1977; Lobel, 1980; Montgomery, 1980a, 1980b). Various micrograzers have been shown by Brawley and Adey (1981) to affect algal community structure in the tropics, but no attempt has been made to study them in West Africa.

Sea-urchins normally feed on anything that is available and broadly suited to their rasping mode of feeding; often they completely remove all algae other than crustose and endolithic species (Lawrence, 1975). From experiments in Ghana there is some information on the influence of *Echinometra lucunter* on algae growing in tidepools (see above), though no quantitative data exist for the feeding activities of the sublittoral urchin *Diadema antillarum*. Around the ledge and cliff areas of raised rocky banks off the Ghanaian coast the abundance of sessile animals and the inconspicuousness of most algae other than crustose forms (see Fig. 9.5) is probably a consequence of the nocturnal grazing activities of this urchin. Though *Diadema* may derive some of its nutritional requirements from lithothamnia the largest feeders of crustose algae in these cliff areas are likely to be parrot fishes. The common occurrence of oysters and barnacles in *Diadema*-grazed areas suggests that the sea-urchin may promote the establishment of these sessile animals which are normally competitively inferior to ground-covering filamentous and fleshy algae. Intense grazing appears to exclude many algae from the rocky banks, and thus limits competitively dominant algae largely to refuge areas which include the cobble areas or low, little-structured rocky banks. In common with other tropical areas, plant–animal interactions seem more important than they do in colder seas, and should prove a fruitful field for future research.

References

Allen, J. R. L. (1965). Late Quaternary Niger delta and adjacent areas: sedimentary environments and lithofacies. *Bull. Am. Ass. Petrol. Geol.*, **49**, 547–600.
Anang, E. R. (1979). The seasonal cycle of the phytoplankton in the coastal water of Ghana. *Hydrobiologia*, **62**, 33–45.
Barbey, C. (1968). Africa in the physical world Seas and coasts in western Africa. In *International Atlas of West Africa*, pp. 1–2, Pl. 1. OAU Scientific, Technical, and Research Commission, Paris.

Bassindale, R. (1961). On the marine fauna of Ghana. *Proc. zool. Soc. Lond.*, **137**, 481–510.

Berrit, G. R. (1969). Les eaux dessalées du Golfe de Guinée. *Proceedings of the Symposium on Oceanography and Fisheries Resources of the Tropical Atlantic, 20–28 Oct. 1966, Abidjan*, pp. 13–22, Unesco.

Bodard, M., and Mollion, J. (1974). La végétation infralittorale de la petite côte sénégalaise. *Bull. Soc. phycol. Fr.*, **19**, 193–221.

Boughey, A. S. (1957). Ecological studies of tropical coast-lines. 1. The Gold Coast, West Africa. *J. Ecol.*, **45**, 665–687.

Brawley, S. H., and Adey, W. H. (1977). Territorial behaviour of threespot damsel fish (*Eupomacentrus planifrons*) increases reef algal biomass and productivity. *Envir. Biol. Fish.*, **2**, 45–51.

Brawley, S. H., and Adey, W. H. (1981). The effect of micrograzers on algal community structure in a coral reef microcosm. *Mar. Biol.*, **61**, 167–177.

Buchanan, J. B. (1958). The bottom fauna communities across the continental shelf off Accra, Ghana (Gold Coast). *Proc. zool. Soc. Lond.*, **130**, 1–56.

Chi-Bonnardel, R. van (1973). *The Atlas of Africa*. Edition Jeune Afrique, Paris.

De May, D., John, D. M., and Lawson, G. W. (1977). A contribution to the littoral ecology of Liberia. *Botanica mar.*, **20**, 41–46.

Ekman, S. (1953). *Zoogeography of the Sea*. Sidgwick and Jackson, London.

Feldmann, J. (1937). Recherches sur la végétation marine de la Méditerranée. La Côte des Albères. *Revue algol.*, **10**, 1–339.

Gauld, D. T., and Buchanan, J. B. (1959). The principal features of rock shore fauna in Ghana. *Oikos*, **10**, 121–132.

Griffiths, J. F. (1972). Climates of West Africa. In *World Survey of Climatology*, Vol. 10. Elsevier, Amsterdam.

Houghton, R. W. (1976). Circulation and hydrographic structure over the Ghana continental shelf during 1974 upwelling. *J. phys. oceanogr.*, **6**, 909–924.

Houghton, R. W., and Mensah, M. A. (1978). Physical aspects and biological consequences of Ghanaian coastal upwelling. In R. Boje and M. Tomozak (Eds.), *Upwelling Ecosystems*, pp. 167–203. Springer-Verlag, Berlin, Heidelberg, New York.

Ingham, M. C. (1970). Coastal upwelling in the northwestern Gulf of Guinea. *Bull. mar. Sci.*, **20**, 1–34.

Jackson, S. P. (1961). *Climatological Atlas of Africa*. Commission for Technical Co–operation in Africa South of the Sahara, Joint Project No. 1, Pretoria.

Jeník, J., and Lawson, G. W. (1967). Observations on water loss of sea weeds in relation to microclimate on a tropical shore (Ghana). *J. Phycol.*, **3**, 113–116.

John, D. M. (1972). The littoral ecology of rocky parts of the north-western shore of the Guinea Coast. *Botanica mar.*, **15**, 199–204.

John, D. M., and Lawson, G. W. (1972a). Additions to the marine algal flora of Ghana I. *Nova Hedwigia*, **21**, 817–841 (1971).

John, D. M., and Lawson, G. W. (1972b). The establishment of a marine algal flora in Togo and Dahomey (Gulf of Guinea). *Botanica mar.*, **15**, 64–73.

John, D. M., and Lawson, G. W. (1974). Observations on the marine algal ecology of Gabon. *Botanica mar.*, **17**, 249–254.

John, D. M., and Lawson, G. W. (1977a). The marine algal flora of the Sierra Leone peninsula. *Botanica mar.*, **20**, 127–135.

John, D. M., and Lawson, G. W. (1977b). The distribution and phytogeographical status of the marine algal flora of Gambia. *Feddes Reprium*, **88**, 287–300.

John, D. M., Lieberman, D., and Lieberman, M. (1977). A quantitative study of the structure and dynamics of benthic subtidal algal vegetation off Ghana (Tropical

West Africa). *J. Ecol.*, **65**, 497–521.

John, D. M., Lieberman, D., Lieberman, M., and Swaine, M. D. (1980). Strategies of data collection and analysis of subtidal vegetation. In J. H. Price, D. E. G. Irvine, and W. F. Farnham (Eds.), *The Shore Environment*, Vol. I. *Methods.* Systematics Association Special Volume No. 17(a), pp. 265–283. Academic Press, London, New York, San Francisco.

John, D. M., and Pople, W. (1973). The fish grazing of rocky shore algae in the Gulf of Guinea. *J. exp. mar. Biol. Ecol.*, **11**, 81–90.

John, D. M., and Price, J. H. (1979). The marine benthos of Antigua (Lesser Antilles). I. Environment, distribution and ecology. *Botanica mar.*, **22**, 313–326.

Kaufman, L. (1977). The three spot damselfish: effects on benthic biota of Caribbean coral reefs. *Proc. Third Int. Coral Reef Symp.*, **1**, 559–564.

Laborel, J. (1974). The West African reef corals: an hypothesis on their origin. *Proc. Second Int. Coral Reef Symp.*, **1**, 425–443.

Lawrence, J. M. (1975). On the relationships between marine plants and sea urchins. *Oceanogr. mar. Biol. ann. Rev.*, **13**, 213–286.

Lawson, G. W. (1955). Rocky shore zonation in the British Cameroons. *J. W. Afr. Sci. Ass.*, **1**, 78–88, 1 Pl.

Lawson, G. W. (1956). Rocky shore zonation on the Gold Coast. *J. Ecol.*, **44**, 153–170, 2 Pl.

Lawson, G. W. (1957a). Some features of the intertidal ecology of Sierra Leone. *J. W. Afr. Sci. Ass.*, **3**, 166–174.

Lawson, G. W. (1957b). Seasonal variation of intertidal zonation on the coast of Ghana in relation to tidal factors. *J. Ecol.*, **45**, 831–860.

Lawson, G. W. (1966). The littoral ecology of West Africa. *Oceanogr. mar. Biol. ann. Rev.*, **4**, 405–448, 2 Pl.

Lawson, G. W. (1978). The distribution of marine algal floras in the tropical and subtropical Atlantic Ocean: a quantitative approach. *Bot. J. Linn. Soc.*, **76**, 177–193.

Lawson, G. W., and John, D. M. (1977). The marine flora of the Cap Blanc peninsula: its distribution and affinities. *Bot. J. Linn. Soc.*, **75**, 99–118.

Lawson, G. W., and John., D. M. (1982). *The Marine Algae and Coastal Environment of Tropical West Africa (Beih. 70 zur Nova Hedwigia)*. J. Cramer, Vaduz.

Lewis, J. R. (1961). The littoral zone on rocky shores — a biological or physical entity? *Oikos*, **12**, 281–301.

Lewis, J. R. (1964). *The Ecology of Rocky Shores*. English Universities Press, London.

Lieberman, M., John, D. M., and Lieberman, D. (1984). Factors influencing algal species assemblages on reef and cobble substrata off Ghana. *J. Exp. Mar. Biol. Ecol.*, **75**, 129–143.

Lieberman, M., John, D. M., and Lieberman, D. (1983). Determinants of algal species composition on reef and cobble substrates off Ghana. *J. Ecol.*, in press.

Lobel, P. S. (1980). Herbivory by damselfishes and their role in coral reef community ecology. *Bull. mar. Sci.*, **30**, 273–289.

Longhurst, A. R. (1958). An ecological survey of the West African marine benthos. *Fishery Publs colon. Off. London*, **11**, 1–102.

Longhurst, A. R. (1962). A review of the oceanography of the Gulf of Guinea. *Bull. Inst. fr. Afr. noire, sér. A*, **24**, 633–663.

Longhurst, A. R. (1964). The coastal oceanography of western Nigeria. *Bull. Inst. fr. Afr. noire, sér. A*, **26**, 337–402.

McMaster, R. L., LaChance, T. P., and Ashraf, A. (1970). Continental shelf geomorphic features off Portuguese Guinea, Guinea, and Sierra Leone (West Africa). *Marine Geol.*, **9**, 203–213.

Marchal, E. (1960). Premières observations sur la repartition des organismes de la zone intercotidale de la région de Konakri (Guinée). *Bull. Inst. fr. Afr. noire*, sér. A, **22**, 137–141.

Martin, L. (1971). The continental margin from Cape Palmas to Lagos: bottom sediments and submarine morphology. In F. M. Delany (Ed.), *The Geology of the East Atlantic Continental Margin*. Part 4, *Africa*, Report No. 70/16, Inst. Geol. Sci., pp. 79–96.

Montgomery, W. L. (1980a). Comparative feeding ecology of two herbivorous damsel-fishes (Pomacentridae: Teleostei) from the Gulf of California, Mexico. *J. exp. mar. Biol. Ecol.*, **47**, 9–24.

Montgomery, W. L (1980b). The impact of non-selective grazing by the giant blue damselfish, *Microspathodon dorsalis*, on algal communities in the Gulf of California, Mexico. *Bull. mar. Sci.*, **30**, 290–303.

Ogden, J. C., and Lobel, P. S. (1978). The role of herbivorous fishes and urchins in coral reef communities. *Envir. Biol. Fish.*, **3**, 49–63.

Pauly, D. (1975). On the ecology of a small West-African lagoon. *Ber. dt. wiss. Komm. Meeresforsch.*, **24**, 46–62.

Pilger, R. (1911). Die Meeresalgen von Kamerun. Nach des Sammlung von C. Ledermann. *Bot. Jb.*, **46**, 294–313, 316–323 (1911–12).

Price, J. H., and John, D. M. (1980). Ascension Island, South Atlantic: a survey of inshore macroorganisms, communities and interactions. *Aquat. Bot.*, **9**, 251–278.

Reyssac, J. (1970). Phytoplancton et production primaire au large de la Côte d'Ivoire. *Bull. Inst. fond. Afr. noire*, sér. A, **32**, 869–981.

Sanusi, S. S. (1980). A study on grazing as a factor influencing the distribution of benthic littoral algae. M.Sc. thesis, University of Ghana, 217 pp.

Sourie, R. (1954a). Contribution à l'étude écologique des côtes rocheuses du Sénégal. *Mém. Inst. fr. Afr. noire*, **38**, 1–342.

Sourie, R. (1954b). Principaux types de zonations verticales des algues sur le littoral rocheux de la presqu'ile du Cap Vert (zone intercotidale). *Rapp. Commun. int. bot. Congr.*, **8**(17), 151–153.

Stephenson, T. A., and Stephenson, A. (1949). The universal features of zonation between tidemarks on rocky coasts. *J. Ecol.*, **38**, 289–305, 1 Pl.

Thompson, B. W. (1965). *The Climate of Africa*. Oxford University Press, Nairobi, London, New York.

Toupet, C. (1968). Major climatic elements. In *International Atlas of West Africa*, pp. 1–3, Pl. 10–13. OAU Scientific, Technical, and Research Commission, Paris.

Watts, J. C. D. (1958). The hydrology of a tropical West African estuary. *Bull. Inst. fr. Afr. noire*, sér. A, **20**, 697–752.

Plant Ecology in West Africa
Edited by G. W. Lawson
© 1986 John Wiley & Sons Ltd

CHAPTER *10*

Montane vegetation

J. K. Morton
Department of Botany, University of Waterloo,
Ontario, Canada

I. Introduction

Mountains are areas of land with narrow summits and considerable local relief, which rise above the surrounding countryside. This distinguishes them from plateaux which are broad areas of relatively level but elevated land. Mountains are higher than hills but there is no clear distinction between them. An arbitrary elevation of about 1000 m has often been used to define the lower limit of mountains, but this is not universally applicable for there are extensive areas of the earth's surface above this elevation which are of low relief and could not be considered as mountainous. Elevation determines the presence of the characteristic plants and animals associated with mountains through its influence on many other factors, including rainfall, cloud, prevailing winds, temperature, and fire.

There are four widely separated areas of high land in West Africa (Fig. 10.1). Running from west to east, these are:

1. the *Guinea Highlands* of Guinea, Sierra Leone, Liberia, and Ivory Coast;
2. the *Togoland Hills* and Atakora Mountains which extend along the Ghana–Togo border and into northern Benin;
3. the *Jos Plateau* of northern Nigeria; and
4. the *Camerouns system* consisting of Cameroun Mountain and the offshore island of Equatorial Guinea (Bioko) and the Adamawa massif extending along the Nigeria–Cameroun border and eastwards into Chad.

There are also many other hills and rocky outcrops of lower elevation scattered across West Africa which, though not montane in their elevation, often provide suitable habitats for montane plants and animals.

II. Environmental conditions

The environmental conditions associated with mountains are very different from those of the lowlands and montane plants must be adapted to these

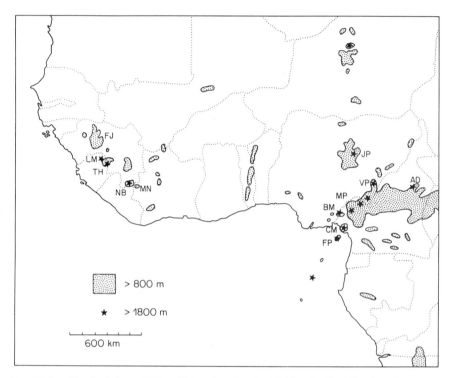

Fig. 10.1.   Physical map of West Africa showing the highlands. The Nigerian/Camerounian peaks are: JP (Jos Plateau); VP (Vogel Peak); MP (Mambila Plateau); AD (Adamawa massif); BM (Bamenda Highlands); CM (Cameroun Mountain); FP (Fernando Po = Bioko). The Guinean Highland peaks are: FJ (Fouta Djalon); LM (Loma Mountains); TH Tingi Hills); NB (Nimba Mountains); MN (Man Massif). (After Cole, 1974).

conditions to survive. Though environmental factors are closely interrelated, we can consider them under the following headings: temperature, rainfall, wind, radiation, evaporation, biotic, and fire.

1.   Temperature

For each 100 m ascent on a mountain the average temperature drops by about 0.6°C. Thus, as Boughey (1953) observed, the temperature on the top of Cameroun Mountain, the highest peak in West Africa, would be a chilly 4°C when in Limbe (= Victoria) at the base it is 32°C. It is for this reason that, even on mountains in the tropics, the temperature falls with increasing elevation until on the highest mountains, such as Kilimanjaro, it is arctic with permanent ice-fields and glaciers. In West Africa only Cameroun Mountain is high enough to experience frequent freezing temperatures. Even here snow does not persist though it occasionally settles on the summit. Of equal importance are the extremes and rapid fluctuations in temperature which mountains in the tropics experience. They heat up very quickly in clear daytime weather and lose heat just as rapidly through the thin atmosphere when clouds shut out the sun or as night approaches. Frequently, daily temperature fluctuations on these mountains are greater than anual fluctuations in many lowland regions of the world.

2. Rainfall

Precipitation is usually much higher on mountains than on the surrounding lowlands, particularly if the prevailing winds blow off the sea or are otherwise laden with moisture. When winds come into contact with a mountain range they are forced to rise. As a result the air is cooled because temerature drops with increasing elevation. If the air is nearly saturated with water vapour, condensation takes place forming mist or clouds and then rain (or hail and snow at very high altitudes). Another factor which can induce rain on mountains is convection currents. Mountains are subject to higher solar radiation. This heats the land which in turn increases the temperature of the air. Warm air rises and is cooled at higher elevations thereby condensing water vapour which is changed into mist, cloud, and rain. This is one of the reasons why banks of cloud are often associated with mountains during the heat of the day, particularly in the tropics where solar radiation is high and the effects of convection currents are more pronounced. There is a further reason why rising air over mountains produces cloud and rain. Air as it rises expands, for atmospheric pressure is lower at higher elevations. When gases, in this case air, expand they cool, thus accentuating the effects of high elevation. The influence of mountains on rainfall in West Africa is pronounced. Debundscha at the foot of Cameroun Mountain has one of the highest recorded average annual rainfalls in the world with 10 149 mm y^{-1} (Letouzey, 1968). The mountainous coastal peninsula of Sierra Leone is ringed with cloud above about 700 m almost every day of the year and rainfall (averaging about 6000 mm y^{-1} at Guma) is much greater than in the lowlands (3500 mm at Freetown 15 km away). In Ghana the Atewa Hills, though they rise to only about 700 m have a much higher rainfall than the neighbouring areas of lowland forest.

3. Wind

It has already been noted how moisture-laden winds, as they blow against mountains, rise and precipitate their water vapour in the form of cloud and rain. However, winds are not always heavily laden with moisture. The harmattan winds which blow across much of West Africa from the Sahara Desert during December, January, and February each year, are very dry and do not carry sufficient moisture for this to be precipitated as rain. Instead the windward sides of mountains, in those parts of West Africa which are affected by the harmattan, are seared by the hot and dry wind. These effects are even greater on mountains than they are in the lowlands over which the winds blow. The resulting effect on the vegetation is pronounced. The Loma Mountains and Tingi Hills of Sierra Leone experience the full force of the harmattan. Both mountain ranges have dense forest on the middle and lower slopes of their south and west sides, but on the slopes exposed to the harmattan (the north and east) the belt of forest is narrow and in some places is broken. Here the lowland savanna connects with the montane grassland so that savanna fires can sweep up the mountain during the dry season. Similar effects of the harmattan winds were observed by Schnell (1961) in the Chaîne de Fon in Guinea. Jeník and Hall (1966) in a study of the effects of the harmattan on the Djebobo massif in the Togoland Hills of Ghana, comment that the intense desiccation and high temperatures of these winds are responsible for the development of extreme local ecoclimatic conditions and the existence of communities of xeromorphic plants resembling those of semi-deserts. They observed that trees are distorted by the dry winds which also cause soil erosion by blowing away the finer particles of soil on the exposed slopes.

4.   Radiation

On mountains solar radiation (insolation) is usually greater than in lowland areas. The reasons for this are twofold. Firstly, at higher altitudes, atmospheric pressure diminishes (at the top of Cameroun Mountain the pressure is only about 60 per cent of that at sea-level) — the air is 'thinner' or more rarefied. As a result the very important effect of air in filtering out the rays of the sun is lessened and solar radiaton is greater, light intensities (on a clear day) are very high and the land heats up rapidly from the increased radiation. All wavebands of radiation are not absorbed at an equal rate by the atmosphere. The effect on ultraviolet radiation is particularly important. At sea-level most is filtered out by the atmosphere, but on mountains the level of ultraviolet radiation is much higher. The effect of this on living organisms is considerable; plants become more stunted and compact, often forming cushions; red pigment (anthocyanin) production is increased, for it helps the plant to filter out the injurious radiation; delicate growing points are protected either by a dense covering of woolly hairs or by layer upon layer of old leaf bases. The second factor resulting in higher solar radiation on mountains is that the atmosphere tends to be clearer at higher elevations. The lowest layers of the atmosphere are usually heavily laden with water vapour, dust, smoke, and other forms of pollution. Mountains usually rise above the lower layers into the clear upper atmosphere. This is often very noticeable in climbing a mountain. A blanket of haze can be seen shrouding the lowlands, while the mountain itself is in crystal-clear air. Water vapour, dust, and smoke prevent solar radiation from reaching the surface of the earth, and the plants and animals which live there. Without this protective blanket, mountains receive much greater levels of radiation on clear days. The intensity of this radiation varies considerably, however, because of the frequent occurrence of mist and cloud on mountains. When present these greatly reduce the level of radiation reaching the surface of the ground (and the plants). In particular they almost completely remove infra-red radiation which is the major source of heat. As pointed out by Hedberg (1964), the radiation climate of mountains is very variable, displaying a fairly regular diurnal cycle, according to the often regular daily cycle of cloudiness. The reverse effect occurs at night when the rarefied atmosphere permits rapid radiation of heat from the land at high elevations, especially on clear nights when there is no protective covering of cloud. This is a further reason why mountains in the tropics experience extremes of temperature, as well as a different quality of radiation, and why these extremes are more diurnal than seasonal. Data collected by Jeník and Hall (1966) in the Togoland Hills of Ghana exemplify this. They recorded diurnal changes in temperature in the grassland from 40 °C during the day to 13 °C at night under a clear sky. Similarly on Cameroun Mountain rapid radiation of heat from the land on a clear night frequently causes temperatures to plummet to below freezing and icicles and frost often form on the vegetation near ground level.

5. Evaporation

During certain seasons evaporation on mountains can be very high due to the combined effects of solar radiation and wind. Jeník and Hall (1966) observed this during the period when the harmattan winds were blowing on the Togoland Hills (Table 10.1). Evaporation was exceptionally high (62.1 ml day^{-1} as measured by a Piche evaporimeter) creating desert-like conditions in the grassland on the top of the hills. This compared with a much lower rate of evaporation (16.9 ml) in grassland at the bottom of the range. As observed by Jeník and Hall, this high level of evaporation appears 'to have no parallel in the published data. Schnell (1952c) records a diurnal total of 27 ml from the Fouta Djalon Mountains and Ross (in Walter, 1951) records 28 ml from the edge of the Namib Desert in Namibia (South-West Africa). The maximum rate per hour (4.4 ml) occurred when the sun's rays were at right angles to the land and the harmattan wind was blowing strongly. Because of these high levels of evaporation, montane plants possess a great variety of xeromorphic adaptations which enable them to reduce water loss and withstand desiccation. Adaptations include evergreen leaves with thick cuticles, deciduous leaves, stomata sunk in deep grooves in the leaf, development of a protective layer of dense woolly hairs, fleshy underground rootstocks (bulbs, tubers, rhizomes, tap-roots, and corms), and concealing the growing points in protective leaf bases or bud scales. Yet other species escape the rigours of the dry season by being

Table 10.1. Microclimatic extremes on the Djebobo massif, Ghana as recorded during the dry harmattan period on 30–31 December 1965. After Jeník and Hall (1966) (temperature in °C, evaporative power of the air in ml). Reproduced by permission of The Journal of Ecology

	Riverain forest	Bottom grassland	Summit steppe	Windward steppe	Leeward savanna
Maximum air temperature	28.6	40.0	35.4	35.0	43.7
Minimum air temperature	12.0	13.2	20.3	20.5	19.6
Diurnal range of air temperature	16.6	26.8	15.1	14.5	24.1
Maximum soil temperature	22.4	32.6	34.7	—	31.0
Minimum soil temperature	19.2	23.2	24.6	—	23.0
Diurnal range of soil temperature	3.2	9.4	10.1	—	8.0
Maximum water temperature	22.8	—	—	—	—
Minimum water temperature	18.0	—	—	—	—
Diurnal range of water temperature	4.8	—	—	—	—
Maximum evaporation per hour	1.0	2.1	4.1	4.4	3.0
Total diurnal evaporation	6.3	16.9	62.1	62.2	35.2

ephemeral, completing their life-cycle and dispersing their seeds during the favourable seasons and surviving the dry season as seeds.

6. Biotic factors

These are similar on mountains to those in lowland vegetation. Large herbivores such as antelope, water-buffalo, and elephant graze on the vegetation and trample the ground, particularly around water-holes and along the often well-established trails which they use. In montane communities biotic factors can be of particular importance in relation to plant dispersal and establishment. A high proportion of the montane flora is adapted to animal dispersal, with either succulent fruits or barbed or sticky seeds or fruits. As a result many montane herbs are characteristically associated with tracks through the forest and grassland where their seeds or fruits are dispersed by passing animals (e.g. species of *Achyranthes, Adenostema, Sanicula, Dichrocephala,* and *Desmodium*). A number of species are dependent on the disturbed habitats created by the trampling of large mammals. For instance on the Loma Mountains the small insectivorous sundew, *Drosera burkeana*, is almost confined to the wet disturbed ground around drinking-holes and water-seepage areas where trampling reduces competition from perennial species and creates an open habitat where the *Drosera* can grow. Man is the major biotic factor affecting the mountains of West Africa, largely through the activities of hunters who use fire to drive out the game. Fires started by man have greatly increased the impact of burning on the montane vegetation throughout West Africa, pushing back the forest edge, destroying enclaves of trees and shrubs and causing a great expansion in the areas of grassland. Jacques-Felix (1968) discusses this in relation to the Adamaoua region of Cameroun where fire and grazing have led to the destruction of woody vegetation, degradation of the grassland and soil, and changed the species composition as many native species have been replaced by alien weeds. The resulting degraded grassland is much poorer in its nutrient content and cannot support as many cattle as the natural montane grassland of this area. Grazing of cattle is affecting the vegetation in several other areas of West Africa, particularly where these lie within the savanna zone. Thus the Fouta Djalon and Jos Plateaux and the Atakora and Mandara Mountains have been extensively affected by grazing. In recent years the Obudu Plateau of Nigeria has been developed for ranching cattle. There can be little doubt that the few remaining areas of montane grassland which are still relatively unaffected by grazing will soon be exploited.

7.   Fire

The montane grasslands are a fire-controlled climax, as are the lowland savannas. The sharp demarcation between the upper edge of the forest and the montane grassland is the result of periodic fires. These sear the edge of the forest and prevent the trees and shrubs from spreading into the grassland. They also prevent seedlings of the forest trees from becoming established in the grassland. Fire is endemic in the montane grassland and fires are started naturally by lightning as well as by man. Mountains are particularly subject to violent thunderstorms. Furthermore these are usually at their most frequent and severe at the end of the dry season when the vegetation is as dry as tinder and is readily ignited. The first storms are often 'dry storms' with frequent lightning but either very localized rain or no rain at all. These are ideal conditions for starting fires. In March 1964 the writer witnessed such an event on the Loma Mountains. For several days storms developed each afternoon with heavy thunder and lightning but no rain. Lightning struck the grassland on the steep slopes some distance below the summit. This started a fire which roared upwards and over the summit. During the next few days it gradually spread across the whole plateau, burning most of the montane grassland but leaving unburnt enclaves in many places. In modern times man has become the major factor starting fires. In the mountains of West Africa this is usually done by hunters in order to drive out game from the cover of rank vegetation, but in areas where the land is cultivated, fire is part of the normal agricultural practice used to clear the land. Schnell (1952b) considers that most of the montane grassland on the Nimba Mountains is the result of the destruction of the montane forest and soils by man-made fires.

The sequence of degradation which he describes is as follows:
1.   Montane forest dominated by *Parinari excelsa*.
2.   Open woodland and secondary forest.
3.   Tall herbaceous and scrub vegetation dominated by *Dissotis* spp. and the grass *Setaria chevalieri*.
4.   Montane grassland dominated by *Loudetia kagarensis* and *Eriosema* sp.

Similarly Keay (1955) and Richards (1963) consider that the lower montane grassland of Cameroun is the result of excessive burning caused by man in the upper zone of montane woodland, little of which now remains intact apart from a few enclaves in gullies or on the lee-side on natural fire breaks.

Most areas of montane grassland in West Africa are either small enclaves on the tops of the higher peaks and ridges or, if more extensive, they are broken up by gallery forest which ascends the gullies, or by forest and thicket which tend to occupy rocky areas. As a result the occasional fires started by lightning or other natural causes were usually unable to spread into neighbouring enclaves of grassland. Burning was very restricted and occurred at intervals of several or even many years. Man has changed this. Hunters go from one grassy enclave to the next and start fires in each so that most burn every year. This change in the frequency of burning has also affected the intensity of the fires. When an area of grassland has not burnt for several years there is a considerable accumulation of dead plant remains which burn fiercely when they are eventually ignited. These occasional fires are thus much more intense than when an area burns every year. The native plants of the montane grassland are adapted to conditions of infrequent but severe burning. Other species, which cannot withstand these conditions, are unable to become established. When the fires are less severe under conditions of annual burning these species are able to invade the montane grassland and compete with the native plants. This is happening in many parts of West Africa resulting in a degradation of the montane grassland and a change in species composition. As previously noted, Jacques-Felix (1968) describes such a situation in the Adamaoua region of Cameroun. Similar events are taking place on the Loma Mountains where lowland savanna species and alien weeds are invading the montane grassland. Another effect of frequent burning has been to push back the edge of the forest and to destroy large areas of thicket and woodland. The trees and shrubs which grow in the forest edge are able to recover from the effects of infrequent fires, putting out new shoots and branches to make up for those damaged by the fire, but annual burning prevents recovery and results in a gradual degradation of the forest edge and recession of the forest. Fire is the single most serious threat to the survival of montane vegetation in West Africa and the unique assemblage of montane plants. Urgent action is now needed to prevent or control man-made fires in these communities.

III.   Types of montane vegetation

The vegetation of mountains is characteristically zoned. Zonation is determined primarily by altitude but is also affected by other factors such as exposure to prevailing winds and rain, steepness of slope, the type of terrain, drainage, etc. Even mountains which occur within the savanna zone usually have forest, either on the lower slopes or in gullies and ravines which provide shelter and moisture for the growth of trees. The presence of the forest is due to the effects of the mountains on local climate in generating higher rainfall. Up to an elevation of 800—1000 m this forest is essentially the same in species composition and structure as lowland rain forest, but higher up it gradually changes into montane or mist forest. The factor determining the presence of this type of forest is the level at which banks of cloud created by a cooling of the air masses as they rise to pass over the high land form on the slopes of the mountain. This is a zone of high rainfall, high humidity, and lower temperatures due to reduced solar radiation resulting from the cloud cover. In the montane forest the trees are characteristically festooned with epiphytes — mosses, club-mosses, ferns, begonias, orchids, peperomias, etc. Lianes are less abundant than in lowland forest and the trees are not usually buttressed. The layering of the forest, so characteristic of the lowland forest, is not well developed, and trees are smaller — 25–30 m high – and the canopy is broken. Gaps in the forest, created by falling trees, are common and they become an impenetrable tangle of rank herbs and herbaceous climbers.

The montane forest extends to about 2000 m where it gradually becomes more open with a fairly even canopy of smaller trees. The ground layer is also more open and not so rank, consisting of small shrubs and herbs. In this zone the branches of the trees are festooned with lichens and the trunks encrusted with lichens and other epiphytes, particularly orchids. This upper montane forest or montane woodland is a much drier zone than the mist forest. It tends to occur above the banks of cloud which daily envelop the mountain slopes and it receives a much lower rainfall. As a result montane woodland is very prone to damage by fire and much has already been destroyed and replaced by grassland.

The forest and woodland usually give way to montane grassland at between 2000 and 3000 m. Two zones are recognizable on the highest mountains of West Africa. There is a lower zone of tall grasses with scattered, often stunted and gnarled trees. It is usually a very colourful zone with many beautiful flowering herbs and shrubs. The grasses are from 1 to 2 m high and are characteristically tufted species in which the individual tufts are widely spaced with bare soil between them. In this bare soil many geophytes grow, usually flowering in the dry season or with the early rains, and also small ephemerals which germinate with the first rains, then flower and fruit rapidly completing their life-cycles before the onset of the dry season. The very narrow transition zone between forest and grassland is one of the richest in species and the most spectacular in colour, for here are to be found a wealth of small trees, shrubs, and rank perennial herbs.

On the highest peaks this tall montane grassland gives way to upper montane grassland at elevations above 3000 m. This zone is characterized by shorter species of grass only about 0.5 m high, which form dense tussocks of tightly packed shoots. Trees are absent and the zone has been compared by some writers to the steppe of more northerly climates. Exposure to wind, tempera-ture-change, and radiation is severe, but rainfall is low because the zone is above the level of many of the rain-producing clouds.

The upper limit of forest on the mountains of West Africa is, under the most favourable conditions, about 3000 m but frequently it is much lower than this where local climate, prevailing hot dry winds, drainage, or soil create conditions less favourable for tree growth. Thus on the Togoland Hills of Ghana the forest gives way to grassland on the summits at under 1000 m. On the south side of Cameroun Mountain, which receives the highest rainfall, forest reaches a little over 3000 m, but in many places on the north side it gives way to montane grassland at 2000 m because of the rain shadow created by the mountain.

Forest often extends into the grassland zone along gullies and ravines where it is less exposed and where moisture is more readily available. These tongues of 'gallery forest' are an important feature of the vegetation of mountains, breaking up the grassland into enclaves and providing refuge for wildlife. They provide a series of highly specialized microclimates where temperatures and humidity are very different from in the neighbouring grassland. Data obtained by Jeník and Hall (1966) from the Togoland Hills illustrate this. Maximum and minimum temperatures in the gallery forest were 28.6 and 12°C respectively, while those in the nearby grassland were 35.4 and 20.3°C. Similarly, evaporation per hour reached 4.1 ml in the grassland but was only 1.0 ml in the forest.

Woody vegetation also occurs within the grassland zone where rocky outcrops given protection to trees and shrubs from drying winds and wind-blown fires. Where rocky boulder-strewn outcrops are extensive, large areas of thicket with scattered trees develop. Parts of the Loma Mountains are domin-ated by this type of vegetation in which the endemic and very beautiful *Dissotis leonensis* is a characteristic species. Similar zones of thicket and· scrub are found on the rocky Idanre Hills of southern Nigeria. These thickets are characterized by shrubs and trees which have adaptations to withstand the long periods of drought that occur in these situations and are accentuated by the steep rocky terrain. Many are deciduous or have thick fleshy evergreen leaves. Succulents such as the tree euphorbias are a feature of this zone.

On the mountains of East Africa there is a well-developed and very distinct zone of bamboo at the upper edge of the forest. This is dominated by *Arundinaria alpina* — a different species of bamboo from the one in the lowlands. This zone is absent on most of the mountains of West Africa but occurs in the Bamenda area of Cameroun above about 2000 m. The zone is floristically very poor because the roots of the bamboo completely dominate the upper layers of the soil and exclude most other plants.

Several workers, notably Keay (1955), Boughey (1953), Richards (1963), and Hall (1973) have described the zonation of vegetation on the mountains of West Africa, usually with particular reference to Cameroun Mountain. The following are the most readily recognizable zones as they apply to the whole of the West African mountains.

1. *Lowland rain forest* up to 800–1000 m (lowland savanna in mountain systems well north of the forest region occurs up to 1000 m).
2. *Montane forest* (mist forest) 800–2000 m.
3. *Montane woodland* 2000–3000 m (the bamboo woodland of the Bamenda area comes within this zone). Severely degraded by burning and often now replaced by the following zone.
4. *Lower montane grassland* 2000–3000 m. Dominated by tall grasses with scattered trees and enclaves of woodland or thicket in gullies and rocky areas.
5. *Upper montane grassland* above 3000 m. Characterized by short grasses and the absense of trees.

IV. The principal montane regions of West Africa

1. The Guinean highlands

The main components of these are the Fouta Djalon Plateau of Guinea, the Loma Mountains and Tingi Hills of Sierra Leone and the Nimba Mountains lying on the Liberia–Guinea–Ivory Coast border. All are part of the ancient and heavily eroded Cambrian and Pre-Cambrian rocks which make up most of the African continent. The rock is mainly granitic with local intrusions of quartzite, gneiss, dolorite, and other igneous rocks.

The *Fouta Djalon* is an extensive plateau of about 1100 m elevation, heavily eroded and covered with a hard crust or 'pan' of iron oxides. As a result the bedrock is only exposed in valleys around the edges of the plateau and on the many relatively low peaks which rise above it. The elevation of the plateau is not high enough to support montane vegetation, but isolated areas of such vegetation are found on the peaks which rise to over 1500 m. The Fouta Djalon lies north of the belt of tropical rain forest and is well within the drier Guinea savanna region. Much of the forest, which at one time covered the sides of the plateau, the valleys, and the peaks, has been destroyed and the vegetation and soil of the main plateau are also heavily degraded by cultivation and burning. Adam (1958) in a description of the soil and vegetation draws attention to the importance of the iron 'carapace' which covers the plateau and produces numerous laterite pans ('borvals') where drainage is impeded. Tree growth is inhibited on the carapace and these areas are covered by a very poor type of grassland in which *Loudetia*, *Anadelphia*, and *Tristachya* are dominant. Cultivation is only possible where there is no carapace or where soil has accumulated on top of it. The Fouta Djalon has a rich endemic and submontane flora though little montane vegetation. The remnants of montane forest indicate that it was dominated by *Parinari excelsa* — a semi-evergreen tree with small xeromorphic leaves. The climate of the region is characterized by the intense dry season during which fires sweep through the grassland, but total rainfall is moderate, averaging about 2000 mm yr^{-1} though higher on some of the peaks.

The *Loma Mountains* and *Tingi Hills* are situated in the interior of Sierra Leone near the source of the River Niger. They are composed of granite, though the main peak of the Lomas (Bintumane) is capped with dolorite — a very hard igneous rock which fractures into enormous cubes and blocks. The Tingis and the other peaks of the Lomas are granite domes, in which the rock often flakes away in large sheets like the scales of an inverted onion. Bintumane rises to 1942 m and Sankanbiriwa, the main peak of the Tingis, to 1717 m. Both massifs rise steeply from the surrounding plains which are covered with a mosaic of savanna (mostly derived) and forest, for they lie at the northern edge of the tropical rain forest region. The lower slopes are heavily forested with lowland rain forest. This gives way at about 800 m to montane forest dominated by *Parinari excelsa* (Table 10.2). The montane forest extends to the edge of the main plateaux of both massifs at about 1400 m and up the gullies as gallery forest to near the summits. Tree ferns (*Cyathea manniana*) dominate many areas of the gallery forest where there are permanent streams. Another tree fern (*C. dregei*) occurs at the edge of the grassland. On the main plateaux and upper slopes the forest gives way abruptly to montane grassland. This zone, which extends to near the summits of the main peaks is dominated by coarse tall grasses (1–2 m high) characteristic of the lowland savanna i.e. species such as *Monocymbium cerasiiforme*, *Hyparrhenia diplandra*, and *H. smithiana*, *Andropogon perligulatus* and *A. shirensis*, *Loudetia kagarensis* and *Anadelphia leptocoma*. However, the montane composition of the vegetation is apparent from the abundance of often very colourful montane herbs and shrubs which occur within this zone, for example *Eriosema spicatum*, *Vernonia nimbaensis*, *Phyllanthus alpestris*, *Melanthera abyssinica* and *Kotschya*

Table 10.2.   Trees and shrubs of the montane forest of the Loma Mountains and Tingi Hills

Tree layer

Parinari excelsa	*Memecylon* spp.
Syzygium staudtii	*Ficus eriobotryoides*
Polyscias fulva	*Ilex mitis*
Nuxia congesta	*Afrasersalisia cerasifera*
Trichoscypha oba	*Olea hochstetteri*

Shrub layer

Psychotria calva	*Trichilia heudelotii*
Clausena anisata	*Cephaëlis peduncularis*
Ouratea squamosa	*Vangueriopsis vanguerioides*
Vincintella passargei	

lutea. Near the summit of both massifs the grassland changes in structure and species composition. The grasses are shorter (under 1 m) and montane species take over, for example, *Pennisetum monostigma*, *Schizachyrium schweinfurthii*, *Eragrostis invalida*, *Loudetia jaegerana*, *Tripogon major*, *Panicum lindleyanum* and *P. ecklonii*, *Elionurus argenteus*, *Sporobolus mauritianus*, and *Andropogon mannii*.

The montane grassland has pronounced seasonal aspects with the tall lush green growth of grass during the rains turning to rich golden brown at the beginning of the dry season (November–April). Many of the herbs and shrubs flower and fruit during this period so that at the end of the rains these grasslands are a blaze of colour, particularly at the edge of the gallery forest. Fire sweeps the grasslands during the dry season and the grasses put out new growth almost immediately, stimulated by the fire and the heavy dew which settles during the cool clear nights. These bright green meadows of very short grass soon come into flower with a wealth of geophytes — orchids, gladioli, legumes, Commelinaceae, numerous Composites, and several sedges (Table 10.3). Cole (1974) in an analysis of life-forms in the plants of this region, notes the dominance of such hemicryptophytes and geophytes in the montane grassland (Fig. 10.2).

These mountains do not have the extensive laterite pans or iron carapaces so characteristic of the Fouta Djalon Plateau. As a result borvals are absent, though depressions in the rock and gently sloping surfaces harbour a similar flora of short grasses and ephemeral herbs of which *Neurotheca loiselioides*, *Nerophila gentianoides*, and *Bulbostylis briziformis* are among the most characteristic and colourful. The steep slopes of exfoliating granite have a distinctive flora of which the two sedges *Afrotrilepis pilosa* and *A. jaegeri* are the most characteristic feature. They form extensive thick mats and cushions on the steep rock slopes and act as sponges, retaining water long into the dry season which supports many other herbs including colourful gladioli and

Table 10.3. Geophytes and hemicryptophytes in the montane grassland of the Loma Mountains and Tingi Hills

Bulbostylis trichobasis	*Cyanotis longifolia*
Cyperus margaritaceus	*Cyanotis caespitosa*
Commelina africana	*Eupatorium africanum*
Vernonia nimbaensis	*Hypoxis angustifolia*
Phyllanthus alpestris	*Helichrysum mechowianum*
Euphorbia depauperata	*Eriosema spicatum*

orchids. The rock in other parts of these mountains is fractured into large boulders and these areas support a xeromorphic scrub and thicket vegetation. Among the characteristic shrubs and small trees are the spectacular *Dissotis leonensis* with its thick fleshy stems and pink blossom, also evergreens with thick leaves such as *Craterospermum laurinum*, *Pavetta lasioclada*, *Eugenia leonensis*, Gaertnera paniculata, and *Harungana madagascariensis*, and deciduous very hairy species such as *Premna hispida* and *Sabicea vogelii*. The boulders in this community protect it from the annual fires which sweep through the montane grassland.

Elevation coupled with their location on the boundary between the savanna and forest regions, determines the climate of these mountains. Jaeger (1969, 1971) carried out detailed climatic studies on the Loma Mountains and Cole (1974) analysed and summarized them. Temperatures decrease at the rate of 7.5°C per 1000 m regardless of season — from 25–35°C maximum and 15–22°C minimum in the savanna at the foot of the mountains to 19–25°C maximum and 12–15°C minimum in the montane grassland. Relative humidity drops in the dry season by 19.4 per cent per 1000 m but increases in the wet season by 9.4 per cent per 1000 m to saturation level almost throughout that season. These conditions reflect the extreme exposure of these mountains to the drying harmattan winds and the increased precipitation which elevation induces during the wet season. No data on annual rainfall have been collected, but Jaeger's observations indicate that rain falls daily on the mountain from the end of July to the beginning of September. Between 23 July and 12 October 1964 1816 mm were received. However at the foot at Sekurela (elevation 600 m) only 1045 mm fell during the same period. The dry season extends for three months (December, January, and February) but morning dew is heavy in situations sheltered from the dry harmattan winds. Though desiccation is severe on the exposed montane grassland during the harmattan, the dry season is a short one with occasional rain falling in the mountains between the wet and dry seasons.

The *Nimba Mountains* lie within the tropical rain forest region and hence are much more heavily forested than the Loma Mountains and Tingi Hills. Forest

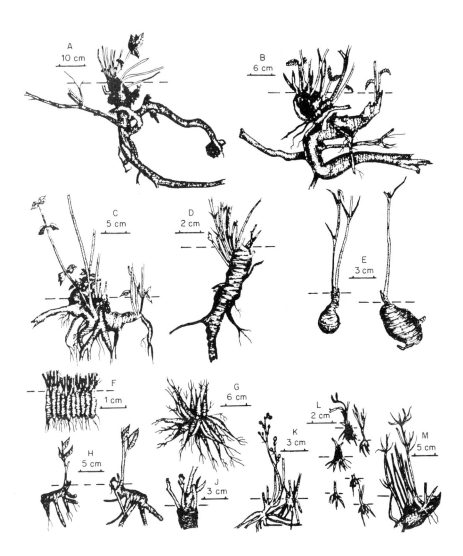

Fig. 10.2. Life-forms of selected montane grassland species on the Loma Mountains, Sierra Leone, at 1680 m. The ground level is indicated by broken lines. (A) *Eriosema parviflorum* subsp. *collinum*; (B) *Kotschya lutea*; (C) *Melanthera abyssinica*; (D) *Euphorbia depauperata*; (E) *Margaretta rosea*; (F) *Bulbostylis trichobasis*; (G) *Cyanotis lourensis* (cf. *C. longiflora* and *C. caespitosa* with smaller tubers); (H) *Vernonia nimbaensis*; (J) Indet. (? *Eupatorium africanum*); (K) *Sopubia simplex*; (L) *Hypoxis angustiflolia*; (M) *Coreopsis camporum*. (After Cole, 1974.) With permission from Cole, *Botanical Journal of the Linnaean Society*, **69** (1974), p. 200. Copyright 1974 by the Linnaean Society of London.

reaches to the tops of even the highest peaks but grassland occurs on some of the ridges and summits. The mountain range is a heavily eroded fragment of the Pre-Cambrian granitic bedrock from which all the mountains of the Guinean system are derived. However, the Nimba Mountains have extensive areas of quartzite and schists and a rich deposit of iron ore which is mined. The range consists of a ridge forming a chain of peaks extending for about 40 km along the Guinea/Ivory Coast border into Liberia. It culminates in Mont Richard-Molard (1752 m) but has many peaks and ridges over 1200 m. The Massif de Man, lying in the western part of Ivory Coast, is an outlier of the Nimba Mountains with peaks reaching to 1340 m. Dense rain forest clothes the lower slopes of the Nimba range. It gives way to montane forest above 800–900 m, and this is dominated by *Parinari excelsa,* with *Syzygium staudtii, Garcinia polyantha, Amanea bracteosa,* and *Gaertnera paniculata* as conspicuous components. Epiphytes are particularly well developed in this zone because of the high humidity from cloud, mist, and rain. The epiphytes include a wealth of ferns and orchids, club-mosses, begonias and peperomias encased in dense growths of bryphytes and lichens (Jaeger and Adam, 1975). Annual rainfall ranges from about 2500 to 4000 mm on the mountains, but more important than the total rainfall is its distribution. Climatic data presented by Adam (1971) show that in 1962 every month had rain on at least one day; December, January, and February were the driest months with one, two, and six days with rain respectively. All the other months of the year had much more frequent rain, with seven months in which rain fell on 20 or more days of the month (Table 10.4).

The montane grassland is confined to the crests of some of the ridges and to steep slopes where the rock is exposed or soil is very thin. It is dominated by *Hyparrhenia subplumosa* and *Loudetia kagarensis* and in wetter places by the sedge *Hypolytrum cacuminum.* As on the Lomas and Tingis the montane grassland supports a rich and colourful flora of herbs and shrubs including many of the same species. On the steep rocky outcrops *Afrotrilepis pilosa* clings to the surface and forms large mats of vegetation with its own characteristic microflora of geophytes and ephemerals including *Disa welwitschii, Polystachya microbambusa, Osbeckia porteresii,* and *Swertia mannii.* Fires sweep through these grassland areas, their frequency having increased since human exploitation of the area began. They are causing some degradation of the vegetation, but the more moderate dry season of these mountains reduces their severity.

Lying midway between the Loma and Nimba Mountains is the Chaîne de Fon with the Pic de Tibé in the north and the Massif du Ziame in the south, the former rising to 1650 m and the latter to 1300 m. Schnell (1961) in an account of the vegetion and flora notes that the range is characterized by an iron pan or carapace on the elevated areas. This has been eroded in the valleys. Upland savanna covers the ridges and many of the peaks, while forest, submontane in

Table 10.4. Rainfall at Buchanan and on Nimba Mountains (After Adam, 1971). Reproduced by permission of the Muséum national d'Histoire naturelle

	Buchanan (0 m)		Nimba Mountains			
			Yekepa (500 m)		Geologist's camp (340 m)	
	Rainfall		Rainfall		Rainfall	
Months in 1962	No. of days	mm	No. of days	mm	No. of days	mm
January	5	78.4	1	23.9	1	15.5
February	6	51.3	8	73.9	6	125.7
March	10	38.8	8	87.5	11	195.1
April	16	182.4	17	202.0	20	267.7
May	22	310.1	16	190.6	15	238.6
June	26	803.5	20	267.9	26	395.7
July	26	778.0	25	170.2	29	407.7
August	14	314.2	24	185.2	29	616.0
September	24	1027.6	28	225.1	28	605.9
October	23	391.5	26	210.0	22	242.4
November	22	294.0	22	206.4	20	362.5
December	7	40.8	2	52.6	2	36.6
Total	201	4310.6	197	1895.3	209	3509.4

the upper reaches, is able to grow in the valleys and gullies where the absence of the iron pan permits the rooting of trees. The vegetation and flora is very similar to that of the Nimba Mountains and *Parinari excelsa* is the characteristic tree in the upper forest.

2. The Togoland Hills and Atakora Mountains

These are a chain of peaks extending from Amedzofe in southern Ghana, diagonally through Togo to the northern tip of Benin. Nowhere do they exceed 1000 m in elevation and as a result they do not support a well-developed montane vegetation, but the vegetation on the upper slopes and plateaux has some of the features of the lower montane forest and grassland of other regions in West Africa, and a distinct though small element of montane species. The southern part of the chain is heavily forested, even though it lies in the 'Dahomey Gap' where the tropical rain forests of the Guinean region are separated from those of the Cameroun–Congo region by a narrow gap of savanna which extends to the coast. Further north the range rises out of the Guinea savanna, but many parts of the slopes and most of the steep valleys are forested. The tops of almost all the peaks and many of the ridges and plateaux are covered with grassland. This is dominated by lowland savanna grasses such as *Loudetia arundinacea*, *Schizachyrium sanguineum*, and *Monocymbium ceresiiforme* and has scattered savanna trees of *Lophira lanceolata*, *Syzygium guineense*, *Cussonia barteri*, and *Hymenocardia acida*, but there are many submontane herbs such as *Haumaniastrum alboviride*, *Otomeria camerooni-ca*, *Eriosema monticolum*, *Helichrysum mechowianum*, *Cyperus margar-itaceus* var. *nduru*, *Sporobolus mauritianus*, *Coreopsis asperata*, *Inula klingii*, and *Senecio lelyi*. The upper reaches of the forest, particularly in the steep-sided ravines have many species typical of the lower montane forest, including trees and shrubs such as *Parinari kirstingii*, *Pachystella brevipes*, *Lecaniodiscus cupanioides*, *Maesa lanceolata*, and *Eugenia leonensis*, and herbs such as *Hypoestes triflora* and *Carex neo-chevalieri*. Jeník and Hall (1966), in a study of climatic conditions during the dry season in the Djebobo part of the range, recorded low night temperatures in the upper forest, with temperatures dropping to 12°C. They suggested that the presence of montane species at these relatively low elevations may be due to the deleterious effect of cold on the competitive power of lowland species, thereby enabling the montane species to survive, particularly in the upper forest where temperatures were consistently lower. Their data (Fig. 10.3) also demonstrated the severity of the harmattan winds on these exposed slopes. Evaporation was very high and simulated desert conditions, hence only trees with thick leaves could survive. Exposure to the hot drying winds on the north-facing slopes resulted in a high incidence of fires and distortion of many of the trees; the strength of the winds being sufficient to create wind-blown soil erosion. Woodland and thicket on the tops of the range were confined to sheltered pockets on the lee-side of the ridges and peaks.

3. The Jos plateau

This is an isolated area of elevated land in northern Nigeria some 250 by 150 km, rising about 400 m above the surrounding lowlands to an elevation of 1200 m. Several ranges of hills rise out of the plateau to nearly twice that height. The geology of the region is varied but consists mainly of granites and basalts. The former produce the rugged peaks and the latter the softer landscape of the main plateau where flat-topped low hills with an ironstone cap are a characteristic feature. The basalts have given rise to the fertile soils which make the Jos Plateau a major agricultural area. Rainfall is higher than in the surrounding lowlands which lie within the drier parts of the Guinea savanna, and reaches 2000 mm yr^{-1} in the wetter south-west. Temperature reflects the higher elevation and is lower than in the surrounding areas with

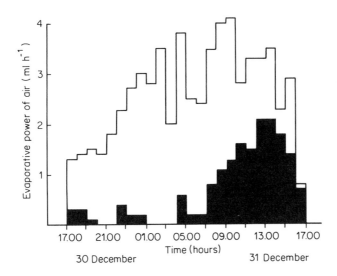

Fig. 10.3. Djebobo massif, Ghana. Diurnal march of the evaporative power of the air at 10 cm above ground in the bottom grassland (black) and the summit tree steppe (white) during harmattan weather on 30–31 December 1965. (After Jeník and Hall, 1966.) Reproduced by permission of The Journal of Ecology.

monthly minima from 15.5 to 18.5 °C and maxima from 27.5 to 30.5 °C.

Though the vegetation of the Jos Plateau is distinct it is not montane. Heavy grazing, shifting cultivation, and mining have destroyed most of the dry forest which originally covered the area. As a result soil and land erosion are a serious problem. Grassland now forms a continuous cover except in the watercourses where riparian forest, or remnants of it, occur in the wetter areas. The flora of the plateau is dominated by lowland savanna species. However, it is enriched by a large number of species characteristic of East and Southern Africa, some of which also occur in Cameroun and other upland areas of West Africa. The presence of this interesting element in the flora of the Jos Plateau is due to the similar climate and elevation of these regions, all of which lie within the same continuous vegetation belt of dry savanna and savanna woodland. Another feature of the flora of the plateau is the abundance of succulents, both the tree-like euphorbias and smaller succulents such as species of *Caralluma*, *Huernia*, *Sarcostemma*, *Kleinia*, and *Aloe*.

4. The Cameroun range

The largest and most important montane system in West Africa is the Cameroun range which culminates at its southern end in the semi-active volcano of Cameroun Mountain (4072 m) and the offshore dormant volcanic island of Equatorial Guinea, Bioko, attaining 2850 m. The range is the result of a volcanic rift in the granite Pre-Cambrian rock which underlies the whole of West Africa. The rift stretches north-east along the Nigeria/Cameroun border and the range gradually decreases in elevation as it extends through the Bambuta Mountains, the Bamenda Highlands, and the Mambila Plateau, as the Adamaoua range which curves eastwards through the interior of the Cameroun Republic and into Chad. A northern arm of the range extends along the Nigeria/Cameroun border towards Lake Chad as the Mandara Mountains. Outliers of the system include the Obudu Plateau and Mount Sankwala on the southern Nigeria/Cameroun border, and the Vogel Peak and Yola Plateau further north. In the southern part of the range, on Cameroun Mountain and Bioko, there is no dry season and diurnal and monthly fluctuations in temperature and humidity, at least on the lower and mid-slopes, are within a narrow range. Near sea-level, temperatures average 22–29 °C and humidity 85–90 per cent. The mountains force moisture-laden winds from the Atlantic Ocean upwards into cooler strata and create exceptionally wet conditions on the windward side. The range cuts across the vegetation belts of West Africa and extends to the edge of the dry Sahel region. Except in the far north, the lower slopes and steep-sided valleys are forested due to the increased rainfall induced by the mountains. Montane grassland occurs on even the most southerly peaks, the montane forest giving way to grassland at about 2800 m.

The volcanic origin of the Cameroun range makes it unique among the mountains of West Africa. Though only Cameroun Mountain is active, extinct volcanic craters, cones, and lava flows are everywhere apparent. Some of the craters are filled with beautiful lakes surrounded by dense forest. Occasionally, poisonous volcanic gases are discharged and produce 'miraculous' kills of fish in these lakes. The porous volcanic rocks weather rapidly and produce fertile soils. However, because most of the rock is so porous, drainage is rapid and the vegetation of many of the elevated areas is subject to severe seasonal drought.

Only on the high mountains of Cameroun, in West Africa, is the full range of montane vegetation zones present. Lowland rain forest extends up the lower slopes to about 1000 m. It is heavily exploited for agriculture and lumber, with slash and burn farming and extensive plantations of oil palm and bananas. It gives way gradually to montane vegetation, the distinctive zones of which are well described by Keay (1953, 1955).

Evergreen montane forest — mist forest or cloud forest — extends to about 2000 m in a belt around the mountains which is enveloped by mist and cloud for long periods and for some part of most days throughout the year. The tree canopy is irregular and broken. It includes such trees as *Polyscias fulva*, *Schefflera abyssinica*, *Ficus chlamydocarpa*, *Entandrophragma angolense*, and *Tabernaemontana brachyantha*. The almost continuous high humidity of this zone results in a profusion of mosses, liverworts, and epiphytes such as orchids, ferns, begonias, peperomias, and club-mosses festooning the branches and trunks of the trees. Also the thick evergreen leaves of the understorey trees and shrubs are covered with epiphyllous liverworts which reach their optimum development in this zone. This also is the zone where tree ferns (*Cyathea manniana*) are locally dominant in many of the gullies.

Above 1800–2000 m the forest becomes drier and more stunted and broken. Hanging lichens begin to make an appearance, festooned from the branches and leaves of the trees. Different species of trees also occur in this zone — *Syzygium staudtii, Schefflera mannii, Nuxia congesta,* and *Pittosporum mannii* — and a dense forb vegetation entangles breaks in the forest, dominated by *Mimulopsis violacea,* various species of *Impatiens* and *Rubus pinnatus.* This upper montane forest zone opens abruptly into grassland at about 2800 m as one scrambles through the tangle of low trees, shrubs, and rank herbs into the open fire climax of the lower montane grassland. This is dominated by tall species of grass 1–1.5 m high of which *Loudetia camerunensis, Andropogon distachyus,* and *Pennisetum monostigma* are the most characteristic. The forest edge has many flowering shrubs and rank herbs which are ablaze with colour at the end of the rains and in the early dry season (October–December). These include species of *Pentas, Lobelia, Pycnostachys, Coreopsis, Anisopappus, Satureja, Indigofera,* and *Senecio.* This tall montane grassland is characterized by scattered gnarled trees of *Agaurea salicifolia* and, in sheltered hollows and gullies, woodland and thicket made up of this species and others among which the commonest are *Lasiosiphon glaucus, Hypericum lanceolatum, Philippia mannii, Adenocarpus mannii,* and *Myrica arborea.* Keay (1953) considers that this type of woodland is the climax vegetation in the lower montane grassland zone, but excessive burning from man-made fires has destroyed the woodland and allowed the grasses to dominate. This woodland is very similar to a well-defined zone on the mountains of East Africa — the Ericaceous zone. Particularly in the damper gullies it is the home of many temperate plants — genera such as *Geranium, Thalictrum, Cerastium, Stellaria, Rumex, Galium, Rubus, Viola, Sanicula, Poa,* and *Plantago.* Few areas remain, either on Cameroun Mountain or in the Bamenda Highlands, where this type of woodland persists over any extensive area. These few remaining areas survive where natural fire breaks provide protection. In the Bamenda area closed communities of the mountain bamboo (*Arundinaria alpina*) occur within this zone. A similar bamboo forest is a characteristic and well-developed feature on the mountains of East and Central Africa.

This lower montane grassland, characterized by its tall grasses and scattered trees, merges into an upper zone of grassland at about 3000 m. Here the grasses are shorter (up to 0.5 m high) and form very tight dense tussocks with cushions of mosses and lichens between them. Trees are absent. Almost all the plants are xerophytes. The grasses have inrolled leaves and the young shoots are protected in the dense tightly packed bases of dead leaves. Other plants include the heather, *Blaeria mannii* — a small shrub with tiny leaves in which the stomata are deeply sunk in the underside of the leaves, the fleshy-leaved *Senecio clarenceanas*, and the very woolly *Helichrysum mannii* and *H. cymosum*. The dominant grasses in this zone are *Pentaschistis pictigluma*, *Deschampsia mildbraedii*, *Koeleria cristata*, and *Andropogon dummeri* with the small ephemeral *Aira caryophyllea* abundant on the bare ground between the grass tussocks.

The summit of Cameroun Mountain (above 3500 m) consists of cindery lava from recent eruptions. Vegetation is sparse and the area has the appearance of a semi-desert. However the 'rock' is encrusted with mosses and lichens which are well adapted to withstand the extreme and prolonged desiccation to which the area is subject. Few vascular plants manage to establish themselves here, but there are scattered clumps of *Pentaschistis pictigluma* and *Senecio clarenceanus* on the fairly recent lava flows and a dense cover of tussocks of *Deschampsia mildbraedii* on older areas.

The environment on the upper slopes and summit of Cameroun Mountain is extreme and most unfavourable to plant growth. Diurnal fluctuations are much greater than seasonal ones. Night temperatures frequently fall to below freezing; frost and hail are common and icicles form on rocks and the vegetation, but the sun rapidly raises temperatures during the day, particularly in sheltered places close to the ground. Solar insolation also fluctuates wildly as passing clouds and banks of mist blot out the sun and cause temperatures to plummet. Humidity also fluctuates greatly with long periods of the year when mist and cloud cover the summit almost continuously and everything is drenched in condensation. Rainfall is low, for the summit is above most of the rain-producing clouds. Also water drains away instantly in the porous lava, and the vegetation and surface of the ground dry out rapidly in the intense solar radiation and wind. Outside the Arctic and Antarctic there can be few more inhospitable places for plant growth than the exposed summit of Cameroun Mountain.

The Cameroun range of mounains emerges from the tropical rain forest region into the Guinea savanna to the south-west of Bamenda and then becomes progressively drier to the north and east. Forest, however, occurs in sheltered valleys and gullies as far north as Vogel Peak where Hepper (1961) comments on the occurrence of fine gallery forest along the streams which flow from the higher ground. The Vogel Peak area, like the other northern parts of the Adamaoua range, experiences a severe and prolonged dry season. Despite this and the continuity of grassland from the lowland savanna to the top of the peaks, montane communities occur as far north as Vogel Peak which rises to about 2040 m. The northern limit of the range, the Mandara Mountains, with peaks rising to 1442 m is not made up of recent volcanic rocks but is granite. It is heavily grazed and burnt and does not support montane vegetation, though it has a number of montane species in its flora.

An outlier of the Cameroun range in southern Nigeria — the Obudu Plateau and Sonkwala Mountains — has been the subject of several studies: Hall (1971), Hall and Medler (1975), Keay (1979), etc. The highest peak is Mount Koloishe, 1920 m. The area rises out of the lowland forest and savanna mosaic and is forested in the valleys and on some of the steep slopes. The top of the plateau consists of low, rolling, grass-covered hills which burn fiercely each year. It is now used as a cattle ranch. The grassland is about 1 m high and dominated by *Loudetia simplex*. Stunted savanna trees (*Terminalia, Vitex, Maytenus,* and *Cussonia*) are scattered through it, together with the shrub *Protea madiensis*. The flora has many species associated with the lower montane grassland of Cameroun. Epiphytes abound on the trees and boulders, both in the grassland and the forest. Rainfall is high on the plateau (4300 mm yr^{-1}) — about twice that in the neighbouring lowlands. However, a pronounced dry season, coupled with annual burning, are the factors determining the presence of the grassland.

Montane forest occurs in the upper valleys and as enclaves in the grassland of the plateau. Many of the trees are common lowland species, for example *Piptadeniastrum africanun, Carapa procera,* and *Sterculia tragacantha,* but montane species also occur, including *Pittosporum mannii, Polyscias fulva, Maesa lanceolata,* and *Nuxia congesta,* along with tree ferns (*Cyanthea camerooniana* and *C. manniana)* and herbs such as *Homalocheilos ramosissimus, Pycnostachys meyeri, Crassocephalum vitellinum,* and *Sonchus angustissimus.* The vegetation and flora are, as one would expect, very similar to those of the nearby Cameroon range.

The mountains of West Africa, though they make up but a small portion of the total land area, are a distinct, very important, and unique ecological region. They are covered with very different types of vegetation from those found in the lowlands and support distinct and in some respects unique floras and faunas. To the biologist they are an open-air laboratory in which he can study the ecological factors which maintain their characteristic plant and animal communities, and the evolutionary processes which have given rise to their unique floras and faunas. Unfortunately these areas are under increasing pressure from human activity as mining, logging, hunting, grazing, and cultivation encroach upon them. It is most important that representative areas of adequate size be preserved for future study and for the valuable resources which they contain. These mountains with their own peculiar floras and faunas are a part of our human heritage and must be preserved for the enrichment and enjoyment of future generations.

References

Adam, J. G. (1958). *Elements pour l'étude de la végétation des hauts plateaux du Fouta Djalon. Part I.* Bureau des sols, Direction des Services Économiques, Gouvernement Général de l'Afrique Occidentale Français, Dakar.

Adam, J. G. (1971). Flore descriptive des Monts Nimba. I. *Mém. Mus. Nat. Hist. Nat.*, n.s. **20**. Paris, 528 pp.

Boughey, A. S. (1953). The vegetation of the mountains of Biafra. *Proc. Linn. Soc. Lond.*, **165**, 144–150.

Cole, N. H. Ayodele (1974). Climate, life forms and species distribution on the Loma Montane grassland, Sierra Leone. *Bot. J. Linn. Soc.*, **69**, 197–211.

Hall, J. B. (1971). Environment and vegetation on Nigeria's highlands. *Vegetatio*, **23**, 339–359.

Hall, J. B., (1973). Vegetational zones on the southern slopes of Mount Cameroon. *Vegetation*, **27**, 49–69.

Hall, J. B., and Medler, J. A. (1975). The botanical exploration of the Obudu Plateau area. *Nigerian Field*, **40**, 101–115.

Hedberg, O. (1964). Features of Afroalpine plant ecology. *Acta Phytogeographica Suecica*, **49**. Uppsala, 144 pp.

Hepper, F. N. (1961). Plants of the 1957–58 West African Expedition: I. *Kew Bull.*, **15** (1), 56–66.

Jacques-Felix, H. (1968). Evolution de la végétation au Cameroun sous l'influence de l'homme. *J. Agric. Trop. Bot. Appl.*, **15**, 350–356.

Jaeger, P. (1969). Première esquisse d'une étude bioclimatique des Monts Loma (Sierra Leone). *Bull. Inst. Fond. Afr. Noire*, **31**, sér. A(1), 1–21.

Jaeger, P. (1971). Le Massif des Monts Loma (Sierra Leone). IV. Observations climatologiques. *Mém. Inst. Fond. Afr. Noire*, **86**, 73–112.

Jaeger, P., and Adam, J. G. (1975). Les forêts de l'étage culminal du Nimba Libérien. *Adansonia*, sér. 2, **15** (2), 177–188.

Jeník, J., and Hall, J. B. (1966). The ecological effects of the Harmattan wind in the Djebobo Massif (Togo Mountains, Ghana). *J. Ecol.*, **54**, 767–779.

Keay, R. W. J. (1955). Montane vegetation and flora in the British Cameroons. *Proc. Linn. Soc. Lond.*, **165** (2), 140–143.

Keay, R. W. J. (1959). *An Outline of Nigerian Vegetation.* Government Printer, Lagos, Nigeria, 55 pp. + map.

Keay, R. W. J. (1979). A botanical study of the Obudu Plateau and Sonkwala Mountains. *Nigerian Field*, December **1976**, 106–119.

Letouzey, R. (1968). Etude phytogéographique du Cameroun. *Ency. Biol.*, Vol. 69. Editions P. Le Chevalier, Paris, 511 pp.

Morton, J. K. (1961). The upland floras of West Africa — their composition, distribution and significance in relation to climatic changes. *Comptes Rendus de la IVe Réunion Plenière de l'AETFAT*, 391–409.

Morton, J. K. (1972). Phytogeography of the West African mountains. In D. H. Valentine (Ed.), *Taxonomy, Phytogeography and Evolution*, pp. 221–236. Academic Press, London.

Richards, P. W. (1964). *The Tropical Rain Forest.* Cambridge University Press, 450 pp.

Richards, P. W. (1963). Ecological notes on West African vegetation. III. The upland forests of Cameroons Mountain. *J. Ecol.*, **51**, 529–554.

Schnell, R. (1952a). Contribution à une étude phytosociologique et phytogéographique de l'Afrique occidentale: les groupements et les unités géobotaniques de la région guinéenne. *Mém. Inst. Fr. Afr. Noire*, **18**, 41–234.

Schnell, R. (1952b). Végétation et flore des Monts Nimba. *Vegetatio*, **3**, 350–406.

Schnell, R. (1952c). Végétation et flore de la région montagneuse du Nimba. *Mém. Inst. Fr. Afr. Noire*, **22**, 1–604.

Schnell, R. (1961). Contribution à l'étude botanique de la Chaîne de Fon (Guinée). *Bull. Jard. Bot. Bruxelles*, **31**, 15–54.

Walter, H. (1951). *Einführung in die Phytologie*. III. *Grundlagen den Pflanzenverbreitung*, Part 1. Standortslehre, Stuttgart.

Plant Ecology in West Africa
Edited by G. W. Lawson
© 1986 John Wiley & Sons Ltd

CHAPTER *11*

Change of vegetation with time

M. Adebisi Sowunmi
Department of Archaeology, University of Ibadan,
Nigeria.

I. Introduction

The vegetation zones in West Africa today have been discussed in Chapter 1. These vegetation types are the major plant cover of climatic zones. It will be shown in this chapter that these plant communities have been subjected to varying degrees of changes through time. Such changes have been brought about by climatic and other environmental oscillations as well as the activities of man and his domesticated animals.

1. Time perspective

It is necessary first to identify a temporal framework in considering what vegetational changes have occurred through time in West Africa.

The radiocarbon dating method was started in 1950 and this date is referred to as 'the Present'. Years before the Present are indicated as BP. It is convenient to start considering vegetational changes in West Africa from the early Tertiary period which commenced about 65 million years ago (i.e. c. 65 m. y. BP).

This choice was made because by the early tertiary period the supercontinent known as Gondwanaland, comprising Africa, Antarctica, Australia, India, and South America had completely broken up, and Africa then became a separate continent (see Fig. 11.1).

2. Environmental background to vegetation changes

The vegetation of any area is an integral, indeed a basic, component of the ecoystem and is sensitive to any changes in the established homeostatis or equilibrium. Consequently, vegetation changes are themselves a response to and a reflection of variations in one or more of the other factors of the

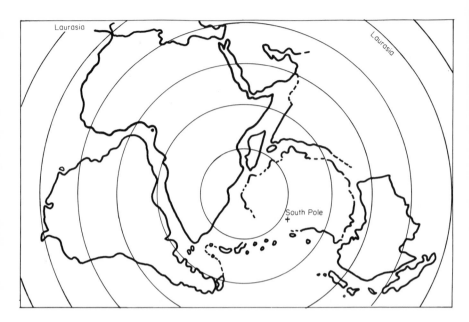

Fig. 11.1. A simplified reconstruction of Gondwanaland. Simplified from du Toit, 1937, Fig. 7. (After Hallam, 1973, Fig. 11.) Reproduced by permission of Oxford University Press.

environment. Various studies, such as geological, geomorphological, and biological, have shown that environments in West Africa, during the time period under review, were not stable but were subjected to crustal movements such as subsidence and uplifting of land masses (epeirogeny), mountain building (orogeny), faulting and volcanicity. In addition, there were phases of erosion and deposition of sediments, as well as fluctuations in the levels of rivers, lakes, and the sea — the Atlantic Ocean. These phenomena were not peculiar to West Africa but were world-wide. The environmental phenomena and changes which seem to have influenced the vegetational history of West Africa will now be considered in some detail as a means of further elucidating some of the processes involved in the vegetation dynamics of the region through time.

(a) Relationship between geomorphology and climate
Before considering the various environmental changes, it is necessary to highlight the relationship between geomorphology and climate in West Africa. The two prevailing wind systems which have a dominant influence on the climate of the region are the Atlantic, moisture-laden south-westerly monsoon and the continental, dry, north-east trade winds. Long-term and periodic

variations both in their vigour and the consequent relative positions of the Intertropical Convergence Zone (ITCZ — i.e. the zone where these two wind systems converge; it is at its maximum northward extent in July and maximum southward extent in January, Fig. 11.2a+b) contribute to climatic changes. There is good reason to believe that at least the south-westerly monsoon wind system has probably blown from the south-west since the Palaeocene period (c. 65 m.y. BP) (Burke, 1972). By that time, the Atlantic Ocean between South America and Africa already had a shape similar to what it is today, though it was smaller.

There is a great deal of evidence from palaeobotanical, palaeontological, and geomorphological studies that from the Neogene (beginning c. 26 m.y. BP) till about 3000 years ago — when human impact on the environment made it difficult to isolate the effects of natural phenomena — the climates in West Africa were characterized mainly by alternating wet and dry phases. As Grove (1958, p. 532) succinctly put it '. . . the relief and soils (of the Sudan and Sahel zones of West Africa) are the outcome of deep weathering, water erosion and the formation of extensive lakes in humid periods; and of wind action and dune-building in the intervening arid periods'. This close relationship between geomorphology and climate also holds true for the more southern, Guinea zones (savanna and forest), except that there is no evidence as yet of the formation of extensive lakes. Furthermore, in the entire region, there was the formation of clayey and alluvial soils as well as pronounced fluviatile discharges in the wet periods with drastic reduction or complete cessation of flow when the climate was dry (Table 11.1). Furthermore, as outlined by Herbrard *et al.* (1971) and as borne out by the results of interdisciplinary studies, arid periods in the tropics are marked by, among other things, a predominance of erosion over soil formation culminating in the formation of extensive and massive deposits of wind-blown sands or dunes referred to as ergs. Wet phases, on the other hand, are characterized by features such as formation of lakes, channelization of a network of rivers, and soil development.

(b) Successive landscapes
During the Palaeocene (c. 65–54 m.y. BP), the sea-level rose and its waters consequently flooded parts of the continent. Such invasion of land by the sea is known as sea transgression. The Atlantic transgressed over parts of the southern sedimentary basin in the area stretching from Ivory Coast to Nigeria (Tastet, 1975), while the 'Saharan sea' (Petters, 1978) transgressed over parts of north-western Nigeria and south-western Niger (cf. Kogbe and Sowunmi, 1975).

The period between the Eocene and Miocene (c. 54–26 m.y. BP) was one during which there were variations in the continental depositional environments. New landscapes emerged due, at least in part, to sea regression, i.e. a lowering of sea-level following a diminution of its water volume, resulting in a

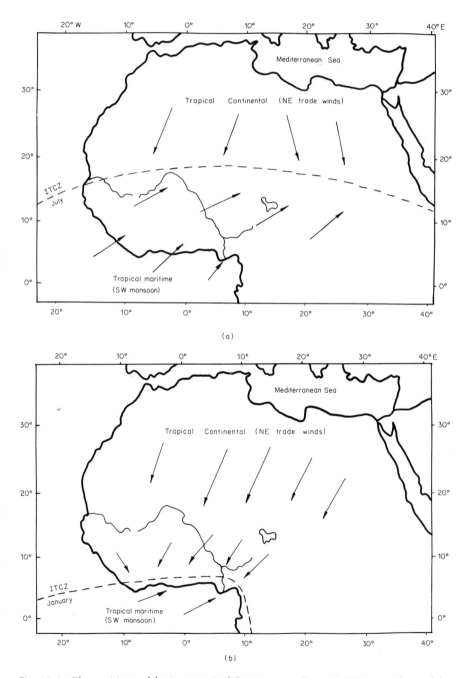

Fig. 11.2. The positions of the Intertropical Convergence Zone (ITCZ) in (a) July, and (b) January; prevailing winds in West Africa are also indicated.

Table 11.1.   Summary of fluctuations in sea-level, relative to today's zero, and inferred concomitant climates in West Africa. Details are provided in the text. '?' indicates where no data were available

Geological period	Time-span (in years bp)	Sea-level	Inferred climate
Late Quaternary	1700 to the Present	Slight fall to reach today's level (sea regression)	Perhaps as today, but drier than at 5,500 years bp
Late Quaternary	2000–1700	Rise in level (sea transgression)	Perhaps as today, but drier than at 5,500 years bp
Late Quaternary	4000–2000	Fall in level (Tafolian regression)	Drier
Late Quaternary	5500–4000	Rise in level (Nouakchottian transgression)	Very Wet
Late Quaternary	12 000–5500	Rise in level (sea transgression)	Wet
Late Quaternary	?–12 000	Rise in level (but to a level below today's zero)	Dry
Late Quaternary	30 000–?	Fall in level (Ogolian II regression)	Dry
Late Quaternary	>40 000–30 000	Rise in level (Inchirian transgression)	Wet
?Middle to ?Late Quaternary	?	Fall in level (Ogolian I regression)	Dry
Middle Quaternary	?–70 000	Rise in level (Aioujian transgression)	Wet
Middle Quaternary	?700 000–?	Fall in level (Akcharian regression)	Dry
Early to Middle Quaternary	1m.–?700 000	Rise in level (Tafaritian transgression)	Wet
Early Quaternary	2m.–1m.	?	?
Miocene and Pliocene	26m.–2m.	?Fall in level	Seasonal, drier than preceding phase
Eocene and Oligocene	54m.–26m.	Fall in level (sea regression)	Seasonal, predominantly dry
Palaeocene	65m.–54m.	Rise in level (sea transgression)	Wet

concomitant exposure of land covered during the preceding transgressive phase or phases. A good illustration of this phenomenon is provided by the palaeogeography of southern Nigeria. The Nigerian coastline was just south of Onitsha (c. 6°N) in the lower Middle Eocene. Thus the terrain that forms the bulge in the lower Niger basin, including the present Niger delta (this bulge lies within c. 4° 20' to 6°N and 5° to 9°W) was submerged by the sea at that time. However, subsequent sea regressions led to the successive emergence of land until the coast advanced seaward beyond its present limit in the Pliocene/ Pleistocene period (Short and Stauble, 1967) (see Fig. 11.3).

Secondly, there were erosional phases as exemplified by the following phenomenon. In West Africa there is a geological formation of clayey sandstones known as the Continental Terminal, and believed to be of Miocene to Pliocene age (Tastet, 1975). At different localities it overlies sediments of ages ranging from Palaeocene, Eocene, and Lower Cambrian to Upper Pre-Cambrian, an evidence of a break or gap in deposition during the time interval between the laying down of the underlying sediments on the one hand and the Continental Terminal on the other. Such a break within geological layers is termed an unconformity and marks a period either of non-deposition or erosion of deposited materials. The unconformity below the Continental Terminal is considerable and is said to indicate a phase of tremendous erosion (cf. Tastet, 1975).

The Miocene and Pliocene, on the other hand, were periods of soil formation or deposition during which latosols — represented by ferric oxide cementation in the lower portion of the coastal part of the Continental Terminal (Tastet, 1975) — as well as clayey sands of the formation itself developed.

The earliest Quaternary deposits (overlying the Continental Terminal) known from the coastal part of the sedimentary basin of the Ivory Coast, Togo, Benin, and Nigeria are the 'Terre de Barre'. These comprise mature clayey sandstone of continental origin. The clay is kaolinite which derives from well-drained latosols and is therefore indicative of a long period of soil formation inland. Their specific age is not known for certain, but by correlation with similar desposits in Senego-Mauritania, they have simply been assigned an early Quaternary 'Quaternaire Ancien' age (Tastet, 1975). On the basis of the homogeneity and fineness of the sediment (more than 50 per cent consists of particles less than 50μm in diameter) as well as the 'rolled' surface of the quartz grains, it has been suggested that they were not likely to have been transported by wind action but rather carried in suspension and deposited by water (Tastet, 1975). This is indicative of high rates of fluvial activity.

Continental deposits of the latter part of the Quaternary in West Africa are both better dated and well defined. Some specific examples of landscapes during both arid and wet phases will next be highlighted, as background to considering the various vegetation communities.

(i) Landscapes of arid phases during the Late Quaternary period.   In the Sudan and Sahelian areas, stretching from Senego-Mauritania through Upper Volta, Niger, and Nigeria to Chad are successive series of extensive and massive dunes or ergs, which formed the landscape during several severely arid phases of the late Quaternary. These dunes are indeed said to extend 'from the Atlantic

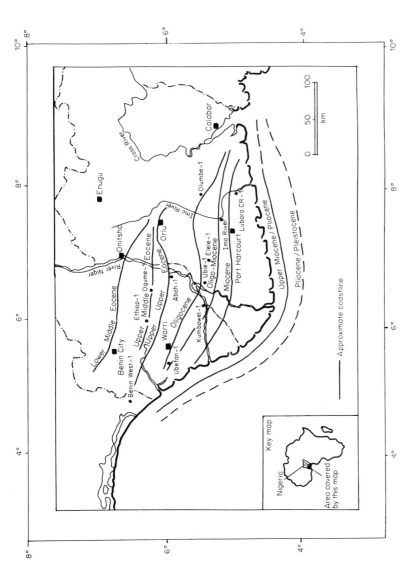

Fig. 11.3. Paleogeography of Tertiary Niger delta — stages of delta growth. (After Short and Stauble, 1967, Fig. 6.) Reproduced by permission of the American Association of Petroleum Geologists.

to Ethiopia' (Chamard and Courel, 1975, p. 63). On the basis of their numerous morphological characteristics it has been possible to correlate these deposits.

In Senego-Mauritania three distinct formations of dunes were recognized, namely, ergs I, II, and III (Leprun, 1971). Erg I is analogous and contemporaneous with the 'ancient erg' of Niger and Upper Volta, which in turn is similar to and contemporaneous with the dead erg of Hausaland (North-east Nigeria) and that of Chad (cf. Leprun, 1971; Chamard and Courel, 1975; Grove, 1958, Figs. 1 and 2). These dunes were formed during an arid period in the late Pleistocene, some time before 40 000 years BP.

Erg II is similar in character to the 'recent erg' of Niger and Upper Volta as well as that of the red dunes of Senego-Mauritania. The deposition of this erg corresponds in time with the lowering of sea-level known as the Ogolian sea regression of Senego-Mauritania dated 20 000–18 000 years BP. This sea regression was world-wide and marked the most severe phase of the last glaciation when sea-level fell to -120 m in the North Atlantic and to -110 m in the equatorial Atlantic (Martin, 1972). It is significant to note that erg II is the largest of the series, suggestive of a greater severity of aridity. During the arid period between 20 000 and 12 000 years BP, as a result of the formation of massive dunes, the desertic zone extended down to 13°N latitude, whereas today the southern most limit of active dune formation is 16°N latitude (Servant and Servant-Hildary, 1972) (see also Fig. 11.4). Furthermore, there is palynological evidence that the Senegal River either completely ceased to flow or was greatly reduced during that period; at least it no longer discharged into the Atlantic (Rossignol-Strick and Duzer, 1979a, b) its mouth having been blocked by dunes. Other rivers in the region were presumably similarly affected. Lake levels in intertropical Africa — from Mauritania to the eastern rift — were very low, between 21 000 and 12 500 years BP, with minimum levels attained from 15 000 to 13 000 years BP, confirming the prevalence of a very arid climate then (Street and Grove, 1976).

The third formation of dunes in Senego-Mauritania, erg III, is dated to between 7000 and 5500 years BP. A similar formation has also been found in Central Niger (Leprun, 1971). There was a re-establishment of arid and desertic conditions in several areas north of 15°N latitude between 4000 and c. 2000 years BP; this period is partly coincident with a lowering of the sea-level by about -3.5 m along the Mauritanian coast, termed the Tafolian regression (Hebrard et al., 1971; cf. Einsele et al., 1977).

Geomorphological evidence of arid phases in the Quaternary period are not confined to the Sudan and Sahelian zones alone, but have also been found in the more southern and moister Guinean zones. However, such geomorphological data as well as C-14 dates are comparatively very few and have come from a limited number of locations only. Particularly scanty are data from lowland forest regions.

A very significant contribution to our knowledge of environmental conditions during the latter part of the Quaternary period in the forest region of West Africa was made by Talbot and Delibrias (1977, 1980), on the basis of their studies of Lake Bosumtwi in Ghana. This is a crater lake, which lies in the lowland rain forest zone of southern Ghana near the ecotone with the northern savanna, and was created by a meteoritic impact that occurred about 1.3 m.y. BP. Records of fluctuations in the level of this lake provide a very reliable indication of environmental conditions both because of its location and because it is fed by over 37 inflowing rivers while it has no outflow (Whyte, 1975). However, there is an indication that it probably overflowed its banks during one or several periods of maximum lake level in the past through the only valley cut through the crater rim (Talbot and Delibrias, 1977).

Four distinct phases of low lake levels or regressions have been identified, namely between 12 000 and 9000 years BP, around 8000 years BP, 4500–4000 years BP, and some time during the last 1000 years (Talbot and Delibrias, 1980). The regression of 4500–4000 years BP was the most drastic, when the lake level fell below what it is today. It is remarkable that these regressions were synchronous with those recorded for other tropical African lakes in the humid, semi-arid, and arid regions in West and East Africa. This synchronaeity of low lake levels is a strong indication that such fluctuations in lake levels reflect climatic changes that occurred throughout the African tropics (cf. Talbot and Delibrias, 1980). The causes of regression in Lake Bosumtwi are complex and as yet incompletely understood. It has been suggested that a probable factor might be a hydrological deficit due to the combined effects of a rise in temperature concomitant with a slight decrease in precipitation (Talbot and Delibrias 1977, 1980). However, the deposition of aeolian sands in South-east Ghana during the period 4500–4000 years BP clearly shows that the regression then was caused by aridity and the increased vigour of north-east trade winds.

A preliminary lithological study of a core from the Niger delta in Nigeria (Sowunmi, 1981a) showed that some time before 35 000 years BP the rivers that drained into the delta carried and subsequently deposited very coarse-grained sand and gravel which were eroded as the streams and rivers in that drainage system were rejuvenated. These coarse materials — found at a depth of −36 m — very probably accumulated under dry conditions during which, in the absence of an effective vegetation cover, the finer components would have been removed by sheet floods and winds (cf. Burke and Durotoye, 1970). Thus an arid phase, prior to this stream rejuvenation, and occurring probably before 40 000 years BP is indicated. It is significant to recall that in the Sudan and Sahelian zones of West Africa, a dune formation, erg I of Senego-Mauritania, is believed to have been built some time before 40 000 years BP. The climate must have been generally dry throughout the West African region, even as far south as the Guinean forest at that time. The phase of gravel deposition mentioned above might have been synchronous with the formation of the older of two

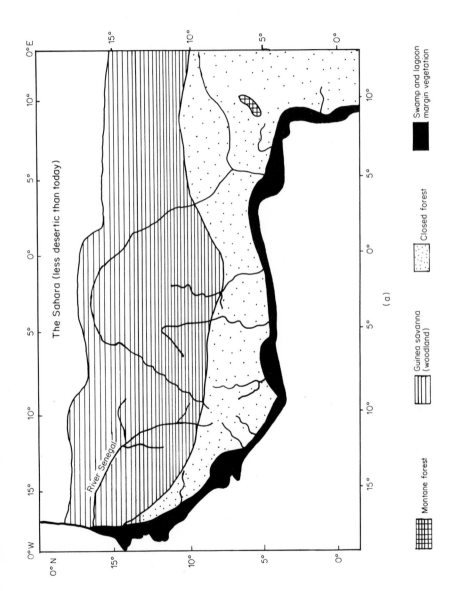

(a)

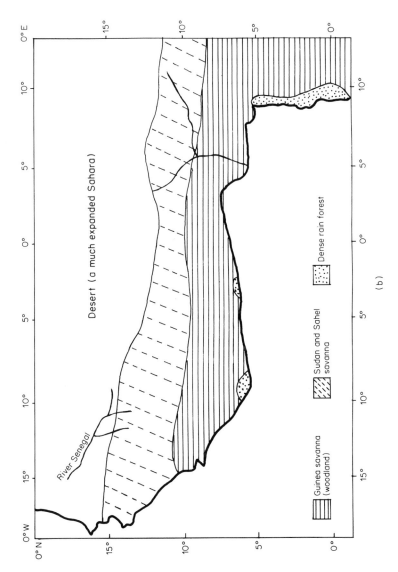

Fig. 11.4. Provisional vegetation maps of West Africa during (a) a wet phase, and (b) a dry phase in the late Quaternary period (based by the author mainly on biogeographic-al, palaeobotanical, and fossil geomorphological evidence).

pediment gravel deposits widely distributed in south-western Nigeria, tentatively dated to c. 60 000 years BP (Burke and Durotoye, 1970). This pediment gravel is extensively cemented by geothite (ferrous hydroxide) and red haematite (ferric oxide) to form an iron, tropical hard pan. The younger, uncemented, pediment gravel deposit is assumed to have been formed c. 20 000 years BP when conditions in the Sudan and Sahel zones throughout West Africa were dry, with the formation of massive dunes. The formation of these pediment gravels in south-western Nigeria resulted from a major climatic change as 'conditions [for their formation] today begin at least 500 km north of south-western Nigeria' (Burke and Durotoye, 1970, p. 93).

Finally, in a part of West Central Cameroun that is at present under a wet tropical climate, there occur canyons whose formation cannot have taken place under the present climatic conditions. A canyon is a deep, steep-sided river valley normally found in arid or semi-arid regions. Their formation can therefore only be attributed to a dry period some time in the past. Furthermore, in the upper part of river terraces, which form the embankment of these canyons, are thick sediments of colluvial clay which seemed to have been weathered *in situ*. The joint occurrence of these canyons and colluvial clay are indicative of the prevalence of arid phases, characterized by low rainfall and weak fluvial activity, during their formation. Two such phases have been dated to c. 21 000 and 2000 years BP (Hurault, 1972). This additional evidence of a dry phase c. 21 000 years BP is very noteworthy and will be referrred to later.

(ii) Landscapes of wet phases during the Tertiary through to the late Quaternary period.   There is widespread and abundant evidence that, in general, arid phases alternated with wet in the African tropics. Thus, for example, in the latter part of the Quaternary, sand-dunes intercalate either with lacustrine deposits as in the Chad basin (Servant and Servant-Hildary, 1972) or with beach fossils from a sea transgression as in Senegal (Grove, 1958), while the most recent dunes, formed 7000–5500 years BP in Senego-Mauritania, are overlain by thin soils over which is a vegetation cover today (Leprun, 1971); furthermore the tropical forest of Zaire is rooted in Pleistocene dunes, whose specific age is unknown.

Environmental conditions during the wet phases were on the whole similar to present-day ones, but certain periods were wetter than now. A further and very important distinction is that the Sahelian zone had more favourable climates with formation of lakes.

In Senego-Mauritania are deposits of either ferruginous cuirass or calcareous concretions, depending on the nature of the parent rock, dated to Pliocene, Plio-Pleistocene and Middle Pleistocene times (Nahon, 1972). These are considered to be indicative of wet phases since, on the one hand, ferruginous cuirass results from weathering of clay and subsequent accumulation of the released iron sesquioxides under warm and wet climatic conditions, and, on the other, accumulation and subsequent mobilization of calcium occurs in areas with adequate soil and subsoil moisture as well as effective plant cover.

Between Upper Cretaceous and Recent times there were several important volcanic eruptions in Cameroun, the lava of which have overlain and thus preserved the various deposits that characterized the landscape immediately prior to each eruption (cf. Letouzey, 1968). The study of these deposits has contributed immensely to our knowledge of the geomorphology of the more southern, inland portion of the Guinean zones. The 'Lignite group' of Nigeria of Middle to Upper Tertiary age is represented in southern Cameroun where it is overlain by basaltic lava. The deposits comprise clay, alternating with gravel and lignite, indicative of the prevalence of wet conditions at the time of deposition. In the Mbos Plain (also in South-west Cameroun) are lacustrine, clayey, and alluvial deposits, intercalated with volcanic lava and dated to Upper Tertiary/Quaternary times. Similarly, finely bedded clay overlain by stratified clayey sandstone and apparently fluviatile gravel occur in a valley near Hangloa (North-central Cameroun). These sediments have been relatively dated on palynological evidence to the Upper Tertiary/Quaternary period.

In the Lower Pleistocene period, c. 1 m.y. BP (0.87–1.37 m.y. BP) the Mammelles volcano in the Dakar region in Senegal erupted. Ash tuffs from this eruption covered the surrounding terrestrial landscape and buried its vegetation (Hebrard, 1974; Cantagrel et al., 1976). Subsequently, processes of soil formation and erosion followed, as evidenced by the diverse types of sediments overlying the volcanic tuff, such as calcareous sandstones, alluvial gravel, and clay and red dunes (cf. Lappartient, 1972).

During the wet phase between c. 40 000 and 30 000 years BP (the Inchirian transgression), lagoons were formed in the littoral parts of Ivory Coast, Togo, Benin, and Nigeria. Clayey mangrove swamps developed in the lower Niger delta extending into the lower parts of present-day freshwater swamp. Lakes were also formed in the present-day arid Chad basin.

After the severe aridity of 30 000–12 000 years BP wet conditions were again restored c. 12 000–4500 BP in West Africa, with local fluctuations back to drier conditions. This period was marked by the formation of lakes in parts of the now semi-arid to arid Sahel zones. The Senegal River flowed into the sea once again. The phase was contemporaneous with part of the Nouakchottian transgression (named after Nouakchott, Mauritania) which was at its maximum c. 5000 years BP. During that maximum phase the sea penetrated inland to a varying extent throughout the West African coast, reaching up to about 250 km in Senegal (Michel and Assemien, 1970), thus mangrove swamp was more extensive then than now.

II. Vegetation communities through time

Having outlined the geomorphological and climatic history of West Africa from the Tertiary through to Recent times, the stage is now set for a discussion on the vegetation communities that characerized the landscape at various points in time. The data for the Tertiary through to the beginning of the late Quaternary period are very sketchy for four main reasons. Firstly, it has not been possible to obtain reliable chronometric dates from most of these older deposits as they lie outside the range of C-14 dating. The oldest C-14 date is currently about 80 000 years BP. The few reliable dates available are those from isolated sediments that occur along with volcanic materials which can be subjected to K–Ar dating methods. Volcanic rocks are particularly suitable for K–Ar dating on account of their being rich in potassium while having a very low quantity of atmospheric argon. The K–Ar method extends the dating of fossils to about 50 m.y. BP, i.e. the Eocene. Secondly, some fossil taxa cannot be related to extant ones, thus limiting the degree of identification. Thirdly, some of the sediments have been eroded away thus creating gaps in the record, for example the large unconformity between the Continental Terminal and the stratum immediately underlying it. Fourthly, data are available for only a small area in the region. In contrast, several late Quaternary sediments have been radiocarbon dated. Identification with extant species is less problematic — even though the high species diversity of the tropics is an additional source of difficulty. Being comparatively much younger, these sediments have been subjected to erosion for a much shorter geological time. Though data for the late Quaternary are more widespread, intensity of research and consequently the amount of unpublished works in the region have been very uneven, those relating to the lowland forest areas being particularly scanty.

   Hitherto, there is very little direct botanical evidence for the Tertiary and Quaternary vegetational history of sub-Saharan West Africa. Indeed, in the recently published and otherwise comprehensive work of Flenley (1979) on the geological history of the equatorial rain forest, there is a complete absence of data from lowland tropical West Africa. The reconstruction of successive vegetations that follows here has been based mainly on the works of the following authors: Germeraad *et al.* (1968), Kogbe and Sowumni (1975) for the Tertiary period in Nigeria, and Letouzey (1968) for the Tertiary period in Cameroun, Letouzey (1968), Cameroun; Sowunmi (1981a, b), Cameroun and Nigeria; Hall *et al.* (1978), Ghana; Assemien (1971), Assemien *et al.* (1970), Fredoux and Tastet (1976), Martin (1972) and Michel and Assemien (1970), Ivory Coast; Rossignol-Strick and Duzer (1979a, b), Sahel and Sudan zones; and Hebrard (1974), Senegal — all for the Quaternary period. In view of the latitudinally zonal character of the major West African vegetation types and the overall striking similarity of the species composition within each type — the varying degrees of endemism notwithstanding — data from the area listed above may be extrapolated to those for which data are not yet available, as appropriate.

   Identification of the components of these past vegetation types has been based for the most part on their fossilized pollen and spores (Fig. 11.5); in a few cases fossil leaves and/or fruits were used. It is pertinent to point out some of the limitations of this approach. Firstly only those plant parts which were in areas where they could become embedded in sediments, under conditions conducive to fossilization, have been so buried. Secondly, subsequent to such a burial, there is differential preservation — a function of the degree of resistance to decay processes — whereby those parts less resistant to decay are destroyed. Thirdly, it is not always possible to identify the pollen, spore, or leaf to species level or refer it to any extant species. Thus the ecological significance of such fossils is greatly diminished. Fourthly, it is assumed that the ecological range of plants has not changed since Tertiary times. Nevertheless, in spite of the fact that these micro- and macrofossils provide only a very incomplete or partial picture of the entire vegetation community at a particular point in time, it is still possible to obtain a reasonable idea of the major characteristics of the vegetation provided the fossils used are those of plants whose ecological range is both known and restricted.

For data on the present-day distributions and natural habitats of the species considered, reference was made to Hutchinson and Dalziel (1954–1968) and Keay (1959b).

1.   Palaeocene (c.65–54 m.y. BP)

On account of a major sea transgression in West Africa during the Palaeocene, the southern parts of the region — at least in the Gulf of Guinea, including the Proto-Niger delta that was formed in the Cretaceous — were submerged by the Atlantic Ocean. In addition, there was an inland sea, the Saharan Sea, that inundated parts of North-western Nigeria and South-west Niger. The exposed continental mass must therefore have been less extensive in area than now.

The present mangrove swamp of the lower Niger delta was not in existence, either floristically or with regard to location. Instead of this there was an estuarine swamp vegetation of which the palm *Nypa* was a component. The southern limit of this swamp in Nigeria was c. 6 °N, (on the basis of the palaeogreography of the Niger delta (see Fig. 11.3), a more northern position that today's mangrove swamp which is c. 4 ° 20'N. *Rhizophora* is today the dominant genus in the Niger delta, and the high, arched stilt roots give the creeks their characteristic look (Fig. 11.6). In view of the absence of *Rhizophora*, the aspects of the swamp vegetation in the Palaeocene must have been very different since *Nypa* is a 'prostrate-stemmed, gregarious palm' (Hutchinson and Dalziel, 1968, p. 165) (Fig. 11.6).

There is neither micro- nor macrofossil evidence of the occurrence of grasses in the tropics as a whole during the Palaeocene. Since savannas, by definition, are characterized by an abundance of grasses which predominate over trees, it can be rightly presumed that there was no savanna vegetation in West Africa at this time.

What then was the vegetation on dry land like? It is difficult to tell. However, it seems there was a tropical forest community as evidenced, on the one hand, by the occurrence of a pollen type referred to *Ptychopetalum* spp. (*Anacolosa* type) (Olacaceae), in Nigeria, and, on the other, two fruits of *Canarium schweinfurthii* in Cameroun. Both lines of evidence are admittedly very tenuous. *Ptychopetalum* is today represented in West Africa by only two species: *P. anceps*, a forest shrub in Sierra Leone, Liberia, Ivory Coast, and Ghana, and *P. petiolatum*, a forest shrub or small tree in southern Nigeria and

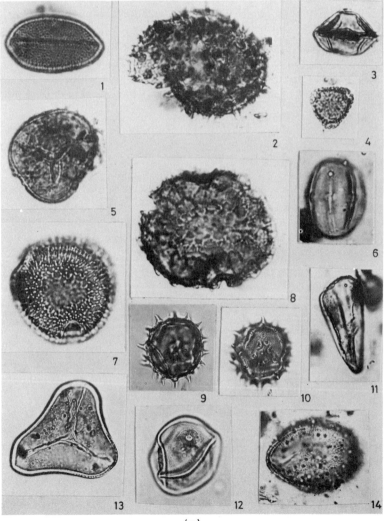

(a)

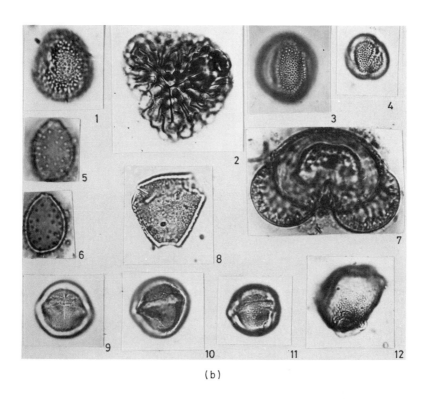

(b)

Fig. 11.5. Some fossil pollen and spores from the Continental Terminal Formation of north-western Nigeria and south-western Niger, and the Niger delta of Nigeria. All × 650 (Photo: Sowunmi.) Reproduced by permission of Mouton Publishers (Division of Walter de Gruyter & Co.).
(a) (1–5) Continental Terminal Formation (Upper Eocene to Miocene): (1) cf. *Kentia* sp.; (2) cf. *Nypa* sp.; (3) *Rhizophora* sp.; (4) *Uapaca* sp.; (5) fern spore. (6–14) Niger delta (late quaternary deposits, 35 000 years BP to the Present): (6) cf. *Butyrospermum paradoxum*; (7) *Canthium* sp. (aff. *C. subcordatum*); (8) *Ceiba pentandra*; (9–10) Compositae (*Tridax procumbens*); (11) *Cyperus conglomerata*; (12) Graminaceae, (13) *Elaeis guineensis*; (14) *Hyphaene thebaica*. (b) Niger delta deposits (contd.): (1) *Ilex* cf. *mitis*; (2) cf. *Lygodium microphyllum*; (3) cf. *Maytenus senegalensis*; (4) *Mitragyna* sp.; (5–6) *Pandanus candelabrum*; (7) *Podocarpus milanjianus*; (8) *Protea* cf. *elliottii*; (9–11) *Rhizophora racemosa*; (12) *Rytigyna senegalensis*.

Cameroun. *Canarium schweinfurthii* is a large forest tree, up to 40 m high, occurring from Senegal through to Cameroun. The spatial extent as well as the density and species diversity of this forest community is not known for certain. However, judging from the paucity of fossil material, it seems the number of species represented was probably rather low.

Fig. 11.6. Upper right and left (1 and 2) characteristic views of the *Rhizophora* swamp forest in the Niger delta today; note the distinctive arched stilt roots of the red mangrove *Rhizophora racemosa*). (Photo: Sowunmi.) (3) *Nypa fruticans* along the Foru River, Northern District, Territory of Papua, New Guinea. (Photo: CSIRO, Division of Land Research and Regional Survey, Canberra, Australia.) (After Tralau, 1964.) Reproduced by permission of K. Svenska Vetenskapsakademien (R. Swedish Academy of Sciences).

2.   Eocene (c. 54–38 m.y. BP)

The same estuarine swamp *Nypa* community of the Palaeocene still persisted along the southern coast. *Rhizophora* was still not represented. In addition, there is evidence of a freshwater swamp or riverine forest in areas not reached by the tidal stream; this community was typified by *Symphonia globulifera*, which rapidly became dominant in the coastal vegetation following its first appearance in the Middle Eocene period.

North-western Nigeria and south-western Niger had a freshwater swamp forest with ferns, *Uapaca* sp. (aff. *U. staudtii*), and the palms *Arega*, *Mauritia*, and *Ravenea*. In the river valleys were *Elaeis guineensis* and the forest tree *Mimusops warneckei*, and perhaps *Uapaca* sp. as well. The pollen grains of *Elaeis guineensis* at this time were monocolpate, i.e. with one elongated aperture, termed a colpus. Pollen grains with a three-slit colpus were found in younger deposits (Miocene) from the Niger delta (Zeven, 1964), indicative of a later evolution of the latter-mentioned type of oil-palm pollen. The floristic composition of these areas in the Eocene is in sharp contrast to what obtains today. These contiguous areas are now in the Sudan savanna zone. There, seasonally flooded grounds in river valleys carry virtually pure stands of the drier-savanna tree *Acacia seyal*, while there are no lowland rain forest trees along the river banks as is the case in the moister southern Guinea savanna. This therefore suggests that there still was no savanna vegetation in Nigera or in the rest of tropical West Africa for that matter. Furthermore, none of the three palm trees (*Arenga*, *Mauritia*, and *Ravenea*) have been recorded as being present in any vegetation zone in Wet Africa today. In fact they have not been found in Miocene deposits, having probably become extinct at the end of the Eocene.

A significant development was the appearance for the first time of Gramineae pollen in the Middle Eocene, albeit in low quantities. This period thus marked the first occurrence of this family in Nigeria, and perhaps in West Africa. However, grasses were still a rather minor component of the vegetation, and their ecological significance then, if any, is difficult to assess.

3. Neogene i.e. Miocene and Pliocene (c. 26–3 m.y. BP)

In southern Nigeria, *Nypa* disappeared at the end of the Eocene and *Rhizophora* occurred for the first time, rather suddenly and in high percentages, in the lowest part of the Miocene deposits. This therefore signified the development of the *Rhizophora* mangrove swamp that now occurs along almost all the Nigerian coastline, as well as most of the rest of the West African coast.

There were also riverine forests, represented by *Ctenolophon englerianus* type pollen. *Ctenolophon englerianus*, a riparian forest tree, up to about 25 m high, today occurs only in southern Nigeria. In addition there were probably freshwater swamp forests, as evidenced by the occurence of the spores of *Lygodium microphyllum*, a climbing fern commonly found in humid marsh and swamp forests of West Africa today, as well as those of the freshwater fern *Ceratopteris*.

Significantly, *Rhizophora* also appeared for the first time in the early Miocene shale of north-western Nigeria and south-western Niger, indicative of the development of a *Rhizophora* mangrove swamp there. However, *Nypa* was still a component of the estuarine swamp, disappearing only after the early Miocene. The freshwater swamp forest, on the other hand, had ferns and the palm *Kentia*. *Kentia* does not at present occur naturally in Africa, but is sometimes cultivated. However, it is still native to northern Australia and New Guinea.

Nypa occurs naturally today only in the mangrove swamps of Indo-Malaysia. It was introduced to two localities in the estuary of the Cross River (south-eastern Nigeria) from Singapore Botanical Gardens in the early part of this century, though it has since become more widespread in that estuary. It was more recently transferred to the Niger delta. *Nypa*–type pollen disappeared in Venezuela at the Eocene–Oligocene boundary at which time there was a widespread increase in the occurance of mottled shales — a lithological phenomenon signalling a change to a climate with a more pronounced seasonality of rainfall. By inference, *Nypa* became extinct in Nigeria probably due to increased aridity concomitantly with the development of a more markedly seasonal climate.

Grasses increased and became both abundant and diversified in the Neogene. Though grasses naturally occur today in the tropics in a wide range of habitats — coastal or inland savanna, river valleys, and montane areas — these are mostly the open type of vegetation. Furthermore, a drier, more seasonal climate is more conducive to the development of extensive grasslands than a wet one. Consequently, it is suggested here that a savanna-type vegetation became established in Nigeria — and the rest of West Africa — only in the Neogene. There is no evidence of the occurrence of any woody species along with these savanna grasses. While the absence of arboreal pollen does not necessarily mean that trees and shrubs did not form part of this grassland vegetation, it does imply that a woody component, if any, was sparse. The grassland is not likely to have been a swampy 'prairie' of the type that exists in the forest zone in South-east Cameroun today, for example, since that vegetation community is essentially composed of sedges bordered by trees, with sometimes only a small percentage of grasses.

The disappearance of *Nypa* concomitantly with the establishment and expansion of grassland by the Miocene strongly suggests that climatic conditions at that time were indeed drier than before. Forests must have diminished, being largely replaced by a grassy savanna. It should be recalled that the period between the Eocene and Miocene was characterized by variations in the depositional environment of West Africa, indicative of climatic fluctuations.

Thus, the period was marked by the emergence of new landscapes due, at least in part, to sea regression and intensive erosion, as well as ferric oxide cementation of deposits — all of which are processes associated with a seasonal, predominantly dry climate.

The disappearance of *Nypa* from southern Nigeria at an earlier time than in the north suggests that drier conditions and savanna vegetation were probably established first in the south and only later further north. Furthermore, according to Germeraad *et al.* (1968), the curve of grass pollen in the Neogene of Nigeria showed major fluctuations which were regarded as probably reflecting 'the shifting boundary between forest and savannah in the lowland'. However, no further details were provided as to the possible geographical extents of either geographical type. In view of the controversy over the status of the savanna, it is necessary to discuss the issue at some length at this juncture. There are two divergent viewpoints on the issue. The savanna is considered a climatic climax by some while others regard it as a plagioclimax, i.e. a climax community maintained by continuous human activity — in this case largely the periodic burning of the natural vegetation and agriculture. While it may be rightly argued that the savanna in Africa today is maintained and indeed extended in area by man-made fires, agriculture, and pastoralism, the explanation for the existence of such a vegetation in the Neogene before the evolution of hominids is rather problematic. Natural fires, caused largely by lightning, and, to lesser extents, volcanic eruptions and the spontaneous combustion of coal deposits or marshes, do occur. Theoretically, natural fires could have raged any time since the evolution of vascular land plants about 400 million years ago. However, there is not much geological evidence of the burning of vegetation in pre-Pleistocene deposits (i.e. deposits older than 2 million years) while Pleistocene fires are attributed to man. Archaeologically, the earliest known use of fire by man dates back to about 500 000 years ago in the Choukoutien cave of Pekin, China. In Africa, man similarly first used fire some time in the Middle Pleistocene (i.e. c. 700 000–65 000 years BP).

Evidence from North America indicates that the extensive grasslands or prairies were not created by fire but, rather, represent a climatic climax — presumably predating colonization of the subcontinent by man. However, a minority viewpoint traces the origin of the prairies to 'fires set by Man . . . for thousands of years' (Stewart, 1956, p. 129).

It should be noted that in the savanna, during the dry season, a great deal of dry and combustible vegetal material accumulates on the floor, and this could be easily ignited naturally. Wet, dense forests on the other hand do not favour fires. It follows, therefore, that both the ignition and raging of natural fires are phenomena that are dependent on the vegetation itself being an open and dry one. In conclusion, it seems that the savanna in West Africa during the Neogene was a climatic climax resulting from a deterioration in climate.

However, natural fires might have also been one of the factors that contributed to the maintenance of the status quo.

4. Upper Tertiary/Quaternary (c. 3–2 m.y. BP)

In North-central Cameroun, deposits of this period yielded a rich assemblage of leaf imprints which are morphologically very much like those of living species. They were identified as being the foliage of sedges, grasses, several Leguminoseae, and the genera *Adenia, Albizia,* and *Ficus.* Similarly, all the spores and pollen recovered could be referred to extant taxa among which are Polypodiaceae (*sensu lato*), Cyathaceae, Schizaceae, as well as Flacourtiaceae, Gramineae, Sapotaceae, *Berlinia, Bombax, Protea* (presumably *P. elliottii* var. *elliottii*), *Rauvolfia,* and *Symphonia.* Such an assemblage has been interpreted as indicative of a wet savanna with a gallery forest that eventually became rather swampy and characterized by *Symphonia.* A similar vegetation community occurs further south today in the southern Guinea savanna zone.

Before considering the remaining part of the Quaternary period, it is necessary to point out that many of the extant species in West Africa have been in existence since the Miocene (Upper Tertiary) period. Thus, with regard to species composition, the vegetation of the region seemed to have retained its basic characters. Nevertheless, palynological evidence has shown that there were subsequently many 'examples of contraction of range, leading in extreme cases to extinction' (Germeraad et al. 1968, p. 284). *Nypa* and *Kentia* are two examples of genera that became extinct after the early Miocene, while *Anacolosa* and *Ptychopetalum* (both Olacaceae) typify those genera that underwent contraction of range, being represented in the region today by only two species. Furthermore, there have been shifts in the spatial extent of the major vegetation zones — the savanna and rain forest in particular.

5. Early to late Quaternary (c. 2m.–65,000 years BP)

The only data available for the pre-late Quaternary period are from Senegal. About a million years ago, the Mamelles volcano of Dakar erupted. The ejected ash tuffs covered the surrounding landscape, burying its vegetation. An examination of the fossils beneath the ash revealed an extremely large number of impressions of grass leaves. Above these were leaf impressions and compressions as well as petrified stems and trunks of the following: *Sansevieria, Adansonia digitata, Cordia rothii, Piliostigma, Crataeva religiosa, Guiera senegalensis, Terminalia, Combretum, Commelina, Schoenefeldia gracilis, Acacia pennata, A. macrostachya, Ficus, Phoenix, Elaeis, Cocos,*

Ziziphus, and *Typha*. This vegetation complex is neither littoral nor halophytic, nor yet that of a desert or forest. It is similar to that found today in tropical areas where the soil is sandy but with moist depressions, and where the rainy season is short, i.e. Sudan and Sahel savanna. Such a vegetation is found in Senegal today, *Elaeis* and *Typha* occurring in inter-dunal moist depressions.

The vegetational history of West Africa during the late Quaternary is largely one of fluctuations in the spatial extent and relative dominance of rain forest *vis-à-vis* the savanna. These fluctuations were mostly due to climatic and environmental changes, and in more recent times due to the activities of man and his domestic animals. There are biogeograpic and palynological indications of such vegetational fluctuations, the former constituting indirect or deductive evidence, while the latter is direct and sometimes also occurs in a dated context.

The biogeographical evidence will be considered first. On the basis both of the present geographical distribution of mammals and the varying degrees of endemism exhibited, Booth (1957) postulated that there were alternating periods of wet and dry climatic conditions in West Africa in Quaternary times. During the dry periods, the high (rain) forests of the Guinea–Congolese regions became drastically reduced in extent and were restricted to three *refugia*, viz. Liberia, Gabon/Cameroun and the Upper Congo basin. He disagreed with others who postulated that the three *refugia* were in Liberia, Ghana, and the Niger delta (Calabar), and argued that the Nigerian forests were destroyed during the severely dry periods. With the subsequent restoration of wetter conditions, the forests — along with their associated mammalian population — presumably spread out again from the refuge sites (cf. Hamilton, 1976). Areas which have many species in common are those across which migrations were likely to have occurred. Such areas include the entire high forest between the Volta and the Niger. On the other hand, those areas with a high degree of endemism, such as Liberia, Cameroun, and the most eastern parts of Nigeria around the Nigerian–Cameroun border, as well as Gabon, are regarded as centres of origin from which the common species spread. On account of their being floristically very rich, Hamilton (1976) also postulated that Cameroun and Gabon, were *refugia* for forest species when Nigerian forests were replaced by savanna. Furthermore, the forests of southern Cameroun are considered by several authors to be the richest, floristically, in Africa (cf. Hamilton, 1976).

The paucity of tropical African forests — especially those in the areas west of the Niger — in plant, bird, and insect species in comparison with some other parts of the humid tropics has been attributed to the reduction of these forests during the Quaternary (cf. Hamilton, 1976). Recent reductions were probably caused by subsequent minor climatic fluctuations, such as a drier period after 7000 years BP, as well as deforestation by man (Hamilton, 1976). It will be shown later on that the human element became a significant factor probably from c. 3000 years BP, by which time some form of cultivation had probably begun.

Booth (1957) provided no dates for the bioclimatic changes he postulated since there is no accurate measure of the evolutionary rate of mammals. The date given by Hamilton (1976) for the last major forest contraction in tropical Africa, i.e. >25 000 to c. 12 000 years BP, was based mainly on geomorphological evidence, from several other authors, or aridity in large parts of the region then, while that for Holocene forest expansion – from c. 12 000 years BP was inferred from comparable palynological evidence from Uganda. The occurrence of forest trees such as *Cola gigantea* and *Holarrhena floribunda* in present-day southern Guinea savanna of Nigeria is considered additional biogeograpical evidence of a former more northern extent of the rain forest in Quaternary and perhaps even Tertiary times (cf. Sowunmi, 1981b). It is pertinent to note here that the presence of relict forest of Guineo-Congolese species in present-day Zimbabwe is considered by some as being probably due to a formerly greater extent of such forest at a time of higher rainfall, the reduction being due both to climatic deterioration and man. Also estimates of the former extent of lowland rain forests beyond their present-day boundaries in some parts of Africa, have been made on the basis of the distribution of apparently favourable climates for forest growth. The estimates are as follows for the West African countries mentioned: Guinea 60–120 km; Ivory Coast (to the west of the Tiassale forest constriction) 100–140 km; Ivory Coast (at Tiassale) 240 km; Ivory Coast (to the east of the Tiassale forest constriction) 30–70 km; Nigeria 20–120 km; Cameroun 180–200 km on average, sometimes more. Wickens (1975) obtained from western Sudan (Jebel Marra volcanic complex) evidence indicative of lowland rain forest extension northwards by about 400 km between 12 000 and 3500 years BP. This was on the basis of the occurrence of leaf impressions of the forest species *Combretummolle*, *Elaeis guineensis*, and *Saba florida* in fossil ash of this estimated age; *E. guineensis* is now extinct in the area while *C. molle* and *S. florida* are among the extant gallery forest species, along with several other forest species of the Congo basin such as *Casearia barteri* and *Trema orientalis*.

Direct and more precise evidence for changes in the vegetation of West Africa in the late Quaternary period has come from palynological studies which will now be outlined. The evidence does corroborate — in a general way — that deduced from biogeographical studies, in that there are clear indications of variations in the extent of representation of forest components relative to those of the savanna. The additional value of the palynological data is that some of the actual components of the vegetation communities at specific time periods are known, as will be shown presently.

The Niger deltaic deposits reflect the continental environments, throughout its drainage basin, that affect the flow of its feeder rivers and the type of alluvial and pollen material they deposit in it. This drainage basin traverses the Sudano-Sahelian semi-arid zone, the Guinean savanna, as well as the humid lowland and montane Guinean forests; thus these deltaic deposits have provided very valuable information on the vegetational history of the greater part of West Africa. However, pollen representation was found to decrease with increasing distance of the pollen source from the delta, hence pollen originating from Nigeria and Cameroun predominated, most especially mangrove and freshwater swamp forests components (Sowunmi, 1981a, b).

6.   The Late Quaternary (c. 65 000 years BP to the Present)

During an unknown period in the late Quaternary but prior to 35 000 years BP, freshwater swamp forest was very extensive, reaching further south beyond its present-day limit into the area that now supports *Rhizophora* swamp. The latter was probably further south still, though there are indications of its occurrence on a localized and minor scale in brackish-water creeks within the fresh-water swamp. Some of the components of the freshwater swamp forest included *Calamus deeratus, Martretia quadricornis, Mitragyna stipulosa, Uapaca heudelotii,* and *U.* cf.*paludosa,* which are also present-day components of this forest.

Inland, the forest was at first a more open one with trees such as *Holoptelea grandis* and *Uncaria africana.* It eventually became a wetter and denser forest with *Bosquiea angolensis, Khaya* cf. *ivorensis, Myrianthus arboreus,* and *Pycnanthus angolensis,* among others. There were also riverine forests, represented by, *inter alia, Bridelia micrantha, Pandanus candelabrum, Raphia vinifera,* and *Syzygium guineense.* Ferns were also present.

The grasses and sedges present most probably originated from the outer fringes of the freshwater swamp forest as their representation through various strata fluctuated in a similar manner to that of other freshwater swamp forest components.

There was no pollen of savanna vegetation, indicative of either a restricted occurrence of this vegetation community or of its southern limit being much further north than subsequently and at the present time.

At a later period than the above, but still earlier than 35 000 years BP the sea transgressed over the lower reaches of the freshwater swamp forest, resulting eventually in a drastic reduction of this community with its grasses and sedges. *Rhizophora* swamp was then established at its present site and this mangrove progressively became more abundant and predominant in the estuarine vegetation. As *Rhizophora* swamp expanded, the freshwater swamp forest became further reduced until it nearly completely disappeared. Further inland, lowland rain forest and ferns were still well represented. Thus the virtual disappearance of freshwater swamp forest was caused not by adverse climatic conditions but edaphic ones, as a result of the inundation of land by sea-water. This period was probably contemporaneous with the Inchirian transgressive phase of Senego-Mauritania, c. 40 000 – 30 000 years BP.

During the period >30 000 years BP to some time well before c. 8000 years BP, which by correlation with the rest of West Africa was probably c. 24 000–18 000 years BP, there were fluctuations in the extent of *Rhizophora* swamp, probably due to changes in the intensity and extent of penetration of the tidal stream. Freshwater swamp forest was less extensive than in the earliest period considered above. There was also a drastic reduction of the lowland rain forest with *Canthium* cf. *subcordatum* and *Holoptelea grandis* being among the few arborescent species represented; both are today found in more open parts of high forest. Furthermore, the pollen of northern Guinea and Sudan savanna such as cf. *Borreria chaetocephala* and *Hyphaene thebaica* were noticed for the first time when the late Quaternary deposits of the Niger delta core were analysed (Sowunmi, 1981a).

Hitherto, freshwater swamp forest and grass pollen had similar patterns of occurrence. But, at this point, while the former remained poorly represented there was a marked increase in the abundance of grasses, strongly suggestive of a new source, additional to the outer fringes of the freshwater swamp forest. The occurrence of dry savanna pollen jointly with a notable increase in grasses cannot be fortuitous, but indicates that the dry savanna was the additional, new source of grass pollen. Thus at this time, the rain forest became drastically reduced and more open, while the savanna became more extensive, replacing large parts of the forest. The vegetation then was probably very much like present-day northern Guinea savanna.

There is further palynological evidence of replacement of forest by savanna. Firstly, at c. 21 000 years BP in North-central Cameroun, the forest disappeared and was replaced by an open savanna (Hurault, 1972). Unfortunately, no names of component species were given. Secondly, during part of the period that witnessed tremendous sea regression, a littoral peat from a level 65 m below the present sea-level (the sea-level then) and dated 23 000 ± 1000 years BP yielded a pollen and spore assemblage that was relatively poor. It comprised essentially grasses and sedges with some ferns. This must have been an open grassland savanna that had replaced the original forest (cf. Assemien et al., 1970; Martin, 1972). Thirdly, a portion of an offshore core from Senego-Mauritania dated c. 24 000–19 000 years BP was deposited when there was no mangrove vegetation and only a minor occurrence of Guinean savanna; the 'littoral dunes (were) covered by Cyperaceae and Gramineae, suggesting greater aridity than today' (Rossignol-Strick and Duzer, 1979a, p.124).

From the foregoing, it is evident that the impact of the arid period c.

24 000–18 000 years BP was more severe in the more western parts of West Africa, for example Ivory Coast and Senegal, than in the area east of the River Niger, i.e. eastern Nigeria and Cameroun. In the latter area there was a *refugia* of a more open forest within a predominantly woodland savanna vegetation. In contrast, in the more western parts, there was a complete destruction of forest and its replacement by a semi-arid predominantly grassy savanna.

Palynological evidence from the same offshore core from Senego-Mauritania referred to above showed that the subsequent period, c. 19 000 to 12 500 years BP was even more arid. There was a disappearance of Sudan and Guinean savanna and a reduction of the Sahelian elements. More of the littoral strip was exposed owing to a major sea regression, resulting in an expansion of the littoral *Sebkhas* with Chenopodiaceae (*Sebkha* is a saline soil with very sparse vegetation). Mediterranean pollen was abundant in this marine core, having been wind-transported from north of the Sahara. This is an indication that the north-east trade winds were very strong and intensive. There is confirmatory evidence from a region 250 km inland from the coast that the Senegalese landscape was desertic during this period (Michel and Assemien, 1970). Following the arid period above, wetter conditions were restored all over West Africa as reflected by palaeobotanical, limnological, and geomorphological evidence.

Another littoral peat from Ivory Coast, taken from a depth of 65 m and dated 11 900 ± 250 years BP yielded a pollen assemblage that emanated from a vegetation complex that was distinctly different from that of the earlier peat just considered (dated 23 000 ± 1000 years BP). There was a particularly rich variety of ferns including the epiphytic *Microgamma ovariense*, *Nephrolepis biserrata*, and *Platycerium stemaria*; *Ceratopteris cornuta* (a fern of open fresh water); and *Cyathea camerooniana* a terrestrial tree fern. These ferns today occur in the humid forests of Ivory Coast, and were probably then in the upper parts of estuaries not reached by the tidal stream. Tree pollen was rare and of limited variety. *Rhizophora racemosa* was the most abundant. Present also were *Elaeis guineensis* and *Pentaclethra macrophylla* pollen. The paucity of tree pollen coupled with the presence of *E. guineensis* — a tree of river valleys and open forests — and *P. macrophylla* — a tree of open forests and derived savanna — suggests that the forest which had just been re-established following its destruction c. 23 000 years BP, was more open than today's (cf. Martin, 1972). This is to be expected as the climate was then not as wet as it is today, the sea-level being 63 m lower than it is at present.

With regard to the more open nature of forests in West Africa c. 12 000–10 000 years BP, there is corroborative evidence from Ghana and from offshore Ivory Coast. Among the plants whose leaf impressions were recovered from raised lacustrine sediments above the present shores of Lake Bosumtwi (Ghana) and dated to 10 000 years BP were the lianes *Dalbergia hostilis, Strychnos* cf. *splendens,* and *Tetracera potatoria.* There were also forest trees such as *Bussea occidentalis, Celtis zenkeri, Triplochiton scleroxylon,* and *Dialium guineense;* the lattermost is today typical of forest that is drier than that which occurs near the lake now. But of the greatest significance is the occurrence of both *Canarium schweinfurthii* and *Musanga cecropioides* — trees that regenerate where there are openings within the forest floor. *Canarium* was in fact the most abundant fossil in the sediments. Since there is as yet no evidence that Stone age man cleared the forest and cultivated *Canarium c.* 10 000 years BP, the other possible explanation for the abundance of *Canarium* and the occurrence of *Musanga* is that the forest itself was a naturally open one or at least one with gaps in it.

The fossil assemblage recovered from an offshore core just south-east of Abidjan (Ivory Coast) and dated 9625 ± 125 years BP represented three types of vegetation community. Firstly, mangrove swamp in which *Rhizophora racemosa* and the salt-water fern *Acrostichum aureum* were relatively abundant. Secondly, a freshwater swamp or riverine formation with *Anthocleista, Crudia klainei,* and *Mitragyna.* Thirdly, and of particular relevance to the question of a drier forest *c.* 12 000–10 000 years BP, a vegetation very characteristic of open, drier forests with species of Apocynaceae, Euphorbiaceae (e.g. *Antidesma* sp.), Tiliaceae (e.g. *Christiana africana*), and Verbenaceae. However, this forest had a much greater abundance of woody species than the earlier formation which first developed on the restoration of wetter conditions *c.* 12 000 years BP. By inference from the above results obtained for Ghana and Ivory Coast, the present-day dense lowland rain forests in West Africa were probably also more open and drier than now, with the possible exception of the *refugia* in Liberia and Gabon/Cameroun about 12 000–10 000 years BP.

The study of both offshore and onshore sediments from Senego-Mauritania, dated 12 500–5500 years BP revealed a striking and rapid change in vegetation from that which prevailed *c.* 19 000–12 500 years BP (q.v.). Both Chenopodiaceae and Mediterranean pollen disappeared while a rich estuarine mangrove developed. The Senegal River resumed its flow and had a gallery forest. There was a maximal development of Guinea savanna, characterized by *Bombax.* The increase of Guinea savanna pollen coupled with a drop in Mediterranean pollen is considered to be indicative of, on the one hand, increased monsoonal rain, and, on the other, a decrease in the intensity of the north-east trade winds. However, further inland, the Saharan desert 'pseudo-steppe' still dominated the landscape. Towards the close of this period, the mangrove swamp, the Guinea, and the Sudan savanna diminished, indicating a trend towards aridity. The Sahelian steppe — hitherto only a narrow zone —

became very extensive, reaching its present-day northern limit (Rossignol-Strick and Duzer, 1979a).

The pollen assemblage in a portion of a boring east of Abidjan and dated 8045 ± 100 years BP showed that there was a mangrove vegetation near the coast. Inland was a well-developed forest vegetation, dominated by *Piptadeniastrum* — a large forest tree — found today throughout the forest zone of Ivory Coast. Other components of the forest vegetation included Sterculiaceae (e.g. *Cola*), Mimosaceae, and Papilionoceae. This forest was richer in arborescent species than were the forests during the preceding 4000 years. The trend towards a richer forest over time reflected a progressively wetter climate as the sea-level gradually rose after the Ogolian regression, in this particular case from −63 m through to −22 m. This increasing richness also indicated that forest species were spreading westwards from the Gabon–Cameroun–eastern Nigeria *refugia*.

The Niger delta core referred to earlier revealed that from c. 7500 to c. 6960 years BP there was a marked rise in *Rhizophora* pollen as well as increases in freshwater swamp and rain forest components, including ferns. But this was immediately followed (c. 6960 to c. 5400 years BP) by a remarkable reduction in the representation of several groups such as ferns, sedges, freshwater swamp, and rain forest elements, whle *Rhizophora* remained dominant, reaching a peak level c. 5300 years BP. Significantly, in the Ivory Coast, Assemien (1971) also noted a similar drastic decrease in freshwater swamp forest at a time when *Rhizophora* became established at Agneby, during the maximum of the Nouakchottian transgression, c. 5500 years BP . *Rhizophora* swamp also developed at Bogue, 250 km inland following an invasion by the sea all over the lower valley of the Senegal (and Gambia). This transgression also affected the more northern parts of West Africa as evidenced by the occurrence of casts of mangrove roots in Nouakchottian shell beds at a level of 2 m above sea level on the Mauritanian coast. It thus seems that in the southern sedimentary basin of West Africa, a major sea transgression reminiscent of that of the Palaeocene period submerged at least parts of the continental land mass, destroying much of the forest vegetation there.

From c. 5500 to c. 3500 years BP there was a general progressive trend towards conditions drier than before. In Senegal, for example, further decrease of Guinean savanna pollen along with the appearance of elements characteristic of the Sahel, such as *Acacia*, showed that the vegetation had taken on its present-day nature. However, *Elaeis guineensis* and *Syzygium guineense* were still maintained along depressed watercourses as of today. The mangroves also decreased. The Niger delta core also registered decreases in rain forest and southern Guinean savanna pollen c. 4200 years BP, coupled with the recurrence of both open forest and northern Guinea savanna pollen. This again is suggestive of a reduction in the extent of rain forest and a more southward expansion of the savanna as occurred during the arid period c. 24 000–18 000 years BP. There was also a reduction of the *Rhizophora* swamp. It is noteworthy that Aeolian sands were deposited in South-east Ghana c. 4500–4000 years BP. Furthermore, Einsele *et al*. (1977) showed that there was probably a lowering of sea-level to −3.5 ± 0.5 m at 4100 years BP. This must have been a dry spell reminiscent of the post-Inchirian regression. This combination of factors clearly demonstrates how changes in the environment are reflected in the accompanying modification of the vegetation communities in the relevant area.

There was a notable increase in rain forest pollen c. 3500–3000 years BP, suggestive of a renewed expansion of forest. *Rhizophora* swamp forest reached the same peak level as at 5300 years BP.

From c. 3000 years BP to the present there were fluctuations in *Rhizophora* pollen of the Niger delta core, but all to levels below those of c. 5300 and 3200 years BP . There has thus been a regional decrease in *Rhizophora* swamp forest since 3000 years BP.

7. Human impact on vegetation

The impact of man on the natural vegetation became noticeable from c. 3000 years BP, as evidenced (in the Niger delta core) by a drop in rain forest pollen concomitantly with a sharp rise in *Uncaria africana* (a tree of open forest) and *Elaeis guineensis*. The latter had been a very minor component of the pollen assemblage (≤1.0 per cent) from > 35 000 years BP, but rose to 4.3 per cent c. 2800 years BP. It is known to regenerate only where abundant sunlight reaches the forest floor, as in a naturally open forest or one wher the gaps are artificially created by man. *Elaeis* is today a prominent constituent of forest regions that had been or are still being cultivated and is therefore a good indicator of the opening up of forest by man. Further confirmation of the inference that in West Africa, at least from c. 3000 years BP, man has been practising agriculture — incipient or otherwise — was provided by the occurrence of pollen of weeds of waste places or cultivated land, such as *Aspilia* sp., *Borreria verticillata*, *Cleome ciliata*, and cf. *Hilleria latifolia* when the oil-palm became more abundant (Sowunmi, 1981a, b; and in press).

Today, there is hardly any patch of natural, undisturbed vegetation in West Africa; such as exist must be in regions inaccessible to man and his domestic animals. Thus, for example, the rain forests have been degraded to secondary, more open forests and/or 'derived savanna' on account of intensive cultivation and annual fires. There is historical evidence of vegetation changes in the Sudan and Sahel zones of Nigeria and Cameroun. The 'Great Forest' covering 6000 m^2 of Bornu was destroyed in less than 50 years after large-scale immigration began. Several other places south of Lake Chad which had a forest vegetation when visited by early explorers are now savanna (cf. David, 1976). A very striking illustration of the degradation of forest to derived savanna due to human activities is provided at Olokemeji, south-western Nigeria, where a forest reserve has been created. This forest reserve lies across the boundary between semi-deciduous forest and derived savanna. At the forest end of the reserve from which both fires and farming have been excluded, the vegetation today is still forest, with trees such as *Cola millenii, Hildegardia barteri*, and *Triplochiton scleroxylon*. But, just literally across the road within this forest zone is an area that is annually burnt and farmed.The vegetation that occurs along with the farms is typically derived savanna with grasses such as *Andropogon* sp. and *Hyparrhenia* sp., and trees such as *Butyrospermum paradoxum, Hymenocardia acida*, and *Parkia clappertoniana*. The effects of fire in derived savanna and the regeneration of forest under fire-protection are further discussed in Chapter 5.

In recent years the role of cattle in bringing about some change in vegetation has become evident in Nigeria. Herds of cattle driven down to the south from the north by Fulani herdsmen have contributed to the spread of some savanna species such as *Acacia* sp., *Dichrostachys glomerata*, and *Piliostigma thonningii* to the forest zone in the south. This has been achieved through the germination of the viable seeds of these and other savanna plants contained in the droppings of the cattle and the subsequent growth to maturity of established seedlings (G. Jackson, pers. comm., 1963; field observation by author).

It seems that during 'some unusually dry years, climatic regions including the desert margins [are] displaced to positions comparable with those they occupied in arid periods of the distant past' (Grove, 1973, p. 39). The vegetation is adversely affected and man has now accentuated the effects of these unfavourable periods. Thus the most recent Sahelian drought had very serious deleterious consequences for man and beast alike — crops failed, people and their animal stock died of hunger, or migrated south.

III. Summary

The changing aspect of West African vegetation from the Tertiary through to the Present has been outlined, primarily on the basis of direct botanical evidence. These vegetation changes have been mainly due to evolution, environmental changes, and, in the last few thousand years, by human activity, as reflected in the plant fossil record. The environmental processes and dynamics during the time period have been considered in order to place the study of past vegetation in the proper perspective.

Highlights of the environmental and vegetation changes through time are summarized under the headings below.

Palaeocene. In place of *Rhizophora* swamp forest of today was a *Nypa* estuarine swamp along a coast located c. 2 °N of present position. Inland vegetation was a rain forest with low species diversity. There was no savanna vegetation. This was a time of a major sea transgression and wet climate.

Eocene. *Nypa* estuarine swamp was at the coast. Freshwater swamp or riverine forests had *Symphonia, Arenga, Mauritia, Ravenea,* and ferns as some of the components. There was still no *Rhizophora* swamp, neither was there savanna vegetation. Grasses appeared at this period but were only minor components of the vegetation. The climate was probably the same as in the Palaeocene. Climatic fluctuations occurred between Eocene and Neogene times.

Neogene. The palms *Arenga, Mauritia, Nypa,* and *Ravenea* became extinct in the region. *Rhizophora* evolved (early Miocene) and mangrove swamp soon replaced *Nypa* swamp. Grasses became more abundant and diversified. Forests diminished drastically. Vegetation was now a predominantly grassy savanna. Fluctuations in relative extents of forest and savanna followed. Species composition of these types of vegetation was essentially like today. Climate oscillated but was probably predominantly drier than before.

Quaternary. Characterized by fluctuations in *Rhizophora,* swamp as well as spatial extent and relative dominance of forest *vis-à-vis* savanna.

Circa 40 000–30 000 years BP. There was extensive fresh water swamp forest. *Rhizophora* swamp forest was farther south of present-day limit. Inland forests were at first drier but became wetter. Sea transgression later led to the establishment of *Rhizophora* swamp at present site.

Circa 30 000–12 500 years BP. Forests were either destroyed or drastically reduced, being restricted to *refugia.* Dry savanna expanded southwards, replacing forest. In more northern parts, desertic landscape replaced Sudan savanna. Ogolian regression occurred and was accompanied by the formation of great ergs in Sudan and Sahel zones; climate was very arid.

Circa 12 500–5000 years BP. Forest was reestablished, progressively becoming denser with increasing species diversity following migration from *refugia.* General northward shift of vegetation zones occurred. Nouakchottian transgression resulted in extension of *Rhizophora* swamp further inland. Climate wetter.

Circa 5000–3500 years BP. *Rhizophora* swamp forest diminished. Inland rain forest was reduced and replaced by savanna. Sea regression occurred. Climate drier.

Circa 3500–3000 years BP. Forest expanded and savanna contracted. Climate was wetter.

Circa 3000–Present. Human disturbance of natural vegetation — especially through burning, farming, and grazing of animals — became noticeable. Northern parts of forest were converted to woodland savanna; Sudan and Sahel zones were degraded.

References and Bibliography

Assemien, P. (1971). Étude Comparative des flores actuelles et quaternaires recent de quelques paysages vegetaux d'Afrique de l'Ouest. Thèse Doct. Univ. Abidjan (AOCNRS), No. 5868, pp. 1–257.

Assemien, P., Filleron, J. C., Martin, L., and Taset, J. (1970). Le quaternaire de la zone littorale de Côte d'Ivoire. *Bull. Assoc. Sénég. Et. Quatern. Ouest Afr. Dakar*, **25**, 65–78.

Booth, A. H. (1957). The Dahomey Gap and the mammalian fauna of the West African forests. *Rév. Zool. Bot. Afr.*, **1**, 305–314.

Burke, K. (1972). Longshore drift, submarine canyons, and submarine fans in development of Niger delta. *Am. Assoc. Petrol. Geol. Bull.*, **56**(10), 1975–1983.

Burke, K., and Durotoye, B. (1970). Late quaternary climatic variation in south-western Nigeria: evidence from pediment deposits. *Bull. Assoc. Sénég. Et. Quatern. Ouest Afr., Dakar*, **25**, 79–96.

Cantagrel, J., Lappartient, J., and Tessier, F. (1976). Nouvelles données geochronologiques sur le volcanisme Ouest-Africain. *Bull. Assoc. Sénég. Et. Quatern. Ouest Afr., Dakar*, **47**, 16–17.

Chamard, C., and Courel, M. F. (1975). Contribution à l'étude geomorphologique du Sahel. Les formes dunaires du Niger occidental et de la Haute-Volta Septentrionale. *Bull. Assoc. Sénég. Et. Quatern. Ouest Afr., Dakar*, **44–45**, 55–66.

David, N. (1976). History of crops and peoples in North Cameroon. In J. R. Harlan, J. M. J. De Wet, and A. B. L. Stemler (Eds.), *Origins of African plant domestication*, pp. 223–267. Mouton, The Hague, Paris.

Einsele, G., Herms, D., and Schwarz, U. (1977). Variation du niveau de la mer sur la plate-forme continenal et la côte Mauritaniene vers la fin de la glaciation de Wurm et a l'Holocène. *Bull. Assoc. Sénég. Et. Quatern. Ouest Afr. Dakar*, **51**, 35–48.

Flenley, J. R. (1979). *The Equatorial Rainforest: A Geological History*. Butterworths, London, Boston.

Fredoux, A., and Tastet, J. (1976). Apport de la palynologie a la connaissance paleogeographique du littoral Ivorien entre 8000 et 12000 ans BP, *7th African Micropalaentological Colloquium*. Ile-Ife, Nigeria.

Germeraad, J. H., Hopping, C. A., and Muller, J. (1968). Palynology of Tertiary sediments from tropical areas. *Rev. Palaeobotan. Palynol.* **6**, 189–348.

Grove, A. T. (1958). The ancient erg of Hausaland and similar formations. *Geog. Journ.*, **124**, 528–533.

Grove, A. T. (1973). Desertification in the African environment. In D. Dalby and R. J. H. Church (Eds.), *Report of the 1973 Symposium: Drought in Africa*, pp. 33–45. Centre for African Studies, School of Oriental and African Studies, London University.

Hall, J. B., Swaine, M. D., and Talbot, M. R. (1978). An early Holocene leaf flora from Lake Bosumtwi, Ghana. *Palaeogeogr. Palaeoclimatol. Palaeoecol.*, **24**, 247–261.

Hallam, A. (1973). *A Revolution in the Earth Sciences: From Continental Drift to Plate Tectonics*, p. 31. Oxford University Press, London.

Hamilton, A. (1976). The significance of patterns of distribution shown by forest plants and animals in tropical Africa for the reconstruction of Upper Pleistocene palaeoenvironments: a review. *Palaeoecology of Africa*, **9**, 63–97.

Harlan, J. R., De Wet, J. M. J., and Stemler, A. B. L. (Eds.) (1976). *Origins of African Plant Domestication*, pp. 3–19. Mouton, The Hague, Paris.

Hebrard, L. (1974). Decouverte de la flore et de la végétation ensevelies sous les cinerites du volean quaternaire des Mammelles de Dakar (Sénégal). *Bull. Assoc. Sénég. Et. Quatern. Ouest Afr. Dakar*, **42–43**, 81–90.

Hebrard, L., Elouard, P., and Faure, H. (1971). Quaternaire du littoral Mauritaniene entre Nouakchott et Nouadhibou (Port Étienne) (180–210 Latitude Nord). Quaternaria, 15(2), 297–304.

Hurault, J. (1972). Phases climatiques de tropicales seches a Banyo (Cameroun, hauts plateaux de l'Adamawa). Palaeoecology of Africa, 6, 93–101.

Hutchinson, J., and Dalziel, J. M. (1954–68). Flora of West Tropical Africa, Vols. I–III. Crown Agents, London.

Keay, R. W. J. (1959a). Derived savanna — derived from what? Bull. Inst. Franc. d'Afrique Noire, sér. A, 21, 427–433.

Keay, R. W. J. (1959b). An Outline of Nigerian Vegetation. Fed. Govt. of Nigeria, Lagos.

Kogbe, C. A., and Sowunmi, M. A. (1975). The age of the Gwandu formation (Continental Terminal) in north-western Nigeria as suggested by sporo-pollinitic analysis. Savanna 4(1), 47–55.

Lappartient, J. (1972). La laterite récente des environs de Dakar (Republique du Sénégal): Resume de these. Palaeoecology of Africa, 6, 84.

Leprun, J. C. (1971). Nouvelles observations sur les formations dunaires Sableuses fixees du Ferlo nord-occidental (Sénégal). Bull. Assoc. Sénég. Et. Quatern. Ouest Afr. Dakar, 31–32, 69–78.

Letouzey, R. (1968). Étude phytogéographique du Cameroun. Encyclopedie Biologique, Vol. LXIX. Paris.

Martin, L. (1972). Variation du niveau de la mer et du climat en Côte d'Ivoire depuis 25,000 ans. Cah. ORSTOM Ser. Geologie, IV(2), 93–103.

Michel, P., and Assemien, P. (1970). Études sedimentologiques et palynologiques des sondages de Bogue (Basse vallee du Sénégal) et leur interpretation morphoclimatique. Rev. Geomorph. dyn., 19(3), 98–113.

Nahon, D. (1972). Les cuirasses ferugineuses et encroutements calcaires de l'ouest du Sénégal et de la Mauritanie. Palaeoecology of Africa, 6, 85.

Petters, S. W. (1978). Stratigraphic evolution of the Benue Trough and its implications for the Upper Cretaceous palaeogeography of West Africa. J. Geol., 86, 311–322.

Rossignol-Strick, M. and Duzer, D. (1979a). West African vegetation and climate since 22,500 BP from deep-sea cores palynology. Pollen et Spores, 21 (1–2), 105–134.

Rossignol-Strick, M. and Duzer, D. (1979b). A late Quaternary continuous climatic record from palynology of three marine cores off Sénégal. Palaeoecology of Africa, 11, 185–188.

Servant, M. and Servant-Hildary, S (1972). Nouvelles données pour une interpretation palaeoclimatique de series continentales du bassin tchadien (Pleistocene recent, Holocene). Palaeoecology of Africa, 6, 87–92.

Short, K. C., and Stauble, A. J. (1967). Outline of geology of Niger delta. Bull. Am. Assoc. Petroleum Geologists, 51(5), 761–779.

Sowunmi, M. A. (1981a). Late Quaternary environmental changes in Nigeria. Pollen et Spores, 23(1), 125–148.

Sowunmi, M. A. (1981b). Aspects of late Quaternary vegetational changes in West Africa. J. Biogeog., 8, 457–474.

Sowunmi, M. A. (in press). Beginnings of agriculture in West Africa: The botanical evidence. Current Anthropology.

Stewart, O. C. (1956). Fire as the first great force employed by Man. In W. J. Thomas, Jr. (Ed.), Man's Role in Changing the Face of the Earth, pp. 115–133, Wenner-Gren Foundation for Anthropological Research, Chicago.

Street, F. A., and Grove, A. T. (1976). Environmental and climatic implications of late Quaternary lake level fluctuations in Africa. Nature, 261, 385–390.

Talbot, M. R., and Delibrias, D. (1977). Holocene variations in the level of lake Bosumtwi, Ghana. *Nature*, **268**, 722–724.

Talbot, M. R., and Delibrias, D. (1980). A new late Pleistocene–Holocene water-level curve for lake Bosumtwi, Ghana. *Earth and Planetary Science Letters*, **47**, 336–344.

Tastet, J. P. (1975). Les Formations sedimentaires, quaternaires a actuelles, du littoral du Dahomey. *Bull. Assoc. Sénég. Et. Quatern. Ouest Afr. Dakar*, **46**, 21–44.

Tralau, H. (1964). The genus *Nypa* van Wurmb. *Kungl. Svenska Vetenkapsakademien-shandling, Fjarde Serien*, **10**, 8.

Whyte, S. A. (1975). Distribution, trophic relationships and breeding habits of the fish populations in a tropical lake basin (Lake Bosumtwi — Ghana). *J. Zool.*, **177**, 22–56.

Wickens, G. E. (1975). Quaternary plant fossils from the Jebel Marra volcanic complex and their palaeocolimatic interpretation. *Palaeogeogr. Palaeoclimatol. Palaeoecol.*, **17**(2), 109–122.

Plant Ecology in West Africa
Edited by G. W. Lawson
© 1986 John Wiley & Sons Ltd

CHAPTER *12*

Conservation and land use

I. Hedberg

Institute of Systematic Botany, University of Uppsala,
Uppsala, Sweden

According to the Unesco map of arid lands of the world (Anon., 1977) almost one-third of the West African countries fall entirely or largely within this region, viz. Mauritania, Senegal, Mali, Upper Volta, and Niger. An additional two, viz. Gambia and Nigeria, fall partly within this zone. A considerable part of the northernmost part of the zone consists of bare desert and subdesert, which is followed by thornlands and wooded steppes (Sahel), in the low-rainfall areas (below 500 mm) with very long dry seasons (seven to eight months). The Sahel is replaced to the south by dry savanna, Sudan type, followed to the south by a more moist type, Guinea savanna.

The remaining part of West Africa could be divided into two distinctly different forest regions (Persson, 1977), one with limited and the other with relatively extensive forest resources. According to various sources most of the first-mentioned region, including Guinea-Bissau, Guinea, Sierra Leone, Togo, and Benin, was once covered by closed forest, the scarce remnants of which now form a belt behind the mangrove forest in Sierra Leone and cover the south-eastern corner of Guinea. The rest of this region comprises a band of forest-savanna mosaic and large areas of Guinea savanna. The second region, which includes Liberia, Ivory Coast, Ghana, Nigeria, and Cameroun, has large areas of moist tropical forests followed to the north by forest–savanna mosaic, and a wide belt of Guinea savanna. A belt of Sudan savanna is found further to the north in Ghana and Nigeria and finally, as mentioned above, the northernmost part of Nigeria reaches into the Sahel.

I. Land use

Existing information about land use in West Africa is summarized in Table 12.1. Obviously the figures given are only crude estimates since few detailed inventories have been undertaken. In spite of considerable changes over the

Table 12.1. Land use in West Africa and percentage of labour force employed in agriculture. Extracted from Persson (1977) and *Encyclopedia of the Third World* (1982)

Country	Land area (1000 ha)	Agricultural area		Forests and woodlands	Other areas	Labour force employed in agriculture
		Arable land and land under permanent crops	Permanent meadows and pastures			
Benin	11 262	1 546	442	2 157	7 117	46
Cameroun	46 944	7 300	8 300	30 000	1 944	82
Cap Verde Islands	403	40	10	1	352	>35
Gambia	1 000	200	400	303	227	79
Ghana	23 002	2 574	11 237	2 447	7 596	54
Guinea	24 586	1 500	3 000	1 046	19 040	82
Guinea-Bissau	2 800	275	1 280	1 000	1 057	>84
IvoryCoast	31 800	8 887	8 000	12 000	3 359	81
Liberia	11 137	366	240	3 622	6 909	71
Mali	122 000	11 600	30 000	4 457	77 943	89
Mauritania	103 070	263	39 250	15 134	48 423	85
Niger	126 670	15 000	3 000	12 000	96 700	91
Nigeria	92 378	21 795	25 000	31 069	14 514	56
Senegal	19 200	5 564	5 700	5 318	3 037	77
SierraLeone	7 174	3 664	2 204	301	1 005	67
Togo	5 600	2 160	200	530	2 710	69
Burkina Faso	27 380	5 377	13 755	4 101	4 187	83

last 10 years recent figures for agricultural areas (Anon., 1982b) agree in most cases very well with those given about 10–20 years ago (Persson, 1977).

In the arid region about 45 per cent of the land area is classified as agricultural or forest land, the remaining part being desert or subdesert. In the forest region 45 per cent is classified as agricultural and 25 per cent as forest land (Persson, 1977). This region is probably the area in Africa at present most intensively utilized by man and the vegetation is changing fast, mainly due to shifting cultivation and commercial logging in the tropical moist forest.

The figures in Table 12.1 indicate that the proportion between arable land and permanent meadows/pastures varies considerably. Though large differences no doubt exist between the countries it must be kept in mind that the figures given probably to some extent mean different things from country to country.

Because of the uncertainty in the information on agricultual land and the large proportion of 'other areas' especially in the dry region, a better illustration of the dependence on agriculture (including cattle-farming) would be to give the amount of people employed in it. In 10 out of the 17 West African countries more than 75 per cent of the labour force is employed in agriculture (Table 12.1). Of those 10 countries no less than 6 fall entirely or partly within the arid region, i.e. where the land is unproductive or marginal for food production.

The vegetation cover in the arid part of West Africa (Mauritania, Senegal, Mali, Upper Volta, and Niger) has for a very long time been exposed to a heavy pressure, in the last 30 years alarmingly aggravated by various phenomena, such as:

1. A growing population in need of an increasing area of cultivated land.
2. An extensive pasture farming resulting in heavy overgrazing.
3. An itinerant agriculture system, which does not take into account the need for the upkeep of soil fertility.
4. A lawless exploitation of the existing forests to meet the needs of the local populations for wood.
5. Efforts to apply new agricultural techniques which are not properly undersood by the peasants.

To the above factors, caused by man, must be added those caused by climatic variations, for example, the periods of prolonged drought which are extremely difficult to overcome. Recent investigations have shown that over a number of years such periods occur at a certain periodicity, about twice in each generation (Rapp, 1977).

The results of these events is that the vegetation cover over large areas is ruined, which leaves the thin cover of soil exposed to degrading forces like rains and winds, and results in increasing soil erosion and desertification.

In some of the countries concerned the continuous degradation of the soil has resulted in a decrease in per capita grain production from the 1950s to the 1970s (Table 12.2) which means that at present some of them have to rely upon food imports and sometimes foreign aid, which is highly unsatisfactory from both economic and social points of view.

In the forest regions the main threat to the vegetation, especially for the tropical moist forest, is the fantastic development of technology, and the uncritical application of modern Western techniques in developing countries. Where 30 years ago the forests were slowly cropped by hand by the local

Table 12.2. Per capita grain production (kg) in some West African countries. Extracted from Youdeowei (1981)

Country	Per capita grain production (kg)		% Change
	1950–52	1973–75	
Mali	267	146	−45
Niger	303	169	−44
Senegal	142	186	+31
Burkina Faso	193	180	− 7

people, large bulldozers now clear hectares of forests very rapidly. This type of harvesting is detrimental also because it opens up roads for the small-scale cultivators through hitherto impenetrable forests. The threat to the tropical moist forest from the rapidly growing need for agricultural land is in fact now reported to be as great as that from commercial harvesting. The rate of destruction has reached an appalling rate, which will result in a major loss of forest up to 1990 and there is a grave risk that there will be virtually no forest left in West Africa by the end of the century if the present rate of exploitation is allowed to continue. The area of tropical moist forest has already declined from over 700 000 km^2 to less than 200 000 km^2 and is likely to be reduced to well below 100 000 km^2 during the 1980s, since in a number of countries, for example Nigeria and Ghana, all the remaining forests have been allocated for concessional harvesting. The very low amount of forests per capita, 0.2 ha in Ghana and 0.07 ha in Nigeria (Persson, 1977), certainly means an enormous pressure on the last remnants of these forests. The situation is equally bad in Ivory Coast where between one-third and one-half of the primary forests have been lost in the last 15 years (Meyers, 1980). The tropical forests once growing over practically the whole of Liberia now cover less than half the country (Verschuren, 1982).

II. Conservation

1. Protection of nature

The idea to set aside areas of special value in order to save them from the impact of man was launched in the latter half of the last century, when the world's first national park was established in the USA. Through the years national parks have been created in most countries in the world and also a large number of nature reserves. Because of the increasing use of the term 'National Park' to designate areas with increasingly different status and objectives, the International Union for the Conservation of Nature (IUCN) has recommended the term 'National Parks' only for areas answering the following characteristics:

> A national park is a relatively large area: (1) where one or several ecosystems are not materially altered by human exploitation and occupation, where plant and animal species, geomorphological sites, and habitats are of special scientific, educative, and recreative interest or which contains a natural landscape of great beauty; (2) where the highest competent authority of the country has taken steps to prevent or eliminate as soon as possible exploitation or occupation in the whole area and to enforce effectively the respect of ecological, geomorphological, or aesthetic features which have led to its establishment; and (3) where

visitors are allowed to enter, under special conditions, for inspirational, educative, cultural, and recreative purposes (Anon. 1980b).

Obviously the national park concept is very rigid. On the theme 'nature reserve' on the other hand, there are no less than about 30 variations in the IUCN list, depending upon the reasons for which the areas have been set aside. Such reasons could be some outstanding ecosystems, features, and/or species of flora and fauna, etc. of national scientific interest.

The first nature reserve in West Africa was established 50 years ago and in the last decades an increasing number of national parks and nature reserves have been set aside (Table 12.3 and Fig. 12.1). Suggestions on areas and species in urgent need of protection is given in Hedberg and Hedberg (1968), together with the first fairly detailed survey of West African vegetation.

In recent years a new category of protected (managed) areas, biosphere reserves, has been created. Unlike the above categories these reserves are established under UNESCO's Man and the Biosphere (MAB) Programme and are areas set aside primarily for education, monitoring, research, and con-servation of ecosystems. Up till now, 10 biosphere reserves have been established in West Africa (Table 12.4 and Fig. 12.1).

A biosphere reserve must have adequate long-term legal protection, should be large enough to be an effective conservation unit, and to accommodate different uses without conflict. Each biosphere reserve will be zoned to provide direction as to its management. Four zones may be deliminated as follows: (a) natural or core zone; (b) manipulative or buffer zone; (c) reclamation or restoration zone; and (d) stable cultural zone. Some biosphere reserves are unique in themselves, others involve national parks or nature reserves, such as, for example, Tai Biosphere Reserve in Ivory Coast (see below) or Mount Nimba Biosphere Reserve in Guinea.

The number or area of national parks and nature reserves in a country is sometimes taken as a measure of degree of success of conservation activities. This is, however, far from the truth, especially in the developing countries, since it is not simply a question of area or number, but essentially a question of management quality and if the areas set aside for protection fulfil the purposes for which they were created. At present it seems impossible to have unequivoc-al information on this point, especially as the situation may change very rapidly. As an example may be mentioned Tai National Park in Ivory Coast, the area of which has been reduced by 25 per cent in the last 10 years due to invasion by shifting cultivators from other parts of the country and from the drought-stricken Mali and Upper Volta (Anon., 1981).

Thus the survey of national parks and nature reserves given in Table 12.3 is to be seen more as an attempt to illustrate the degree of conservation awareness than as proof of an overall protection of certain areas or species.

Table 12.3. List of national parks (A) and various types of reserves (B) in West Africa. FR= Faunal Reserve; Fo FR=Forest and Faunal Reserve; GR=Game Reserve; GS=Game Sanctuary; NR=Nature Reserve; R=Reserve; SNR=Strict Nature Reserve. Extracted from *UN List of National Parks and Protected Areas* (Anon., 1980d). Numbers in parentheses refer to location of national parks in Fig. 11.1

Country	Area (1000 ha)	Date
Benin		
A		
(1) 'W'	502 050	1954
(2) Boucle de la Pendjari	275 500	1961
B		
Djona FR	225 000	NA
Pendjari FR	200 000	NA
Atakora FR	175 000	NA
Cameroun		
A		
(3) Bouba Ndjidah	220 000	1968
(4) Benoue	180 000	1968
(5) Waza	170 000	1968
(6) Kalamaloue	4 500	1972
(7)Mozogo–Cokoro	1 400	1968
B		
Dja FR	500 000	1950
Faro FoFr	330 000	1932
Campo FoFr	300 000	1932
Douala–Edea FoFR	160 000	1932
Korup GR	87 000	NA
Lobeke Lake NR	43 000	1974
Bafia	42 000	NA
Lake Ossa	4 000	NA
Pangar–Djerem GR	300 000	1968
Kimbi River GR	5 625	1964
Gambia		
A		
Baboon Island	2 000	NA
Ghana		
A		
(8) Mole	492 000	1971
(9) Digya	312 354	1971
(10) Bui	207 200	1971
(11) Nini–Suhien	16 278	1976
(12) Bia	7 700	1977
B		
Kogyae SNR	32 375	1971

Country	Area (1000 ha)	Date
Guinea		
B		
Mount Nimba SNR	13 000	1944
Ivory Coast		
A		
(13) La Comoe	1 150 000	1968
(14) Tai	330 000	1972
(15) Marahoue	101 000	1968
Mount Sangbe	100 000	1975
(16) Mont Peko	34 000	1968
(17) Banco	30 000	1953
B		
Mont Bandama NR	123 000	1973
Asagny Fauna NR	30 000	1960
Mount Nimba SNR	5 000	1944
N'Zo partial FR	73 000	1972
Mali		
A		
(18) Boucle de Baoulé	350 000	1954
B		
Fina FR	136 000	NA
Asongo–Menaka FR	1 750 000	NA
Elephant R	1 200 000	NA
Badinko FR	193 000	NA
Kongosombougou FR	92 000	NA
Kenie–Baoule FR	67 800	NA
Sounsan FR	37 600	NA
Mauritania		
A		
(19) Banc d'Arquin	1 200 000	1976
B		
Iles Mauritaniennes SNR	10 000	NA
Niger		
A		
(1) 'W'	334 375	1954
B		
Tamou NR	142 640	1962
Gadabedji FR	76 000	1955
Nigeria		
A		
(20) Kainji Lake	530 900	1975
B		
Lake Chad GS	704 480	1978
Borgu GR	350 000	1966
Yankari GR	225 285	1955
Chineurme Duguma	35 431	1978

Country	Area (1000 ha)	Date
Senegal		
A		
(21) Niokolo–Koba	913 000	1954
(22) Delta du Saloum	73 000	1976
(23) Dioudi	16 000	1971
(24) Basse-Casmance	5 000	1970
(25) Langue de Barbarie	2 000	1976
(26) Iles de la Madeleine	500	1976
(27) Kalissaye	250	1978
B		
Ferlo-Sud	633 700	1972
Ferlo-Nord	487 000	1973
Togo		
A		
(28) Fazao–Malfacassa	200 000	1950
(29) Keran	109 200	1950
B		
Koue	40 000	NA
Kamassi	17 000	NA
Keran NR	6 700	1975
Togodo FR	35 000	1952
Fosse Aux Lions FR	9 000	1950
Burkina Faso		
A		
(1) 'W'	190 000	1953
(30) Po	155 000	1976
(31) Deux Bales	115 000	NA
(32) Arly	76 000	1954
B		
Singou	192 800	1955
Bontioli FR	12 700	NA
Sahel	1 600 000	1970
Pama	233 500	NA
Arly FR	130 000	1954
Kourtiagou	51 000	NA
Nabere FR	36 500	1957

NA = not available

2. Reasons for protecting vegetation and species

In view of the food and/or fuel problems facing the people in large areas of West Africa any arguments for conservation steps aiming at protection of vegetation or of individual species appear almost futile. For countries in the forest region in urgent need of export income it is an easy way to exploit the

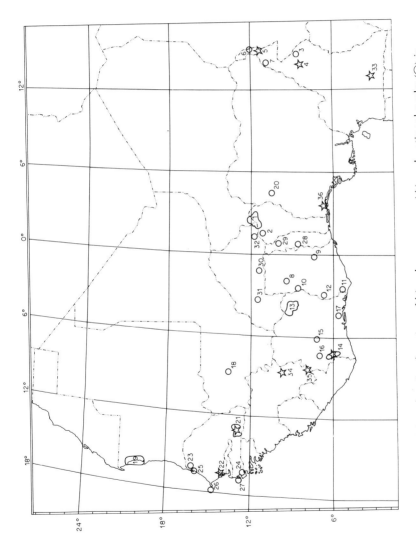

Fig. 12.1. Map showing location of biosphere reserves (☆) and national parks (○) in West Africa (see Tables 12.3 and 12.4). In a few cases a biosphere reserve coincides with a national park, viz. nos. 4, 5, 14, 21, and 22.

Table 12.4.   List of biosphere reserves in West Africa. Extracted from *United Nations List of National Parks and Protected Areas* (Anon., 1980d). Numbers in parentheses refer to location of biosphere reserves in Fig. 12.1

Country	Area	Date
Cameroun		
(33) Dja	500 000	NA
(4) Benoue	180 000	NA
(5) Waza	170 000	1979
Guinea		
(34) Massif du Ziama	116 170	1980
(35) Mount Nimba	17 130	1980
Ivory Coast		
(14) Tai	330 000	1977
Nigeria		
(36) Omo	460	1977
Senegal		
(21) Niokolo Koba	913 000	NA
(22) Sine Saloum delta	180 000	1980
Sambia Dia classified forest	756	1979

seemingly inexhaustible tropical forests. The mere fact that the forest area in eastern Liberia and western Ivory Coast is exceptionally rich in endemic species of both plants and animals (Upper Guinea refuge), or that part of the forests in Cameroun extending into eastern Nigeria (Cameroun–Gabon refuge) is believed to harbour the richest flora in Africa, are good reasons for conservation only among scientists, but poor arguments for hungry people. Short-term economic benefit or today's need for arable land, fuel, etc. naturally weigh heavily against any conservation attempts.

There are, however, very important reasons for protecting natural vegetation and preventing species from becoming extinct, including those from practical and economic points of view, as is shown by the examples given under the headings below.

(a) Catchment areas
The important role of mountain forests as catchments has been known for many years. Permanent flow of streams serving large human habitations can often only be secured if the catchments are kept forested. Deforestation of mountain slopes may also result in disastrous floods in adjacent lowlands.

(b) Soil stabilization
Especially in dry areas an unbroken vegetation cover is essential for preventing soil erosion. Apart from grasses other wild species may also be of great use for

soil conservation and stabilization in semi-arid regions. Any economically useful species that could thrive on marginal lands with poor soil and hot and dry climate in West Africa would thus be extremely valuable. The jojoba plant (*Simmondsia chinensis*), native to north-western Mexico and the south-western United States, has been tried with success in Sudan. This plant not only protects land from loss to deserts but also produces a high-value saleable oil in an environment where little else can even grow. Plans are now being elaborated to grow this plant in other extremely dry areas in Africa. There is, however, also an African plant, the Yeheb nut (*Cordeauxia edulia*) growing in Ethiopia and Somalia which is an important item of the Somali's diet and which seems to have the same outstanding drought resistance. This species, which unlike the jojoba plant yields edible nuts, was threatened by extinction a few years ago, due among other things to browsing by goats, but steps have been taken to safeguard its survival. A project has been drafted in Somalia for a close study of the properties of the Yeheb nut, but from the facts already known it seems highly probable that the plant could be used with success in other countries with similar climatic conditions and thus be of great importance as a soil binder and food plant in the marginal lands of West Africa.

(c) New food plants
Many of the plants used by the local population in forest areas are not cultivated but still growing wild in the forest. No cultivation is needed since the plants are being collected or the fruits picked where the plants grow. If the forest is removed the plants will disappear which means that species, which could have been of great value as food plants in West Africa and also in other parts of the world may be gone for ever. A vast number of useful plants in West Africa have already been recorded (Busson, 1965; Dalziel, 1937), but few of them have been subjected to detailed investigations in order to establish their usefulness in a wider context.

(d) Natural products
It is a well-known fact that medicinal plants play an extremely important role in developing countries and that substances from such plants have also proved useful against various diseases in developed countries. A recent example is furnished by the rose periwinkle (*Catharanthus roseus*), originally from Madagascar and used there to treat diabetes. Investigations carried out some 10 years ago showed that the plant contains more than 70 different alkaloids, two of which, vincristine and vinblastine, are now used as very efficient remedies against leukemia. Whereas in 1960 a sufferer from this disease had only one chance in five of recovery, the prospect has now risen to four chances in five, thanks to these drugs. Another example is the West African species of *Fagara* (*Zanthoxylum*) which show promising results in problems related to sickle-cell anaemia.

In the search for sweet substances, which could be used as substitutes for sugar for diabetics, some West African species, for example *Synsepalum dulcificum* and *Thaumatococcus daniellii*, have been found to contain interesting compounds and in various parts of Africa a member of the spurge genus, *Euphorbia tircualli*, is subject to investigations since to yields hydrocarbons which could be used as an alternative to petrol in diesel engines.

It should also be remembered that apart from plant substances directly used for medicinal and other purposes, the chemical clue to a large number of synthetics used today was derived from plants. It may very well be questioned whether a considerable number of manufactured compounds could have been produced at all without the leads provided by the natural substances occurring in plants. A good example is the synthetic pyrethroid insecticides with very favourable properties which have recently been developed (Elliott and Jones, 1978). Obviously the starting-point for the research leading to those synthetics was the substances (natural pyrethrins) found in *Chrysanthemum cinerariaefolium*, the flowers of which have been used for more than 100 years as a source of outstandingly safe insecticides.

(e) Ecological equilibrium

Awareness of the delicate balance between various species within an ecosystem is also very important, since disturbances caused in one connection might interfere quite unexpectedly with others. A good example is given by the production of durian fruits (*Durio zibethinus*) in an area in Malaysia. These fruits are very much appreciated by the local population because of their taste and as a source of income, but in recent years the fruiting has been continuously decreasing and the fruits have consequently become very expensive as fewer and fewer of them can be found.

The reason for this seemingly unexplainable drop in production became evident through a study of the bats living in the area. These bats live in limestone caves, are pollen-eaters and feed mainly on *Sonneratia* (mangrove trees), but prefer durian pollen when those trees are in flower thereby carrying out pollination of the flowers. The limestone is now being used for cement and the sonneratias are rapidly disappearing as the mangrove is turned into shrimp farms. Thus the fate of the bats in this area is decimation turning eventually into extinction, and in that way also the situation for the *Durio* is becoming precarious. The fruits are very important to the local people and could also become a valuable export product, but in order to prevent extinction of the tree, careful planning must be carried out both regarding the limestone mining far away in one direction and the creation of shrimp farms equally far away in the other direction. Such examples concerning interaction between animal and plant species could certainly also be found in the West African tropical moist forest.

The few examples given above may suffice to illustrate the fact that conservation of species should not be considered a sophisticated demand raised by the developed countries but a matter of deep concern also for developing countries. In this, as in other respects, part of the tragedy is that so many species are likely to be lost before their possible value to society is known. Many potentially useful species have probably already been eradicated and others are on the verge of extinction. In view of the fact that the tropical moist forest is believed to harbour 3 million species of plants, animals, and micro-organisms, two-thirds of which are still unknown to science, efforts to conserve as much as possible of it can hardly be too strongly advocated.

III. Conservation and development programmes

The only way species can be saved in their natural habitats is to protect the whole ecosystem in which they live, and this can be done only if suggested conservation measures have a reasonable appeal to the local population. It is thus quite clear that the only way of conserving natural vegetation and saving plants in developing countries like those in West Africa is by integrating conservation programmes in the economic development plans.

1. Dry regions

For the very dry areas of West Africa the aims of conservation efforts should in the first place be directed towards halting the desertification process which, as mentioned above, has been going on for a considerable time. Recent studies show that the net primary productivity is generally higher in natural vegetation than under agriculture since the natural communities maintain themselves without any expenditure of energy by man (Woodwell, 1977). In many places in West Africa the original vegetation was undoubtedly more productive than the poor crops (including domestic stock) produced afterwards. Traditional pastoralism or farming certainly yield less in many areas than sensible cropping through the years. To improve the present situation some of the most urgent steps would be as follows.

1. To set aside some areas hitherto used for grazing for the restoration of soil cover by keeping the herds out.
2. To protect and help to re-establish the vegetation cover by preventing forest clearing.
3. To allocate certain forest areas for tree-felling to cover the need for wood for various purposes.
4. To enforce by law replanting of forest, after cutting, especially in marginal areas.
5. To create reserves to protect the flora and fauna.

Great diffficulties will cerainly be encountered in persuading the local population to accept steps aiming towards maintenance or restoration of the natural vegetation cover. Legislation will be needed to a very large extent, but does not mean, unfortunately, that conservation programmes will be easy or even possible to accomplish. One very important subject in this connection is education of and information to the local population.

As previously pointed out, any action, however laudable, aiming towards protection of vegetation is doomed to fail unless the local people can be convinced of the benefit to themselves. Another equally important issue is the allocation of funds for aid progammes in these areas, where some of the poorest people of the earth are found. Priority should therfore be given by national and international aid agencies for a sound development, the most important part of which is the ecological foundation. The continuous devastation of the dry areas *can* be stopped if steps be taken that are adapted to the ecological and social conditions in each area.

The above-mentioned steps clearly aim directly at the benefit of the local population and should thus be possible to carry out. Careful planning is necessary and assistance from abroad might be desirable as in the case of the RURGS (Récherche pour Utilization Rationelle du Gibier au Sahel) project in Mali, which is carried out in collaboration between the Republic of Mali and the Netherlands. The project was initiated as a consequence of the severe drought which threatened Mali in the early 1970s and led to the destruction of the natural vegetation so essential for the cattle and the wild fauna of Sahel. In order to justify the project it was underlined that game is an invaluable source of meat and other products for the local people, and furthermore that it would be valuable from a touristic point of view and thus for the economy of the country. It was also stated that the animals at the same time form a major factor against the desertification since it is impossible to keep them without protecting their habitat. Finally, it was stressed that the wild fauna are a priceless part of the heritage of the people of Mali.

Founded on these statements the project was launched during the spring of 1977, starting with a detailed study of the area concerned, i.e. Boucle du Baoulé National Park with adjacent reserves. The initial study concerned climate, geology, soil, vegetation, animal ecology, permanent settlements, nomads, poaching, etc. An interim report was delivered in 1980 (Anon., 1980a) and the final report on the project is anticipated some time after 1982. The very thorough preinvestmnt studies, the multipurpose approach, and the fact that suggested arrangements for the future would be clearly directed towards the immediate benefit of the people living in the area should ease the way for the execution. It should also be of great value to other countries within the arid region when it comes to conservation and land-use planning.

2. Forest regions

Conservation steps in the forest regions of West Africa will probably be much more difficult to encourage, partly because of the seemingly very good supply of wood for local consumption, partly because of the short-term economic benefit obtained from the commercial logging carried out by multinational organizations. As mentioned above, the forests are being cleared at an alarming rate, and though replanting is now being undertaken to an increasing degree this is of little value when it comes to the extinction of plant and animal species and the destruction of valuable natural ecosystems. If successful, forest tree plantation will yield wood and serve as catchment areas, but little is as yet known about long-term results, and they will never give back the precious indigenous trees or other plant species which might have been useful as food or medicinal plants, etc.

To a biologist it is sometimes quite frightening to read reports on ongoing projects such as the CELLUCAM (Cellulose du Cameroun) project in Cameroun. In a Swedish journal (Anon., 1982c) an account was given of this commercial project, about 60 per cent of which is owned by Cameroun, the remaining 40 per cent being shared by foreign companies, institutions, and private persons. The project is reported to be completely unique since 'for the first time all tree species in the tropical moist forest are converted into high-quality, bleached cellulose on a large scale'. The traditional method by which a few trees per hectare with especially valuable timber were cut for export has been abandoned and large areas are virtually stripped of forest. Nothing is mentioned about any aspects of conservation.

Within the activities of CELLUCAM, areas such as mountains and slopes, swamps, and small streams are left untouched, which according to the report means that about 50 per cent of the original forest still remains. Though the erosion risk is touched upon one gets the impression that the reason for leaving the areas mentioned is mostly due to difficulties in exploiting them. The stripped areas are being replanted with high-yielding exotic species such as pine (*Pinus caribbea*) and *Eucalyptus* (*E. deglupta*).

In the article it is also mentioned that Cameroun is one of the richest and most spectacular countries in West Africa, the south and south-western parts 'covered by almost untouched tropical rain forest in which elephants, gorillas, and other animals are living almost undisturbed'. But for how long? It is not too far-fetched to assume that many of the animal and plant species in the areas so far untouched will disappear in the near future. Who is deciding which areas should be left, who will see to the protection of such areas? And even if they could be protected, will the plants and animals survive in an environment which has been more or less efficiently reduced and cut off from adjacent areas? Nothing is mentioned about such aspects and the reader is left with the impression that the management is not concerned.

The overall problem in this connection is of course how to develop tropical forests for the diverse resources they offer to the community at large, now and for ever. Broad-scale elimination will probably have far-reaching repercussions apart from degradation of other ecosystem functions, deterioration of soils, etc. and the extinction of species could well reach hundreds of thousands or even millions. Far-reaching consequences even for areas a long distance from the tropics have been foreseen because of climatic changes caused by changes in the amount of atmospheric carbon dioxide.

There exist, however, programmes for development of the tropical rain forest on a sustained yield basis, for example the Tai Forest project in Ivory Coast, which has been launched to guide and direct the development of the south-western part of the country. Like southern Cameroun this part of Ivory Coast is covered by extensive tracts of forest, which until the mid 1960s were largely uninhabited. At that time the Ivory Coast government decided to develop this region, resulting in a massive investment in various projects. However, it was soon realized that because of its extensive, untouched tropical forest, this area would be an area where scientific research could help to shape and guide future economic development. Thus in 1973 the Ministry of Scientific Research set up the Tai project, the general objective of which was to improve the scientific basis for rational management of the forest ecosystem (Guédé and Guillaumet, 1976; Kanga, 1976; Dosso *et al.*, 1981).

The project includes studies on changes in the flora and various aspects of plant life, dynamics of animal populations, the impact of man, and changes in the forest landscape, etc. Like RURGS this is obviously a project which provides a good opportunity to demonstrate that the needs of development can be compatible with those of conservation and that the aims of develoment are best served by a planned, rational use of the natural resources, in this case the tropical rain forest ecosystem.

Finally a small project from Guinea-Bissau will be mentioned. A far-reaching deforestation is taking place in this country, especially in the north-eastern corner, mainly because of shifting cultivation and the need for household fuel (Anon., 1982a). The primary aim of the project is to diminish the burning and restore the tree cover so that the wood needed by the local population will be available within a reasonable distance. Information and education is an essential part of the programme, and the attempts to protect the forest are linked with efforts to improve agricultural methods. Hence the forest project is associated with a development plan for the north-western fourth of the country where about half of the population lives. An important aspect of the scheme is to involve the local people in the planning. If this project turns out to be successful it will also provide valuable experience for future projects in other countries with similar problems.

IV. Conclusions

The need for protection and conservation of biological resources has long been known by botanists and zoologists round the world. Their call for action towards the increasing exploitation of those resources was first met with doubts and suspicion but has in recent years made most governments aware of the necessity of urgent actions. The planning and execution of such actions sets immense problems, especially in tropical countries like those in West Africa, where today's need for food and fuel must be met before any plans for tomorrow will receive any support.

Even if the situation is sometimes described as desperate it is vital that the efforts to save as much as possible of the remaining natural resouces are never allowed to fade. Increased activities in launching projects like those in Mali, Ivory Coast, and Guinea-Bissau are of the utmost importance, not only for the countries themselves but for the whole region.

Regional and/or international collaboration, often supported by the International Union for the Conservation of Nature (IUCN), World Wildlife Fund (WWF), United Nations Development Programme (UNDP), and other organizations can obviously given promising results. In West Africa this is well illustrated by Liberia, still not mentioned in the 1982 *UN List of National Parks and Equivalent Reserves* (Anon., 1980b). According to recent reports (Verschuren 1982, *in litt.*) the IUCN/WWF mission in 1978–79 (Anon., 1979) in combination with an understanding government is now resulting in remarkable progress in conservation work.

Though conservation, especially in the tropics, is of global importance and thus of international concern, the national responsibility must not be disregarded. What has hitherto too often been neglected, probably because it does not give visible and immediate economic or practical results, is the building up of a scientific infrastructure in tropical countries. Scientific training of West African students in various fields, for example botany, zoology, ecology, and development planning, as well as training of wildlife specialists is of the utmost importance, since lack of national expertise is still a pronounced obstacle in conservation efforts.

Furthermore, public education programmes are badly needed to ease the way for immediate actions as well as to ensure the long-term future of nature conservation. The creating of wildlife associations might be very useful in this connection since by such organizations it is possible to arrange education at all levels.

More money for field assistance and for the training of park wardens would also be rewarding since they are essential for the proper management of protected areas and could also play an important role in education of the local population.

These are some of the activities needed in order to make practical designs to fit country development plans which also take into account the vital need for conservation of natural resources. It is to be hoped that by such national activities, together with international collaboration, brightening prospects for a sound development in the whole of West Africa will be achieved.

References

Anon. (1977). A new map of the world distribution of arid regions. *Nature and Resources*, **13**(3), 2–3 and inside cover.

Anon. (1979). Liberia — dramatic about-turn. *IUCN Bull., n.s.*, **10**(5), 33.

Anon., (1980a). Propositions interimaires pour la conservation et l'utilization de la vie sauvage pour la region de la Boucle du Bauolé en rélation avec les autres utilisation du terrain. Bamako (mimeographed report), 131 pp.

Anon. (1980b). *United Nations List of National Parks and Equivalent Reserves*. IUCN, Gland, Switzerland, 121 pp.

Anon. (1981). S.O.S.: Save the Thai. *IUCN Bull., n.s.*. **12**(3–4), 10–11.

Anon. (1982a). Byskogsprojekt i Guinea-Bissau. *Skogen*, **6–7**, 14–15. Stockholm (in Swedish).

Anon. (1982b). *Encyclopedia of the Third World* Vols. I–III. London, 2125 pp.

Anon. (1982c). Svenske Per bland regnskog, gorillor och pygmeer. *Skogen*, **6–7**, 20–21. Stockholm (in Swedish).

Anon. (1982d). *United Nations List of National Parks and Protected Areas*. IUCN, Gland, Switzerland.

Busson, F. (1965). *Plantes alimentaires de l'Ouest Africain*. L'imprimerie Leconte, Marseilles.

Dalziel, J. M. (1937). The useful plants of West Tropical Africa. In J. Huchinson and J. M. Dalziel (Eds.), *Flora of West Tropical Africa*, appendix. Crown Agents, London.

Dosso, H., Guillaumet, J-L., and Hadley, M. (1981). The Tai Project: Land use problems in a tropical rain forest. *Ambio*, **10**(2–3), 120–125.

Elliott, M., and Janes, N. F. (1978). Synthetic pyrethroides — a new class insecticide. *Chem. Soc. Rev.*, **7**, 473–505.

Guédé, J. L., and Guillaumet, J-L. (1976). The Ivory Coast Tai Forest Project. *Nature and Resources*, **12**(2), 2–5.

Hedberg, I., and Hedberg, O. (Eds.) (1968). *Conservation of Vegetation in Africa South of the Sahara*. Acta Phytogeogr. Suez., **54**, Uppsala, pp. 49–58, 65–105, 115–121.

Kanga, N'Guessan (1976). The Ivory Coast Ministry of Scientific Research. *Nature and Resources*, **12**(2), 6–8.

Meyers, N. (1980). Forest refugia and conservation in Africa — with some appraisal of survival prospects for tropical moist forests throughout the biome. In G. T. Prance (Ed.), *Biological Diversification in the Tropics*. Columbia University Press, New York.

Persson, R. (1977). Forest resources of Africa. Part II. *Dept. of Forest Survey, Research Notes*, 22. Royal College of Forestry, Stockholm, 224 pp.

Rapp, A. (1977). Öknens tillväxt är människornas verk. *Forskning och Framsteg*, 1977(8), 2–12 (in Swedish).

Verschuren, J. (1982). Hope for Liberia. *Oryx*, **19**, 421–427.

Woodwell, G. M. (1977). The challenge of endangered species. In G. T. Prance and T. C. Elias (Eds.), *Extinction is Forever*, pp. 5–10. The New York Botanical Garden, New York.

Yourdeowei, A. (1981). The biology of the environment of Africa: arid and semi-arid zones. In E. S. Ayensu and G. J. Marton Lefevre (Eds.), *International Biosciences Networks*. African Biosciences Network, pp. 37–52. Washington DC.

Plant Ecology in West Africa
Edited by G. W. Lawson
© 1986 John Wiley & Sons Ltd

Postscript —

Ecological Change and Development

G. W. Lawson
Department of Biological Sciences, Bayero
University, Kano, Nigeria

It has been somewhere stated that the way to write a scientific paper is first to say what you are going to say, then say it, and finally say what you have said! If such advice applies also to scientific books I should at this point perhaps be summarizing all that has been set out in the preceding papers. The density of factual information in them clearly precludes such a course of action, but nevertheless it might be possible to disentangle a few threads that are implied through the body of this work or that may be extrapolated from it speculatively.

One feature which keeps recurring in many chapters either overtly or by implication is the importance of the time factor in ecology and it will be convenient to use the time dimension as a framework to examine some of these threads.

In the first place most of the authors spend a good deal of effort looking into the past and reviewing the subjects they have taken up in their respective chapters. What emerges clearly is the way in which individual research projects focus pin-points of bright light on to small fragments of the scientific landscape and the way in which such points as they accumulate give each other mutual aid in exposing to view the broader features of the picture.

Perhaps the most striking feature of the time factor in ecological change is the way that things have acceleratd in recent years. What we may conclude from a study of palaeoecology such as that given in this volume by Dr Sowunmi is that change in the late Pleistocene period was then a slow process occurring over hundreds and thousands of years. When the climate was favourable the forests flourished and advanced gradually northwards: when it became unfavourable they were rolled back towards the sea and clung precariously to a few favoured sites, only to advance again when conditions improved. Such 'marching and countermarching' (to use Moreau's, 1963, phrase) of vegetation across West Africa must have taken place several times over many thousands of years in what geologists — for whom countless ages are, as they were for the psalmist, like an evening gone — are pleased to regard as the relatively recent past. During such movements a few new species of plants originated; a few became extinct.

When mankind came on the scene he was just a part of these great ecosystems. With his small numbers and primitive tools his effect in them would be negligible except in so far as when he mastered fire he might have had an appreciable effect on the creation of the savannas as they now occur. Some ecologists believe that Africa may have been once covered, in the areas not occupied by rain forest, by a drier type of forest but that fire, manipulated by man, may have caused the elimination of fire-tender species and their replacement by the mostly fire-tolerant ones which now occupy the savanna regions (Aubréville, 1949; see also Lawson, 1966, but note remarks in Chapter 5 of the present volume for the alternative view). Even if that is so, savanna has remained for a very long period a relatively stable self-sustaining type of ecosystem.

It is only in the present century and especially within the last 30 years or so that we have seen a spectacularly accelerated rate of change in these old-established vegetation types of Africa. Most evident has been the increasing rapidity of the destruction of the forest. Estimates of this rate are not easy to make, but if the total rain forest regions of the world are considered there seems to be fairly general agreement that they are being destroyed at present at the alarming rate of about 15–20 ha *per minute* which amounts to about 3 per cent of the total forest area per year. If, as seems likely, the average rate of destruction of forest in West Africa does not differ significantly from this it may be expected that all forest in this region may be destroyed in about 30 years, i.e. well within the lifetime of many who will read these pages.

Changes in the forest are easily observed and appreciated; those in savanna perhaps more subtle and less acknowledged. But the explosive growth of human populations in recent years has meant much greater pressure on savanna too. One of the more obvious manifestations of this has been the exploitation of savanna trees for firewood. Though kerosene may be cheaper, the cost of installation of a stove to utilize this fuel is prohibitive for poor people who must perforce buy the more expensive wood to do their cooking. It has been estimated that on average one person in Africa may consume approximately 1 ton of firewood per annum (Anon., 1980). This is roughly the amount of wood produced in much of the savanna per hectare in one year. Taking Nigeria as an example, and bearing in mind that not all of Nigeria is savanna nor all that savanna is necessarily exploitable for firewood and also that not everyone in Nigeria uses firewood for fuel, it may be noted that the total land surface of Nigeria is something over 91 million ha and the human population presently estimated at about that same figure. This suggests that the annual production of firewood might not be able to support a much greater rise in population and that almost certainly in certain areas — around big towns such as Kaduna and Kano in the north for example — the amount of wood being produced is less than the rate of exploitation. Professor Gillet in his chapter emphasizes the importance of trees in combating desertification in the Sahel zone and it cannot be doubted that his words apply also to much of the Sudan zone. It follows that the over-exploitation of savanna for firewood where the removal of wood exceeds the net annual production must be a factor in the possible desertification of the northern savannas.

It is perhaps not out of place here to mention here the steady rise that has taken place in the carbon dioxide content of the atmosphere over recent years. At one time it was thought that this was due to the increasing use of fossil fuels, but it is now believed that only about half of the observed increased in CO_2 is attributable to this cause, the remaining 50 per cent being due to the destruction of the world's forests and the release of the massive amounts of carbon previously stored in their wood. The increased CO_2 in the atmosphere may bring about the well-known 'greenhouse effect' causing a rise in air temperatures, and it has been estimated that only a few degrees rise in temperature might have profound ecological effects such as turning drier regions of the earth into desert, to say nothing of ultimately melting the polar ice-caps thereby raising the general level of the sea and drowning many low-lying coastal areas (Olson *et al.*, 1978).

But these are problems that lie perhaps beyond the scope of this book. What may be legitimately considered are the problems that may be expected to arise in the more immediate future and the possible ways of dealing with them. As already indicated above, the destruction of forest has been proceeding apace and this trend will undoubtedly continue into the future. In some areas reafforestation has been carried out but mainly by fast-growing species with soft wood such as *Gmelinia*, though hardwoods such as teak are occasionally planted. Because of such policies and due to the very long time taken by hardwoods to mature it must be supposed that tropical hardwoods will become increasingly scarce and therefore more valuable in the future.

What is to be most fervently hoped is that sufficiently large areas of untouched primary forest are set aside, before it is too late, as nature reserves to remain inviolable for all time. Frankel and Soule (1981) have emphasized that conservation is not merely preservation of the status quo but must meet the requirement that sufficiently large populations be maintained to allow adaptation and evolution to continue in a natural way. Such action is necessary, however, not only to ensure that a bank of insufficiently studied wild species which may have possible economic uses in the future is maintained, but also to preserve, in working order as it were, intact examples of this anciently evolved and intricately balanced ecosystem for further study and as reference points from which the effective use of exploited forest areas may be measured (Lawson, 1970). It is, of course, necessary to have the strict nature reserves surrounded by a buffer zone of forest in which experimentation and limited disturbance may take place.

With regard to the savanna regions, preservation of examples of these ecosystems are in some ways more certain than for forest as it is in such country that the main game reserves are located. As long as it is regarded as important to maintain them for recreational purposes the vegetation is relatively safe since it must perforce be allowed to exist as a habitat in which the desired animals may live in a natural condition. But outside these reserves the savannas are coming under increasing pressure. Not only is the accelerating demand for firewood for the expanding human population, as already indicated above, a threat to the denudation of woodland and the subsequent risk of desertification, but the impact of numerous agricultural development schemes, however laudable in intention entails other risks to the environment that should not be forgotten in the race for increased productivity. In the northern savanna regions, for example, there are now many irrigation schemes in progress which entail the damming of rivers and the construction of channels for the distribution of water over agricultural land. At the same time there is greatly increasing use of fertilizers for agricultural production. Such changes may have many effects. On the other hand, there is the increased hazard to human health brought about by the presence of permanent bodies of water which encourage the multiplication of the vectors of water-borne diseases such as bilharzia and onchocerciasis. On the other hand there is the potential pollution effect of chemical fertilizers seeping into the man-made lakes and the rivers which are used as a water supply for both humans and animals. Such effects are already begining to become noticeable in some areas, for example in Lake Tiga in northern Nigeria which serves as a source of water for irrigation projects and also as a supply of drinking-water for Kano. Clearly, governments will have to take with account in future the ecological impact of such agricultural and other development schemes by first of all monitoring the environment and then by taking appropriate action on the facts that are thereby revealed, and the information which is available. Such action may not always be politically expedient. The current belief that all development is positive and to the benefit of the nation concerned does not always hold good. The long-term effects may not always be so beneficial to the total environment and therefore to the quality of life. Scientists can only point out such things; it is for those in authority to heed and take appropriate action.

References

Anon. (1980). *Firewood Crops*. Report of an *ad hoc* panel of the Advisory Committee on Technology Innovation Board on Science and Technology, for the International Development Committee on International Relations. National Academy of Science, Washington DC, xi + 237 pp.

Aubréville, A. (1949). *Contribution à la paléohistoire des forêts de l'Afrique tropicale,* Soc. d'Edition Geog., Maritimes et Coloniale, Paris.

Frankel, O.H., and Soulé, M. E. (1981). *Conservation and Evolution.* Cambridge University Press, London.

Lawson, G. W. (1966). *Plant Life in West Africa.* Oxford University Press, London, viii + 150 pp.

Lawson, G. W. (1970). *Ecology and Conservation in Ghana.* Ghana Universities Press, Accra, 21 pp.

Moreau, E. E. (1963). Vicissitudes of the African biomes in the late Pleistocene. *Proc. Zool. Soc. Lon.,* **141**(2), 395–421.

Olson, J. S., Pfuderer, H. A., and Chan, Y. H. (1978). Changes in the global carbon cycle and the biosphere. *Environmental Sciences Division Publication No. 1050. Oak Ridge National Laboratory.* xv + 169 pp.

Index

333